PROPERTY OF
USBI

D1473417

PROPERTY OF
[CCB]

ENGINEERING MANAGEMENT

McGRAW-HILL SERIES in Industrial Engineering and Management Science

CONSULTING EDITOR

JAMES L. RIGGS,
Department of Industrial Engineering, Oregon State University

ENGINEERING MANAGEMENT

DAVID I. CLELAND

Professor of Systems Management Engineering
School of Engineering
University of Pittsburgh

DUNDAR F. KOCAOGLU

Associate Professor of Industrial Engineering
Coordinator, Engineering Management Program
School of Engineering
University of Pittsburgh

McGRAW-HILL BOOK COMPANY

New York | St. Louis | San Francisco | Auckland
Bogotá | Hamburg | Johannesburg | London | Madrid
Mexico | Montreal | New Delhi | Panama | Paris
São Paulo | Singapore | Sydney | Tokyo | Toronto

ENGINEERING MANAGEMENT

Copyright © 1981 by McGraw-Hill, Inc.
All rights reserved.
Printed in the United States of America.
No part of this publication may be reproduced,
stored in a retrieval system, or transmitted,
in any form or by any means, electronic,
mechanical, photocopying, recording, or otherwise,
without the prior written permission of the publisher.

1 2 3 4 5 6 7 8 9 0   FGFG   8 9 8 7 6 5 4 3 2 1

This book was set in Helvetica by A Graphic Method Inc.
The editors were Julienne V. Brown and J. W. Maisel;
the designer was Nicholas Krenitsky;
the production supervisor was Charles Hess.
The drawings were done by A Graphic Method Inc.
Fairfield Graphics was printer and binder.

Library of Congress Cataloging in Publication Data

Cleland, David I
    Engineering management.

    (McGraw-Hill series in industrial engineering
    and management science)
    Includes index.
    1. Engineering—Management.   I. Kocaoglu,
Dundar F., joint author. II. Title.
TA190.C58   620'.0068   80-20369
ISBN 0-07-011316-5

CONTENTS

PREFACE

This book had its beginning in September 1975 when we started an Engineering Management Program at the University of Pittsburgh. Our study of the educational needs of engineers who assume management responsibilities in engineering organizations led us to the conclusion that there were no management books written from an engineering viewpoint. As we studied the market opportunity for engineering management curricula, we became convinced that this was an emerging area of management education.

In recent years more and more engineering schools have started offering management courses in their curricula thereby influencing some practicing engineers to return to school because a suitable educational opportunity is available. It is often many years before an engineer discovers that his career interests lie more in the area of becoming a manager of some technical activity than remaining as a technologist in the engineering discipline. (For ease and simplicity the pronoun "he" will be used throughout the text to denote third person singular when reference is made to an engineer, male or female.) The engineer who senses a turning point in his career seeks an educational opportunity that can be terminal in its own right, usually one that can be pursued on a part-time basis, and one that strengthens his management credentials. The management position the engineer hopes to hold requires a comfortable knowledge of a technology and a philosophical understanding of the management process.

We have not used the term "student" in the restrictive sense of a person pursuing a program of instruction at an educational institution. Rather we use it in a much broader sense to describe that individual who wishes to improve his knowledge of engineering management through a self-improvement program, either by enrolling in formal university courses or following a professional reading program.

This book is written for use in graduate-level engineering management programs. At the undergraduate level, it should have wide appeal for students in engineering schools where management courses are offered for senior undergraduate students in the latter part of their engineering education. Continuing education programs should also find the book appropriate as a text. The chapters have questions at the end that can serve to stimulate discussion in the classroom ambience.

The book should also appeal to the practicing engineering manager, who can use it as a primer in the management of organizations. Many engineers do not have the opportunity to pursue graduate education but desire to improve their engineering management credentials. This book should help in this respect. To this end we have placed at the end of the chapters questions that should be useful to the engineering manager as a self-examination of his management process and style.

We believe that an engineer who assumes management responsibilities needs to develop a philosophy of engineering management. The term "philosophy" is used in the sense of a body of thought that underlies the engineering management discipline. Three key attributes of the philosophy are used in developing this book. First is knowledge—a familiarity with management theory applied to an engineering milieu; second is skill in using management processes and techniques; and third is the development of a set of attitudes—values and aspirations—that can facilitate the leadership role of the engineering manager.

An engineering manager who aspires to be effective in managing an engineering organization needs to develop knowledge, skills, and attitudes in the context of a management philosophy. We present a format for the book around this theme; the format is simple and adaptable to many engineering management situations. The book consists of four parts. Part One, Engineering and Management, introduces the idea of the similarities and differences of engineering and management. This part deals with the phenomenon of the engineer who starts as a technologist and later becomes a manager of technology. This individual's need for management education is described; current approaches to satisfying this need through a form of educational effort is presented. Finally, an engineering management process consisting of distinct functions is suggested in the context of the engineer's needs for management knowledge, skills, and attitudes.

Part Two, Engineering Organization, deals with the broad role of organizing engineering resources to support engineering objectives. Modern organizational approaches such as project management are a springboard for presenting the concept of the project life cycle as a focus for managing an engineering project. The resource common to all organization—the human subsystem—is then discussed, followed by some thoughts on communication. Finally, this section gives an insight into how the engineering manager can deal with people and organizational change in his environment.

Part Three, Management Science in Engineering, explains how quantitative techniques can complement the role of judgment in making decisions. This section starts off with the concepts and philosophies of management science and continues with the mathematical methods used in decision making. Chapters 7 to 9 lead to each other in an increasing order of mathematical emphasis, going from generalities of the management science approach to the specific use of linear programming in decision making. Risk and uncertainty in engineering management decisions are discussed in Chapter 10. The methods developed in these chapters are applied to project selection, control, and evaluation decisions in Chapter 11.

Part Four, The Broader Systems in Engineering Management, wraps the book up. This section takes the reader into a systems context of thinking through technological planning and forecasting engineering and the environment, and engineering and the law. The theme of this part should encourage the engineer to look beyond his engineering discipline to those systems of the world which impinge on the practice of engineering management in today's complex and interdependent environment.

The book integrates the qualitative concepts and quantitative techniques of management. This approach should appeal to the engineering manager who feels a responsibility of integrating the engineering discipline into the social and economics considerations of managing technical systems. The qualitative portions of the book should bring the management issues into perspective. The quantitative portions should complement the mathematical and statistical background of the engineer.

A recent study conducted by the authors showed that the required courses in graduate level engineering management programs were made up of 23 percent management theory and concepts, 34 percent management science methods, 17 percent functional areas, 18 percent finance and economy, and 8 percent term projects. In this book, after the overview of the first chapter, we cover management and organization theory in Part Two, present management science in Part Three, and pull together three major interdisciplinary areas in Part Four. Finance and economy are not included because of their adequate coverage in several excellent books available on those topics.

This book is addressed to three major groups of readers:

1 Advanced engineering undergraduate students taking management courses in their senior years

2 Graduate students in engineering management programs

3 Practicing engineering managers not enrolled in a formal educational program

The needs of these groups are not the same. Each can get a different degree of benefit from each chapter of the book. The accompanying table indicates the intended primary group for which each chapter has been developed.

x   There is a real need for a book on engineering management that captures the attention of the engineering community. We believe that engineering managers who have previously used general management texts will welcome a management book oriented to their particular needs—the management of an engineering organization.

Engineering management continues to evolve and develop. The need for engineers to become professionally trained engineering managers who are generalists is becoming clear. We believe that our efforts in writing this book are a proper step in that direction.

We are indebted to Elizabeth Delisi and Claire Zubritzky who so ably prepared the manuscript. Our thanks to Ronald Leslie of the Litigation Section of Rockwell International, who provided a chapter giving insight into the legal framework of the engineering management process. We are grateful to Jephthah Abara who assisted in preparing the chapters on project life cycle and project control, and provided valuable assistance in the refinement of the manuscript. Finally our deep appreciation to Al Holzman, Chairman of the Department of Industrial Engineering, and Max Williams, Dean of the School of Engineering, who provided us with the "systems environment" to pursue writing and research.

David I. Cleland
Dundar F. Kocaoglu

Chapter	Advanced Engineering Undergraduate Student		Graduate Student in Engineering Management Program		Practicing Engineering Manager	
	Concept	Methodology	Concept	Methodology	Concept	Methodology
1 Why Engineering Management?	X		X		X	
2 Project Management	X	X	X	X	X	X
3 Engineering Project Life Cycle	X	X	X	X	X	X
4 The Human Subsystem in Engineering	X		X		X	
5 Communication in Engineering Management	X		X		X	
6 Managing Change in Engineering Organizations	X		X		X	
7 Management Science Concepts and Philosophy	X	X	X	X	X	X
8 Mathematical Models		X		X		
9 Linear Programming				X		
10 Decision-Making Methodologies	X	X	X	X	X	X
11 Project Selection, Control, and Evaluation		X		X		X
12 Technological Planning and Forecasting	X	X	X	X	X	X
13 The Engineer and the Evironment	X		X		X	
14 Law and the Engineer	X		X		X	

PART ONE
ENGINEERING AND MANAGEMENT

ONE ENGIN
ENGINEERIN
AN
MANAGEME
P/
PART ONE ENGIN
ONE ENGINEERIN
ENGINEERING AN
AND MANAGEME
MANAGEMENT P/

WHY ENGINEERING MANAGEMENT? 1

Nothing is more important than an idea whose time has come. Engineering management has existed for many years as a vital function in technology-based organizations. Recently it has started to gain recognition as a discipline in its own right, different from engineering specialties and also different from general management. Considerable attention is now being directed toward educating engineering managers, who play a dual role as the linkage between management and technical expertise. Such managers allocate resources, work through people, and make and implement decisions while simultaneously formulating technical strategies.

We believe that the time has come for the identification of engineering management as a distinct professional discipline. Individuals who manage engineering organizations find it necessary to understand both the technology involved and the management process through which that technology is applied. Management literature contains very little material that deals with the dual role of the engineering manager—a technologist and a manager. This is somewhat of an enigma, for engineers have made many significant contributions to general management theory and practice.

The pioneer of scientific management, Frederick W. Taylor, who introduced basic analytic methods into factory management early in this century, was an engineer. Taylor's classic statement, "knowing exactly what you want to do, and then seeing that they do it in the best and cheapest way," introduced the era of scientific management at the shop level. Henri Fayol, another engineer, is credited with being the father of modern management theory. Fayol's book, *General and Industrial Management*,[1] is truly the outstanding classic in management theory. Any individual who calls himself a manager should read this book.[2]

---

[1]Henri Fayol, *General and Industrial Management*, Pitman, London, 1949.
[2]Harold Koontz and Cyril O'Donnell, in their *Principles of Management,* 3rd ed., McGraw-Hill, New York, 1964, make this claim. They state: "A study of Fayol's monograph, with its practical and clear approach to the job of the manager and its perception of the universality of management principles, discloses an extraordinary insight into the basic problems of modern, business management" (p. 17). Few management scholars would question their judgment of Henri Fayol.

4

The more recent contributions of engineers and scientists to management have been in the management science–operations-research area. Perhaps this has been so because the mathematical approach to management complements the quantitative background of the engineer.

Yet when one reviews some of the extraordinary contributions to management theory and practice that engineers have made, one is struck by an irregularity: while engineers and scientists have made many contributions to management theory, there have been only a few books written in recent years on engineering management.

It can be argued that the engineer is in somewhat of an identity crisis, not knowing whether to be a technologist for technology's sake or to serve more in a bridging role between science and social demands. Industries develop as the result of a demand when resources to satisfy it are available. The most significant danger for an industry is its preoccupation with its product rather than the business that it is in. Development of engineering in its modern shape as a discipline in the nineteenth century was a direct response to demand for an interface between two isolated segments of society: one represented by society's need for products, the other by the availability of scientific methods and discoveries. Based on this starting point, it is possible to define the *business* that engineering is in as the interaction between those two segments.

This definition necessarily imposes upon engineers the responsibility of integrating and managing all aspects of technical, social, and economic systems. It is questionable, however, as to what degree they are prepared to assume this responsibility. In the present educational system, all engineering students are required to get a broadly based, often highly theoretical background in basic scientific areas. The bright students are then encouraged to undertake a more analytical and abstract study at graduate level. Because of the limited nature of course work in creative applications of technology to realistic problem solving, the graduates usually have little idea of the real needs of the society that they are expected to serve in the capacity implicit in the fundamental purpose of engineering. According to Karger and Murdick, this is a form of myopia, probably gained from the engineering curriculum which is "jammed with mathematics, science, and engineering courses. . . . Engineering curricula seldom require the student to study economics and business management, yet such knowledge is at the heart of 'making science practical for mankind.'"[3]

As a possible solution for this myopia, Karger and Murdick suggest: "Require all engineers to take the equivalent of a three-credit-hour course covering all aspects of new-product problems and

[3]Delmar W. Karger and Robert C. Murdick, "Engineering Myopia," *IEEE Transactions on Engineering Management*, February 1977, p. 24.

a three-credit-hour course covering all aspects of engineering management".[4]

Although the present engineering education is satisfactory by way of providing technical skills and knowledge, its ability to produce students who have an adequate appreciation of engineering management has to be questioned.

To paraphrase J. W. Forrester,[5] who has spoken of the need for the engineer to integrate technology, economics, management, human behavior, and marketing in his work, engineering education gives the student only scientific tools, with just a slight touch on the other subjects in an informal fashion. The engineer has to make decisions based on inadequate information, trying to optimize resources in an uncertain environment, yet at school he primarily deals with essentially deterministic systems. As a manager of socio-technological systems, the engineer has to have conceptual skills and leadership qualities, yet very little of his formal education is directed toward developing them. In analyzing future skills and knowledge requirements for engineers, Forrester recommends that attention should be focused on educating a new breed of engineer, the enterprise engineer, whom he describes as a leader, a designer, a synthesizer and a doer, who understands theory as a guide to practice, and concerns himself with human organization because the pace and success of technology are becoming more dependent on interaction with the social system and less on scientific discovery.

Traditionally, young engineers have considered management as being too intuitive, not based on hard facts and analytical approaches. This perception has led to an almost artificial line of demarcation between the two disciplines. In reality, engineering and management are logical complements of each other. No major technological problem can be solved in a vacuum without explicit recognition of the nontechnical factors such as the social, legal, and economic conditions. Conversely, in a complex society, no major problem can be solved outside the technological system. The engineer, therefore, frequently has to make decisions which go far beyond the boundaries of his technical specialty.

Because of the unique problems and cultural implications of engineering environment, the management of technologies requires a solid engineering background. It would be a naive expectation to assume that nonengineers could become successful engineering managers. On the other hand, it is also naive to assume that successful engineers necessarily become successful engineering managers. No data are available to support either of these assumptions.

An intuitively realistic hypothesis about the engineering-

---

[4]Ibid., p. 25.
[5]J. W. Forrester, *Engineering Education and Engineering Practice in the Year 2000,* paper presented at National Academy of Engineering, University of Michigan, Ann Arbor, Sept. 21, 1967.

6 management interaction could be made as follows: "Successful performance in an engineering specialty is a necessary but insufficient condition for successful engineering management."

This book is addressed to engineers moving into engineering-management positions. Management science techniques and management-theory concepts are blended to give the engineer a balanced exposure to the qualitative and quantitative aspects of management. A rational basis for decision making is developed throughout the book. The emphasis is on the formulation of engineering management problems and the interpretation of their results. The organizational problems of planning, controlling, and scheduling are evaluated in terms of engineering resources and interactions among those resources. The objective of the book is to channel the rational decision-making capability of engineers toward the analysis of engineering management problems. The project orientation of the engineering environment is used as the focal point of the book. The concepts and methods are developed around it.

ENGINEERING TO ENGINEERING MANAGEMENT

Engineering and management have the same basic philosophies, even though the specific nature of the problems are different in the two disciplines. Both engineers and managers are trained to be decision makers in a complex environment. They both allocate resources for the operation of existing systems or for the development of new systems. Both have to recognize, identify, and evaluate the interfaces among system components. Engineering and management decisions differ only because of their emphasis on different subsystems. The engineer is primarily interested in the materials subsystem, dealing with the methods and processes for the allocation of material-related resources to the design, development, and operation of engineering systems. He goes through a rational thought process in his decisions, justifying each step with facts, measurements, and functional relationships that can be proved. His statements are precise and measurable; his thought process is traceable and his decision steps are reproducible for another engineer to accept or refuse the recommendations. He frequently works with relatively well-defined problems in an environment where uncertainties are reduced significantly by the information available about the behavior and properties of materials with which he deals.

A manager's primary emphasis, on the other hand, is on the allocation of human and nonhuman resources to perform the tasks demanded of his organization. His problems are usually more open-ended and less well defined than the engineer's problems. He works in a considerably more uncertain environment compared to an engineer because of the unknowns involved in human behavior and the

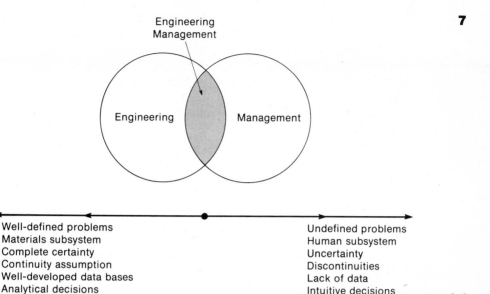

Well-defined problems   Undefined problems
Materials subsystem   Human subsystem
Complete certainty   Uncertainty
Continuity assumption   Discontinuities
Well-developed data bases   Lack of data
Analytical decisions   Intuitive decisions

1.1

Engineering specialty–engineering-management interaction.

external conditions that affect his decisions. He operates in a rapidly changing, discontinuous environment where precise functional relationships are difficult to develop, yet he has to make rational decisions and justify the steps of his thought process. Figure 1.1 depicts the interaction between engineering and management and the overlap between the two fields.

ASSUMING MANAGEMENT RESPONSIBILITIES

The engineer, with his analytical background and training in rational decision making, is a logical candidate for management responsibilities. In fact, according to a recent survey conducted in several technology-based corporations in Pittsburgh over 45 percent of engineers in the 25 to 45 age interval were found to have some form of engineering management responsibility, ranging from staff positions with indirect supervision to the position as head of engineering.[6] According to the Engineering Manpower Commission, the corresponding figure for all engineers in the United States is as high as 82 percent.[7] Other surveys indicate that 40 percent of industrial executives and 34 percent of all top corporate managers in the United States have engineering backgrounds.[8] Change affecting the career

[6]This survey was conducted by the engineering management program at the University of Pittsburgh.
[7]*The Engineer as a Manager,* Engineering Manpower Bulletin 25, September 1973.
[8]Personal correspondence with Dr. R. A. Brown, Industrial Engineering Department, University of Alabama in Huntsville.

8

of an engineer includes those changes which threaten to make his engineering credentials obsolete as well as those that redirect his career. A report of the Carnegie Foundation noted: "The most important fact brought out by . . . surveys over a period of thirty years is that more than 60 percent of those persons who earned [engineering] degrees in the United States, either became managers of some kind within ten to fifteen years or left the engineering profession entirely to enter various kinds of business ventures.[9]

Management opportunities become available to engineers at a relatively early point in their careers. During the 25- to 30-year age period the young engineer has duties that are largely technical in nature. As the age level increases, administrative duties expand rapidly. The percentage of engineers in team supervision is highest in the 31- to 35-year age group. The proportion of project managers peaks at 36 to 40 years, while that of division managers is highest from 46 to 60. By the time engineers are 41, two-thirds or more will have taken on management duties of an increasingly responsible nature. Figure 1.2 summarizes these statistics based on a survey conducted by the Engineers Joint Council.[10]

Engineers reach a decision usually between 3 and 7 years after graduating with an engineering degree. They find themselves in need of choosing between a technical specialty, a technical management position, or a nontechnical management job. There are two reasons for this situation:

1 Engineers, by the nature of their profession, are thrust into a management ambience by assignment to an engineering project almost immediately upon graduation. In an engineering project, they have to deal not only with the materials subsystem but also with the relations among the members of their project team as well as the other specialists from functional groups. They must consider the financial impact of their technical decisions. In a short time they discover that they must work through others, such as draftsmen, designers, and other engineers either directly or indirectly.

2 The reward system in industry, by and large, seems to favor individuals moving through the management, rather than the technical specialty ladder. The dual-ladder concept has been tried in several large companies, but the successful implementation of that concept has been very limited.

[9]Russell L. Ackoff, *Redesigning the Future: A Systems Approach to Societal Problems*, Wiley, New York, 1974, p. 80
[10]*The Engineer as a Manager*, Engineering Manpower Bulletin 25, September 1973.

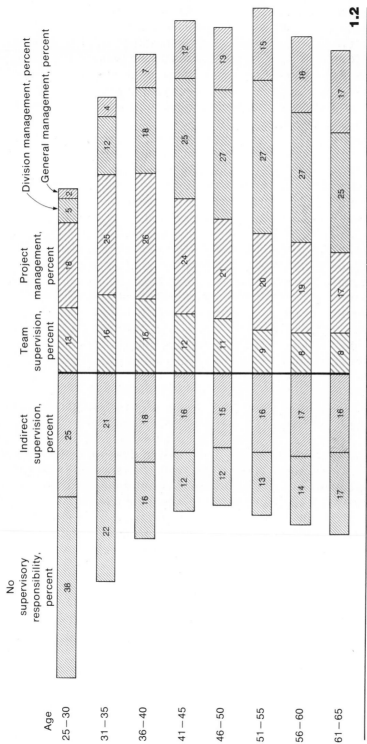

1.2

Engineering management responsibility as a function of age. (*From The Engineer as a Manager, Engineering Manpower Bulletin 25, September 1973.*) Percentages do not add to 100 in some cases because numbers have been rounded off.

10   Even though the engineers are required to make this important choice at an early point in their career path, they realize that they have not been prepared to make management decisions while simultaneously formulating technical strategies. The two roles that they have to play represent different underlying cultures even though both are based on the premise of rational decision making. The engineering education has been very successful in preparing engineers for solving technical problems, but it is not directed toward the management aspects of the engineering profession.

MANAGEMENT KNOWLEDGE REQUIREMENTS
The next excursion for the engineering manager out of his technical discipline is into the world of broad social, economic, regulatory, and environmental considerations. These considerations are not new, but they are certainly accelerating in their impact on engineering decisions. One only has to read of the products that are being recalled by manufacturers for repair or the number of class-action suits in our courts to see the growing impact of social and regulatory considerations. To illustrate the regulatory dimension, an article in a recent trade journal pointed out that over 200 new federal regulatory agencies or departments have been created in the last 10 years. But there is a bright spot in all this—twenty-one such regulatory agencies have been eliminated!

We are not taking issue with this regulatory aspect of the industrial world. Our point in bringing this up is merely to state the emphasis being placed on regulation by our government and the impact this has in influencing the decision aspects for the engineering manager.

One cannot reasonably expect the engineering manager to be expert in all these matters of regulatory, social, and environmental considerations. But he should know enough to know when he needs help. Then he can seek the professional assistance needed to advise him on the impact an engineering management decision can have on the total picture, for the present and for the future.

An example of the complexities of decisions in a major technological project is the multitude of issues surrounding the supersonic transport (SST). When the United States decided to abolish the development efforts in the early 1970s, it was not because of a lack of technical expertise in the field. Rather, the decision was based on the seemingly nontechnical considerations dealing with the economics of the SST, the environmental factors, the legal aspects, the political implications, and so forth. Even if it was a technological triumph, the American SST did not become a final product because of these peripheral issues. The engineering decisions dealing with the materials subsystem and the technological requirements were not adequate to bring the project to a successful completion.

THE ENGINEERING MANAGER'S JOB

The nature of an engineering manager's job and the responsibilities involved center around the process of *making and executing* engineering and engineering management decisions. Routine decisions are never easy, but they are far easier than those complex strategic decisions whose impact on the engineering project may be felt several years into the future. These decisions, growing out of the planning process, are facilitated by the use of modern quantitative management techniques. Such techniques have come into favor as ways to better evaluate the risk and uncertainty of engineering management decisions. The engineering manager requires enough sophistication in these techniques to realize that quantitative approaches to management can be a valuable aid to complement his judgment.

We must accept also that an engineering manager's role requires an adequate working knowledge to permit him to manage a team of professionals and at the same time serve in a fiduciary function for his portion of the company's capital investment and expenditures.

All this leads us to the conclusion that engineering managers wear two hats: the technical hat and the management hat. It should be abundantly clear that education and training beyond the technical training usually received is required. This, coupled with the engineering manager's need to remain abreast of the state of the art, is an obvious challenge.

EDUCATIONAL NEEDS

To what extent does traditional engineering education prepare the engineer for assuming managerial responsibilities? Very little, we believe. The classical engineering subjects do little to develop managerial knowledge and skills. The needs an engineering manager would have to satisfy in order to be a successful manager include the following:

First, and of primary importance to every step of the engineering process, are the economic considerations that enter into the decisions involving the application of technologies. A knowledge of engineering economy is an essential ingredient for engineers and those who aspire to effectively manage engineers.

Next is the evaluation and selection of alternatives in engineering design and resource allocation decisions. When an engineering manager has to determine which engineering project to select and how to distribute his human and nonhuman resources to the various activities in the project, he needs analytical techniques to make the best decisions. The conditions under which the decisions are made are not purely rational, since the engineering manager cannot always afford the luxury of evaluating all alternatives to reach the so-called best or optimum decision. However, the opposite of rational

12   decision making is intuition and the seat-of-the-pants type of management. Intuitive decisions have a justifiable basis if cast in the context of the manager's personal belief, experience, and feel of the situation. Such decisions cannot be subjected to an analytical evaluation, nor can they be traced back systematically. Therefore, even when the engineering manager operates under the "bounded rationality principle" of Herbert Simon,[11] he needs the concepts, tools, and methods of rational decision making. These tools are available in the body of knowledge known as management science.

Another consideration in educating the engineering manager is to develop the necessary skills and attitudes to work harmoniously and cooperatively with people in the organization—subordinates, superiors, peers, and associates. There are no peopleless organizations! An engineering manager must develop the knowledge, skills, and attitudes conducive to being recognized as a leader in his organization. This is a departure for the engineering manager from the relatively predictable world of engineering to the unpredictable world of human behavior. People working together to accomplish a goal require a form of organization—either formal or informal. Today's engineering departments tend to be formal organizations; the person who manages such an organization should have a basic working knowledge of organization theory as it relates to his environment. The engineering manager need not become an expert in organization theory, but he should understand the basic principles of organization and some of the alternative ways of organizing engineering resources, such as the use of the matrix-organization approach.

The need for the engineer to know something of organization theory, particularly the idea of the matrix organization, can be illustrated by describing the new organizational concept in engineering management instituted by General Motors, viz., the *project center*.[12] This concept in management was devised to coordinate the efforts of development engineering in the five automobile divisions in General Motors. A project center consists of engineers temporarily assigned from the divisions to work on a new engineering design. If a major new effort is planned—a body changeover, for example—a project center is formed across the automobile divisions to operate for the duration of the change.

Project centers work on engineering equipment common to all divisions, such as frames, steering gear, electrical systems, and so forth. The project center complements but does not replace the lead-division concept, where one division has primary responsibility for

---

[11]This principle is discussed in detail in D. W. Miller and M. K. Starr, *Executive Decisions and Operations Research*, Prentice-Hall, Englewood Cliffs, N.J., 1960.
[12]See Charles G. Burck, "How G. M. Turned Itself Around," *Fortune*, Jan. 16, 1978, pp. 87–100.

taking an innovation into production. The interpersonal relations this **13** new project-center concept facilitates are noted by one General Motors executive in this fashion:

"We become masters of diplomacy," says Edward Mertz, assistant chief engineer at Pontiac, who was manager of the now-disbanded A-body project center. "It's impossible to work closely on a design without influencing it somewhat. But the [project] center can't force a common part on a division." Indeed, many of G. M.'s engineers feel the project-center innovation has actually helped enhance the divisions' individuality, by freeing some of them to work on divisional projects.[13]

In actuality, General Motors has adopted what has become known as the "project management" approach to organizing and managing engineering projects.[14] From an organizational viewpoint this technique is usually called the "matrix organization"; it results when teams of people drawn from different disciplines are formed to work on an engineering problem. This approach can help engineering managers delegate authority, thus freeing them from involvement in detailed decision making.

It seems quite appropriate to assume that a majority of engineers in the future will continue to integrate technical management responsibilities at some stage in their career, and the proportion of their time spent on such duties will increase as they grow older. Engineers, seeing the need to acquire management knowledge and skills, have created a demand for engineering management programs in universities. Such curricula are usually offered in engineering schools for engineers moving into management positions while maintaining their identity within their technical specialties. One can assume that these programs will continue to provide the future engineer-manager with the required educational opportunities. A brief review of a typical engineering management program gives an insight into the educational experience these engineering managers-to-be hope to acquire.

TYPICAL ENGINEER MANAGEMENT PROGRAM
The typical engineering management program is designed to develop within the student a broad knowledge of the management of technical and scientific organizations. The plan of instruction in such programs has been designed to provide the candidate with the qualitative and quantitative skills necessary to hold management positions in

---

[13]Ibid, p. 96.
[14]Project management is more fully discussed in Chap. 2. The reader who wishes to study even more of the details of project management is invited to peruse D. I. Cleland and W. R. King, *Systems Analysis and Project Management,* 2d ed., McGraw-Hill, New York, 1975.

14   engineering and scientific organizations. These quantitative and qualitative aspects of managing technical and scientific organizations are typically integrated within a *systems approach*. The program of instruction offers both theoretical and applied subjects with sufficient flexibility to appeal to the student who has had experience in working in the real world.

Based on the experience at the University of Pittsburgh, a profile can be developed of the typical engineering management student. Such a student is about 30 years old, with a B.S. in engineering and an average experience of approximately 8 years. He works in an organization that is based on science and technology, either as a technical specialist or at the entry level of technical management. If he holds a nonsupervisory position, his objective is to prepare himself for management responsibilities through the engineering management program. If he is already a technical manager, he wants to broaden his skills for his present position.

In general, he looks for well-planned, structured guidance in the courses he takes. He sees the value of blending the management concepts and management science methodologies in his education and emphasizes the need for relevance in the courses. The engineering management student typically describes himself as a results-oriented, analytical person. He is a successful engineer and sees himself getting immersed increasingly in his field of specialization while management opportunities start developing around him. He feels that he has already proved himself as a successful technologist and there is not too much more to look forward to if he continues as an engineering specialist. He feels he can have upward mobility through the engineering management route, but he knows that he is not qualified to tackle the management problems because he does not have the skills, knowledge, and attitudes necessary for engineering management responsibilities. He has chosen the engineering management program to prepare himself for the challenges of the role that the engineering manager plays by combining management and technical expertise. He feels that such a program is directly addressed to the needs of a technical-scientific manager.

THE GROWTH OF ENGINEERING
MANAGEMENT PROGRAMS
Engineering management programs have grown in popularity significantly in the last several years. In 1980 there were more than seventy engineering management programs at graduate level around the country, and the number was still growing rapidly, as shown in Fig. 1.3. Professional societies have also recognized this trend in engineering management programs and have responded accordingly. The American Society for Engineering Education (ASEE) now has an engineering management division addressed specifically to the educational aspects of engineering management. The Institute of Manage-

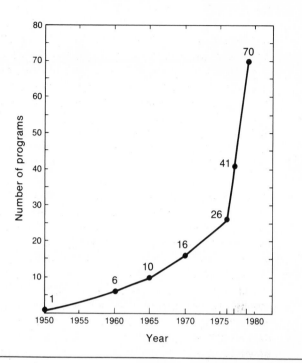

1.3

Growth of graduate-level engineering management programs. (*From Dundar F. Kocaoglu, "Master's Degree Programs in Engineering Management,"* Engineering Education, *ASEE, January 1980.*)

ment Sciences (TIMS) has a division called the College on Engineering Management, dealing with decision-making methodologies for both technical and management decisions. The American Society of Engineering Management (ASEM) is a professional society that brings together the practitioners and academicians to share their views on engineering management.

If the increasing popularity of engineering management programs in the United States colleges and universities is any indication, then engineering management is indeed a profession whose time has come.

ENGINEERING MANAGEMENT PROCESS

An engineering manager who aspires to improve his management performance must develop knowledge, skills, and attitudes in the context of a management philosophy. Such a philosophy must be built on a framework of management theory. We suggest as a starting point the acquisition of knowledge of the major activities that an engineering manager engages in as he manages an engineering organization.

16   These management activities, or as they are more commonly called, functions, were first put forth in their modern sense by Henri Fayol. Fayol described these activities in terms of planning, organization, command, coordination, and control. His definition of managerial activities is itself a primer in management theory. According to Fayol:

> To manage is to forecast and plan, to organize, to command, to coordinate and to control. To foresee and provide means examining the future and drawing up the plan of action. To organize means building up the dual structure, material and human, of the undertaking. To command means maintaining activity among the personnel. To coordinate means binding together, unifying and harmonizing all activity and effort. To control means seeing that everything occurs in conformity with established rule and expressed command.[15]

If we briefly examine management functions in the circumstances of an engineering environment, we can appreciate the particular challenges facing the engineering manager. We have changed these management functions somewhat from the concepts put forth by Fayol.

PLANNING
Planning is the process of examining the potential of future environments, establishing goals and objectives which accommodate the future, and designing a strategy outlining how goals and objectives will be attained. Planning has a time dimension: short-term, usually the forthcoming year; and long-term, those periods beyond a 1-year time. An engineering manager has to be concerned with both long and short terms when planning for his technical discipline. Planning in the short term is necessary to provide engineering design support to emerging projects and to ongoing products that are being manufactured. In addition to this, an engineering base technology has to be maintained and a discipline provided in the state of the art. The pace of technology in modern industrial organizations requires that engineering contribute to the strategic direction of the organization. Thus engineering managers need to be concerned with developing long-range or strategic plans for their discipline.

Adequate planning of engineering effort is one of the most difficult of the functions that face the engineering manager. Perhaps this is why this function is often neglected.

If the chief executives have not endorsed an active strategic planning program within the corporation, the engineering manager's job becomes more difficult. Even with financial and moral support

---

[15]Fayol, op. cit., pp. 5–6.

from key corporate executives, long-term engineering problems can **17**
seem almost insoluble. New technologies and complex, unique
devices that can threaten the technical competitiveness of a product
are appearing almost routinely.

Even the planning for support of ongoing products is
complex in today's industrial enterprise. The simple task of predicting engineering design hours is a case in point. Past experience is
hardly adequate. Estimating personnel, schedule, and product performance is becoming more difficult for the engineering manager. Today, it is common to find professional planners on the staff of
section-level engineering managers.

Planning involves answering the question "What are we
aiming for and why?" It is an activity where problems will become the
rule for those who must make decisions on how engineering
resources will be used in the future.

Technical goals and objectives are established by a careful
examination of the technological capabilities of the engineering organization. They specify the direction in which the organization will
move. The value structures of the managers, the needs for engineering services and the opportunities, and the outside environment play
important roles in the development of objectives. Once the objectives
are determined, the strategies are developed to achieve them within
the limitations imposed upon the engineering organization by the external conditions. Thus, the strategies have to be formulated according to the best estimates available for the future. These estimates can
be obtained through technological forecasting. The purpose of technological forecasting is to determine where it is possible to go from
here and what the dangers and opportunities are. As a result of
technological forecasting, the organization may revise its objectives
and develop a more refined concept of where it intends to go.

Several methods can be used for the purpose of technological forecasting. Time-series analysis is one of the methods. It
involves statistical analysis of trends in technologies observed in the
past. Other methods take future discontinuities into account by
using expert opinion in determining possible breakthroughs, potential directional changes in needs and requirements, and technological capabilities required for the developing needs. Such
methods as morphological analysis, cross-impact analysis, and Delphi method are useful for probing into the future to identify new
technology prospects and to estimate their timing. Another approach that has proved to be successful in technological forecasting
is the development of scenarios[16] in systematic brainstorming sessions. The scenarios describe possible future environments by ask-

---

[16]The concept of scenarios has been developed by Herman Kahn in *The Year 2000*, Macmillan, New
York, 1967.

18 ing "what if" questions about the various political sequences of events.

When developed properly, technological forecasts present a blueprint of the future with respect to technical requirements and their proper timing.

Once the ideas about the future are developed and the technical objectives are refined, the engineering manager formulates strategies to accomplish those objectives in the possible futures that have been predicted. Each strategy represents a different resource profile. Depending upon the resource availability and the estimated outcome of each strategy, he then selects the best alternative to move his organization from where it is to where it should be. Management science techniques described in Chaps. 7 through 11 can be used in selecting the best alternative and allocating the available resources in the best fashion.

ORGANIZING

The management function of organizing is concerned with the question "What's involved and why?" This function deals with how people and nonhuman resources are acquired and utilized in the engineering activity. The common denominator of the human element in an engineering organization is the pattern of authority and responsibility that exists—both formal and informal. In the formal sense, authority and responsibility are dependent on how the people are aligned in some organizational structure. The organization chart usually used to portray this alignment is often referred to as a *model* of the organization. The engineer-reader probably thinks of the term "model" in a different sense—perhaps in a mathematical sense. A model, however, is some degree of representation of the real world. An organization chart is a representation of the real world of how human and nonhuman resources are aligned to do something.

Engineers and those who manage engineers often view the subject of organization with distaste. This is exemplified by the comments one often hears along the lines: "We have a dynamic organization—every few weeks we 'reorganize' things!" There is much truth in such a statement; in an engineering department people are shifted around to support workforce requirements of the engineering projects. Skill-level requirements change during the life cycle of these projects; the ability to change the assignment of people from one project to another with minimum disruption of efficiency is always needed. Little wonder that engineers view organization with misgiving.

An organization should be viewed as a dynamic, goal-seeking, purposeful system. Organizations are not static—i.e., without motion. They change as the demand for the organization's services changes. The changing nature of modern engineering organizations conflicts with the needs of people who seek security and stability in their work environment; extreme care is required to avoid conflict between engineering managers and the people who provide engineering support. Successful leadership—personal leadership—is required of those managers who aspire to effective management of engineering activities.

MOTIVATING

A leader is someone who leads people. While this statement is obvious and oversimplified, it is basically true. People are led by ideas and actions; usually there is some system of rewards and punishments that serves to motivate people—or demotivate them. Engineers and scientists are by education and training inclined to act rationally and objectively about their engineering work. In their interpersonal relations and in their motivational factors, however, they are probably no more rational and objective than anyone else.

Motivating is that portion of a manager's job that deals directly with the human element—the face-to-face personal leadership role the manager must exercise. People are motivated by a wide variety of factors and forces. "Coercion," "conniving," "compensation," and "human relations" are some of the terms that come to mind when the subject of motivation comes up. Motivation will be dealt with in much greater detail later in the book. For the present, we need to stress the fact that motivation is one of the key management functions about which the engineering manager must be concerned. The reader should be able to recollect situations in his career where an engineering manager's personal style of leadership was instrumental in bringing out the best in people. Conversely, the reader can probably also recall managers whose style of leadership was detrimental to motivating people.

The matter of professional obsolescence poses a particular challenge to the engineering manager. Not only do many engineers gravitate to management positions as they grow older, but they are subject to obsolescence on the job. The job performance of engineers tends to peak in their middle to late thirties. According to Dalton and Thompson, "the years of high performance are starting and ending sooner."[17] The aging of technically educated and trained people in the workforce today is unique. Paraphrasing Dalton and Thompson, "no generation of managers has had to deal with this phenomenon before."[18]

Motivation deals with the question, "What encourages people to do their best work?" The challenge to the engineering man-

[17]Reprinted by permission of the *Harvard Business Review*. Quoted from "Accelerating Obsolescence of Older Engineers," by Gene W. Dalton and Paul H. Thompson (September-October 1971), p. 57. Copyright © 1971 by the President and fellows of Harvard College; all rights reserved.
[18]Ibid., p. 58.

20   agers is clear. They will have to find ways to motivate technical people.

DIRECTING

Directing is the management function that is concerned with giving guidance or instruction to those people with whom one shares the running of the organization. Directing is like the management function of motivating that was discussed in the previous section. Motivating might be thought of as a *humanistic* approach to giving direction. Both directing and motivating have strong face-to-face leadership overtones. Both are involved in the day-to-day activities associated with influencing, guiding, and supervising in accomplishing the engineering objectives established through planning. Carrying out the directing function can be as straightforward as the giving of a specific order to an individual. Giving an unobtrusive suggestion to a person who is being counseled is another aspect of directing. As a distinct function of an engineering manager, directing is concerned with the making and execution of decisions—the matter of "Who decides what and when."

The directing function is carried out by an engineering manager who is authorized by organizational policies to make a decision, e.g., approve an engineering change for a development project. A project manager may not have the authority to approve an engineering change without coordinating the change with several other people in charge of such functions as contracting and cost estimating, as well as with the engineering manager. In this situation the project manager, although lacking the formal authority to make the engineering decision, would be responsible for seeing that the decision was made by the person so empowered.

The directing function provides guidance to the people with whom one works, either as a decision maker per se, or through an administrative mechanism to bring about a decision. Indecisiveness—the inability to make a decision—is a difficult situation in any organization. Some managers are indecisive; the people with whom such managers work can become frustrated waiting for a decision. A famous World War II Air Force general is reputed to have had a motto posted in his office which proclaimed: "A bad decision is better than no decision at all!"

The directing function highlights the need for the engineering manager to realize that he is the one who must assume responsibility to see that engineering decisions are made in a timely and relevant fashion.

CONTROLLING

Controlling is the key management function that sees "that everything occurs in conformity with established rule and expressed command." This definition of control is credited to Henri Fayol. We do not believe that anyone has said it any better. Controlling the activities of an engineering organization requires that objectives, goals, policies, procedures, feedback, information, and performance standards be in existence. The engineer will readily relate to the notion of control in a mechanical feedback control system. Control of engineering activities is not as precise; the fundamental similarity is the same, however.

Control is too often associated with the idea of the negative constraint of people—a restriction of the personal liberties of people in their workplace. This association is most unfortunate. Control addresses the question "Who judges results and by what standards?" We believe that professional people are anxious to do good work; they need to know how they are doing. Control looks to the development of performance standards for people and organizations and provides feedback so that progress toward objectives and goals can be effectively sustained.

An engineering project manager constantly evaluates the performance of his team members and measures the achievement level of the project for which he is responsible. These measurements have to be compared with standards to determine the need for corrective action in order to reach the cost, time, and technical performance objectives of the project. The standards are established in the initial stage of the project in form of budgets, technical targets, and time schedules. The budget is the cost profile for the engineering services. It is based on the best estimates of the costs available. Technical targets are the specific results expected to come out of the engineering efforts. In an engineering project for the design of an industrial plant, the technical targets could be the feasibility studies, the plant requirements, and the design of the structural, mechanical, and electric parts. These are broadly defined in the beginning of the project and refined as the project proceeds.

The time schedule shows the relationship among various engineering activities and the time estimates for each one. The tasks to be completed before another one can start are identified and listed in a systematic way. Graphical methods can be used for this purpose. The simplest method is the Gantt chart, shown in the simplified example in Fig. 1.4. A more sophisticated approach is the use of network techniques discussed in Chap. 11. The procedure for these techniques involves the identification of individual tasks, the precedence relationships among them, and the time estimates and the development of a network such as the one shown in Fig. 1.5. Since the duration of each task (structural design, etc.) is already estimated, the engineering project manager responsible for the project compares the actual time spent on each task with the estimated time. If the project is falling behind schedule, he either shifts his resources toward the more time-consuming tasks or revises the initial estimates.

Activity	Time schedule
	January February March April May...................
Feasibility studies	
Plant requirements	
Structural design	
Mechanical design	
Electrical design	

1.4

Gantt chart.

      The engineering project manager's responsibilities in controlling the team can be defined as:

1   Setting the time, cost, performance standards

2   Measuring the time, cost, and performance characteristics in the project

3   Comparing the measured results with the standards set in step 1

A simple project network.

1.5

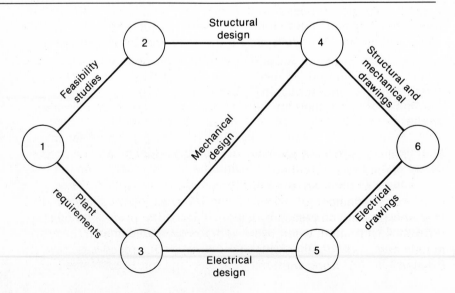

4   Taking corrective action to bring the actual per-   **23**
formance close to the projected performance

This is a continuous process. After step 4, the project manager
goes back to step 1 and follows the same steps.
          The management functions (planning, organizing, motivat-
ing, directing, and controlling) are interrelated and inseparable. For
example, to plan one must:

Organize the people and other resources that will be used
during the planning process.

Motivate the people to take time from their current work to
plan the utilization of resources in the future.

Facilitate the making of planning decisions and provide
leadership for those people who do the planning.

Develop performance standards in the planning process
which can be used to compare actual and planned results
during the control process.

When one carries out the organizing function one must:

Develop an organization plan for how authority, responsibility, and accountability are to be allocated among the people in the organization.

Establish resources, such as salary-incentive programs, off-
duty educational programs, opportunities for attending at
professional meetings, and so forth, to help motivate professionals in the organization.

Set up a decision process to determine who has what authority to make which decisions. This includes providing the
leadership to see that decisions are made on a timely basis.

Institute an information system for providing intelligence on
where engineering projects stand in relation to their objectives, budgets, and schedules. Such information provides
the basis for controlling the future application of resources
to these projects.

So it goes on, with each engineering management function related in
some manner to each other.
          What does an engineering manager really do? The engi-
neering management job consists of a few things that can be simple
in statement but complex in execution: (1) Create an intellectual and
physical environment whereby the members of the engineering community can work together with economic, social, and psychological
satisfaction; (2) work with the members in establishing and ac-

24   complishing organizational and personal objectives, goals, and strategies; (3) periodically review and appraise personal and organizational performance; (4) create a workable set of policies and procedures that will help the members of the organization work effectively toward objectives and goals; (5) keep senior executives informed of how things are going in the technical community; and (6) formulate technical strategies.

To accomplish all this, the engineering manager can draw for assistance from a profusion of management knowledge that has been developed in other organizational contexts. This book will do this for the engineering manager through providing a primer on the application of management thought and practice to the management of resources in the technical community.

SUMMARY

The pacing factor in the application of technology is management. Professional engineers who have a solid foundation in the application of basic engineering sciences require a sound foundation in management knowledge, skills, and attitudes to take on the responsibility of managing engineering organizations.

The engineering manager's role is the link between management and technical expertise. The matching of technical resources with the needs of engineering technical professionals is a major challenge to engineering managers. The underlying concepts of engineering and management require change in the way the engineering manager views the technical environment. Passing from the role of engineer to engineering manager requires the individual to develop new knowledge, skills, and attitudes. Practicing engineering managers will find it useful to develop for themselves a personal philosophy of management that inherently contains an approach to carrying out the management functions of planning, organizing, motivating, directing, and controlling.

DISCUSSION QUESTIONS

1 Why is successful performance in an engineering specialty a necessary condition for successful engineering management? Why is it not a sufficient condition?

2 Engineers have traditionally been associated with technical decisions. Develop the evolution of engineers' involvement with management roles.

3 Identify some of the areas that belong to the "intersection" or overlap of engineering and management.

4 As implied in Question 1, what additional skills must an engineer develop in order to effectively perform managerial functions?

5 Try to relate the five functions of the management process to a real life situation, in the technical or nontechnical world.

6 Examine the six responsibilities of an engineering manager identified in this chapter. Comment and elaborate on each to make sure you fully grasp its implications.

7 What attitudinal changes does an engineering manager have to go through to accept his responsibilities as a technical strategist?

8 How has the role of the engineer in contemporary organizations changed in recent years?

9 What is the difference between engineering knowledge and engineering management knowledge?

10 How can knowledge of the matrix organizational form help the practicing engineering manager?

11 Is the project-center concept being used by General Motors likely to spread to other organizations?

12 What are some of the factors that cause obsolescence in engineers? In engineering managers?

KEY QUESTIONS FOR THE ENGINEERING MANAGER

1 What management philosophies have been developed in my organization for managing technical activities?

2 What other disciplines besides engineering do I integrate in my role as an engineering manager?

3 Does the role of engineering manager in my organization require a solid engineering background? Do the persons who occupy such positions have such backgrounds?

4 What key decisions do I make as an engineering manager? As an engineer?

5 Does the uncertainty of my engineering manager's role make me uncomfortable? Can I live with such uncertainty?

6 Have I identified the engineers in this organization who have the potential to become engineering managers? Have I helped these individuals to start a self-improvement program to develop this management potential?

7 What assistance have I given to the engineers in this organization to develop an awareness of the broad social, economic, and regulatory world in which they practice engineering?

26

8   What do I see as some of my personal needs for engineering management education? What is my strategy for satisfying these needs?

9   Can the concept of the project center as used by General Motors be used in my engineering organization?

10   What is my proficiency level in the matter of engineering management knowledge, skills, and attitudes?

11   Am I perceived by my people as a leader?

12   What strategies have I designed to reduce engineer obsolescence in this organization?

BIBLIOGRAPHY

Ayers, Robert U.: *Technological Forecasting and Long-Range Planning,* McGraw-Hill, New York, 1979.

Blanchard, Benjamin S.: *Engineering Organization and Management,* Prentice-Hall, Englewood Cliffs, N.J., 1976.

Burck, Charles G.: "How G.M. Turned Itself Around," *Fortune,* Jan. 16, 1978, pp. 87–100.

Cleland, David I., and William R. King: *"Systems Analysis and Project Management,* 2d ed., McGraw-Hill, New York, 1975.

David, Edward E., Jr.: "The Schizophrenia of Engineering Education," *Research Management,* vol. 9, no. 6, 1966.

De Bond, Edward (ed.): *Eureka—An Illustrated History of Inventions,* Holt, New York, 1974.

The Engineer as a Manager, Engineering Manpower Bulletin 25, September 1973.

Fayol, Henri: *General and Industrial Management,* Pitman, London, 1949.

Forrester, J.W.: *Engineering Education and Engineering Practice in the Year 2000,* National Academy of Engineering, University of Michigan, Ann Arbor, Sept. 21, 1967.

George, Claude S., Jr: *The History of Management Thought,* Prentice-Hall, Englewood Cliffs, N.J., 1968.

Kahn, Herman: *The Year 2000,* Macmillan, New York, 1967.

Karger, Delmar W., and Robert C. Murdick: "Engineering Myopia," *IEEE Transactions on Engineering Management,* February 1977, p. 24.

Kocaoglu, Dundar F; "Masters Degree Programs in Engineering Management," *Engineering Education,* ASEE, January, 1980   **27**

Koontz, Harold, and Cyril O'Donnell: *Principles of Management,* 3d ed., McGraw-Hill, New York, 1964.

Miller, D. W., and M. K. Starr: *Executive Decisions and Operations Research,* Prentice-Hall, Englewood Cliffs, N.J. 1960.

Roman, Daniel O.: *Research and Development Management: The Economics and Administration of Technology,* Appleton-Century-Crofts, Englewood Cliffs, N.J., 1968.

Shepherd, Clovis R.: "Orientation of Scientists and Engineers," *Pacific Sociological Review,* Fall 1961.

Wolek, Francis W.: "Engineering Roles in Development Projects," *IEEE Transactions on Engineering Management,* vol. EM-19, no. 2, May 1972.

PART TWO

ENGINEERING OR ORGANIZATION

PART TWO ENGI
TWO ENGINEERI
ENGINEERING OR
ORGANIZATION
PART TWO ENGI
TWO ENGINEERI

PROJECT MANAGEMENT 2

This chapter will explore the organizational concepts of project management and their relationship to the engineering community and the larger organizational context in which projects are managed.[1] Project management is so interwoven into the engineering management process that a separate chapter on this topic is required to highlight its importance to engineering management. Much of the material that is presented elsewhere in this book fits in some way into the conceptual framework of project management. Therefore, it is presented as a management technique complementary to the engineering management process.

In the last 20 years, specialized management techniques have become more sophisticated in order to manage ad hoc activities in organizations. These ad hoc activities, typically defined as *projects*, consist of a combination of human and nonhuman resources pulled together to achieve a specified purpose in the organization. Today *project management* is practiced in a wide variety of industrial environments; it is also used in educational, military, ecclesiastical, and governmental organizations. While the practice of project management doubtlessly originated in antiquity, the development of a conceptual framework of project management received much of its momentum from the work of the aerospace and construction industries. In its earliest application, project management took the form of an organizational arrangement consisting of integral teams of people working on a common organizational purpose. Such teams provided a focal point to pull together the organizational resources to be applied to a particular project. In 1964, John Mee described this arrangement as a "matrix" type of organization which entailed an organizational system designed as a "web of relationships rather than a line and staff relationship of work performance."[2]

To complement the organizational aspects of project management, specialized techniques and methodologies have emerged to facilitate the scheduling and budgetary activities of a project. Program evaluation review technique (PERT), progress performance

---

[1]Portions of this chapter have been adapted from D. I. Cleland and W. R. King, *Systems Analysis and Project Management*, 2d ed., McGraw-Hill, New York, 1976, and "Defining a Project Management System," *Project Management Quarterly*, December 1977.
[2]John F. Mee, "Matrix Organization," *Business Horizons*, Summer 1964.

32

reporting, project planning with precedence lead-lag factors, network analysis, and milestone charting are a few of the techniques and methodologies that have been developed to facilitate the planning and control of projects; many of these techniques have been described in terms of a project management system.

Within the engineering community, project management evolved out of the specialized activity of *project engineering*. Individuals who occupy a position as a project engineer work with a group of other engineers who support the project. In such situations questions of authority and responsibility are often raised when the project engineer becomes dependent on other people over whom he has limited control. Project management is inherent in engineering management. Those persons who aspire to engineering management positions should develop a working knowledge of project management.

Project management is usually viewed as a technique for aligning organizational resources around the specific task to be done. For example, an engineering department that is designing a hydroelectric turbine generator system for an electric utility would organize a team of engineers to carry out the systems engineering work required for the generator system. A project manager would be appointed who would organize and develop a team of technologists to work on the project. By working with resource managers in the organization, an interdisciplinary team would begin the design of the system. A matrix type of organization would come into existence.

THE MATRIX ORGANIZATION

The matrix organization is the offspring of the marriage of the traditional functional organization and the interdisciplinary team formed to manage the project. Before describing the matrix organization in more detail, we should examine the nature of the most common type of organization structure in use in industry.

THE FUNCTIONAL TYPE OF ORGANIZATION
The most common form of organizational structure in American industry is the functional type, where personnel and resources having a common function are integrated as an entity in the organization. A functional organizational form is illustrated in Fig. 2.1. The general manager is responsible for establishing organizational objectives such as sales volume, profitability, growth, etc. These are translated into specific goals in each of the functional departments. The marketing department translates the goals into terms of a product most likely to achieve overall organizational objectives. The engineering department develops the hardware and software to support the organizational objectives; manufacturing is responsible for the actual production of the products that have been

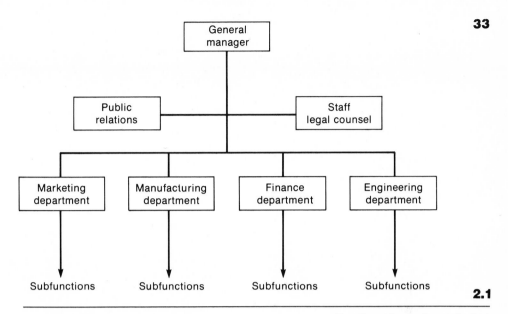

2.1

Functional form of organization.

engineered and for the volume prescribed by the marketing department. The finance department translates organizational goals into financial measurements through establishing accounting procedures
and standards and measuring the profitability of the organization.

The functional form of organization is normally used where
standard products are found. In such an organization, contact
with the customer is limited to the marketing and sales organizations. Coordination between the functional departments is typically
done by the general manager or the department heads themselves, as
needed.

A department manager takes the lead to coordinate activities when that department's work is critical to the solution of the current organizational problem. Such coordination is usually done on an
informal basis.

The functional organization form is oriented toward established products and is not easily adapted to change. Consider the
challenge faced by a manager who wishes to design, develop, and introduce a complex new product, particularly one that requires ad-
vancing the state of the art, is sophisticated, and is of key importance
to the company. Such a product would require a substantial commitment of personnel and other resources in order to assure its successful implementation. An organizational form has to be established that
is built around the product to be designed and developed. Such an
undertaking requires an approach that can bring a focus to the
product design and development activity. The general manager

34   must determine how to build the organization for the design, development, and introduction of the product in such a way that organizational resources can be integrated effectively without detracting from the ongoing production of established products.

One solution that has been found successful in many diverse organizational contexts has been to establish a project, appoint a project manager, and build an organization centered around the design and development of the product. The resulting organizational form is a departure from the traditional functional organization. Such an organization is a matrix type.

NATURE OF THE MATRIX ORGANIZATION
The management of an engineering project requires an organizational arrangement that consists of a project team superimposed on the existing vertical structure of the organization. This can be illustrated by using the design of an industrial plant as a case in point, as shown in Fig. 2.2.

Matrix organization for engineering design of an industrial plant.

2.2

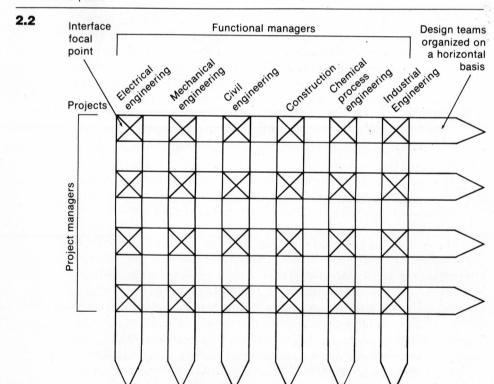

In this figure, the engineering personnel assigned to the design teams come from the different disciplines supporting the project. The interface of the vertical structure and the design team provides a focal point for pulling together the specialized engineering effort to support a project. The figure shows how these engineering disciplines and the engineering project teams interface in the matrix organization. "Interface" is used in the sense of coming together at common boundaries of interest in the management of the project activities.

The management of the engineering project is handled through designating an individual as a *project manager* to provide an accountable focal point for pulling all needed organizational specialized interests into focus on the project. The focus of the specialized support on the project comes together in terms of an *authority-responsibility matrix*—hence the term "matrix organization."

The organizational model reflected in Fig. 2.2 is used to represent this authority-responsibility matrix. In this figure the specialized functional groups—the engineering disciplines—interface with the projects as organizational resources are applied to these projects.

Each project is headed by a project manager, who is a manager in the truest sense of the word. He is responsible for the planning, organizing, motivating, directing, and controlling of the organizational resources to be applied on a project. In such a role the project manager works with the specialized interest groups—the functional managers. A horizontal organization comes into being in such an organizational setting; the basis is thus formed for the matrix organization.

One characteristic of the matrix organization is that each person has two bosses—a project manager and a functional manager. On the surface this appears to be a violation of the management principle of *unity of command*, i.e., an individual should receive orders from only one superior. Consistent with the principle of unity of command is another well-known management principle: *parity of authority and responsibility.* This principle holds that authority and responsibility must be equal for a manager. When viewed in terms of the matrix organization, these principles have a particular significance. In such an organization the functional manager shares his authority with the project manager. The project manager is responsible for the completion of the project on time, within budget, and satisfying the objectives of the project. The functional manager has the responsibility of providing the specialized resources to support the project; the functional manager also tends to retain administration over the people who are assigned to him but who are working on the project.

36

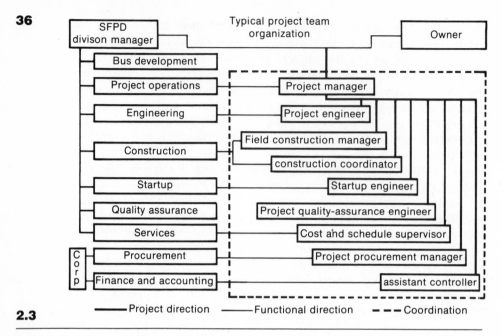

2.3

Typical project team organization. (*From Frederick A. Hollenbach, "The Organization and Controls of Project Management," in* Realities in Project Management, *Proceedings of the Ninth Annual International Seminar/Symposium, Project Management Institute, Chicago Oct. 22–26, 1977, p. 15.*)

EXAMPLES OF THE MATRIX ORGANIZATION

In the Bechtel Power Corporation a matrix organization form has emerged. This emergence has come about in response to the need to designate an individual as a leader of a key group of project people on complex projects who work on a particular construction effort. Such a key group of people includes a project engineer, field construction manager, construction coordinator, procurement manager, cost-schedule supervisor, quality assurance engineer, controller, and startup engineer. Figure 2.3 shows a typical project team organization in the Bechtel Power Corporation.

In the organizational model shown in Fig. 2.3 the project manager provides direction, integration, and synergy of the project team: "He represents the overall project and this is the focal point for the client and the project organization. He is an extension of the division manager for a project. He is the direct communications link between the client and the corporation, and establishes the appropriate channels of communication between the client's organization and both engineering and construction."[3]

[3]Frederick A. Hollenbach, "The Organization and Controls of Project Management," in *Realities in Project Management.* Proceedings of the Ninth Annual International Seminar/Symposium, Project Management Institute, Chicago, Oct. 22–26, 1977, p. 15.

In such an organizational setting the project manager has **37** three major goals relative to the project: to keep the project within budget, to keep it on schedule, and to ensure that it is acceptable to the client and to the corporation.

In Kaiser Engineers, the project manager's role is described so that he has responsibility for all facets of a major project. This responsibility includes the general facets of a turnkey project—engineering, design, procurement, and construction—as well as economics, market analysis, financing, and other such services. It is particularly significant to note the broad organizational context in which this company views its project managers:

The present-day project manager performs his duties in an ever more complex environment. He must be a true professional. He must address many new dimensions in today's projects, such as the public awareness and governmental relations factors. We are all familiar with the delays which are caused today by resistance by special interest groups, involvement in the courts, the multiplicity of new regulations and guidelines, and the complications of international politics and government-to-government relationships. Today, these factors take on the same degree of importance to the success of the program as many of the fundamental technical skills.[4]

Other uses of matrix organization that are not necessarily project-related are found in many contemporary organizations. Companies have experimented with variations of the matrix organization in dealing with different types of complex business situations. The typical diversified international corporation uses a matrix organization for the integration of product, function, and geographical activities to maintain a balance of management in the corporation's international operations.

Some companies use a variation of the matrix organization in which primary operational responsibilities are divided between "business managers," who are usually profit-center managers, and "resource managers," who provide the logistic support needed to carry on the business. Under this arrangement peer managers share strategic and operational decisions; reporting relationships for subordinates includes dual reporting to both the business manager and the resource manager.

It is not our intent to pursue the matrix-organization concept beyond its use in project-oriented activities.[5]

---

[4]V.E. Cole et al., "Managing the Project," in *Realities in Project Management,* Proceedings of the Ninth Annual International Seminar/Symposium, Project Management Institute, Chicago, Oct. 22–26, 1977, p. 57.

[5]For a good description of the matrix organization in a generic context, see Allen R. Janger, *Matrix Organization of Complex Businesses,* The Conference Board, New York, 1979.

38   AUTHORITY AND RESPONSIBILITY IN THE MATRIX ORGANIZATION
The confluence of the stream of authority and responsibility relationships within the matrix organization raises such issues as "who
works for whom?" More specifically, the workings of the matrix organization pose questions such as:

1   *How* will the separate engineering disciplines be
integrated into a systems approach?

2   *Who* will be responsible for the assignment of professionals on a particular engineering project?

3   *Where* will the work be done?

4   *How much* funds are available to be expended on
the project?

5   *What* is to be done in the project by way of attaining project objectives?

6   *How well* has the work been accomplished on the
parts of the project? On the entire project?

7   *When* is the work to be accomplished?

8   *Why* is the work being accomplished?

Obtaining the answers to these kinds of questions in an
authority-responsibility context has to be done from the perspective
of *both* the project manager and the functional manager. These two
managers play complementary roles as they manage resources supporting the projects. Their complementary roles are depicted in Table
2.1.

Complementary roles of project manager and
functional manager in the matrix organization

2.1	PROJECT MANAGER	FUNCTIONAL MANAGER
	1 *What* is to be done in the project by way of attaining project objectives?	1 *How* will the engineering disciplines be integrated into the project?
	2 *How well* has the work been accomplished on the project?	2 *Who* will be responsible for assignment of professionals on the project?
	3 *When* is the work to be accomplished?	3 *Where* will the work be done?
	4 *Why* is the work being accomplished?	4 *How much* funds are available for work assigned?
	5 *How much* funds are available to support the project?	5 *How well* has the functional work been accomplished?

Authority and responsibility
relationships

WHAT, WHEN, HOW MUCH	WHO, HOW		**2.2**
PROJECT MANAGER	DESIGN MANAGER	DESIGN ENGINEER	

PROJECT MANAGER	DESIGN MANAGER	DESIGN ENGINEER

Lead scoping activity for proposed projects
  Inside
  Outside
Represent project objectives to contributing functions
  Cost
  Schedule
  Performance
Lead project teams
  Project Definition
  Project Schedules
  Guidance & Direction
Ensure allocation of supporting resources
Integrate and communicate project information
  Project Planning Council
  Product Board
  Outside
Track and assess progress against plan
Resolve conflicts
Communicate
	Participate in development of project plans
	Participate in determination of project resource needs
	Define design workload
	Assign design personnel, consistent with project needs
	Maintain technical excellence of resources
	Recruit, train, and manage people in the organization
	Assess quality of design activities
	Provide technical guidance and direction
	Communicate
		Prepare detailed plans and schedules for design tasks consistent with overall project plan, including initial definition of support requirements
		Execute design tasks
		Represent design engineering on project teams, in project reviews, for cost estimates, etc.
		Communicate

In Table 2.2 the relative roles of a project manager, a design manager, and a design engineer are shown. This example is taken from the engineering division of a large home-appliance manufacturer. The development of Table 2.2 within this company required many meetings of the people affected by the changing patterns of authority and responsibility portrayed in the table. Note the emphasis placed on *communication* as a responsibility of all participants in this organization.

Within an engineering organization, a project engineer works with a team of people from the engineering subspecialties involved. The design and development of a product is basically a technical process in which engineering has the leadership role, a role that is illustrated in Fig. 2.4, which shows an aerospace program manager's organizational relationship with project engineers in the design and development process. In this context, project engineering directs

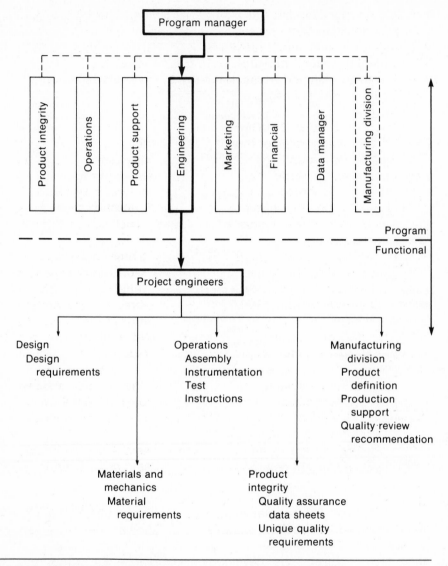

2.4

Role of project entineer. (*Design/Development Process Guide, Subsystem Guide 3-3, Pratt & Whitney Aircraft Group, Government Products Division, United Technologies Corporation, p. 9.*)

all technical aspects of the design and development process and is the program management arm of the engineering department.

There are few things that the project managers can do alone. They must rely on the support of many engineering specialists to bring their knowledge and skills to the project work. The specialists will usually have an administrative reporting relationship to their

immediate supervisor, such as a section manager. The section man- **41**
ager provides administrative support to the engineers, submits merit
evaluations on them, and is responsible for their career development.
The project manager provides an organizational focal point for
overall project planning, direction, and control. This focal point is
based on the work packages of the project.

THE WORK-PACKAGE CONCEPT

The complementary roles of the project manager and the functional
manager center around the work to be accomplished—the work
package. A *work package* is an integral subelement of the project; a
subsystem; a subproduct. Figure 2.5 shows the integral work
packages with the component parts of a hydroelectric plant. Each
part, e.g., intakes, power house, dam, control panel, can be thought
of as a work package.

A work package grows out of a work-breakdown analysis
that is performed on the project. When the work-breakdown analysis
is completed and the work packages are identified, a work-breakdown structure comes into existence. A work-breakdown structure
can be represented by a pyramid similar to that used for describing
the traditional organizational structure. Stated another way, the
work-breakdown structure represents the breakdown of the project
objectives.

The work package that results from the work-breakdown
analysis provides the basis for managing the project. Figure 2.6 illustrates that the basis of the matrix organization is the confluence of
the project and the functional efforts.

In the context of an engineering project, the work-breakdown analysis and the resulting work packages provide a model of

Hydroelectric plant showing integral work
packages.

2.5

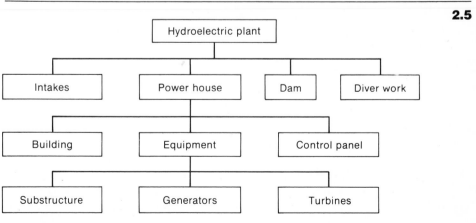

42

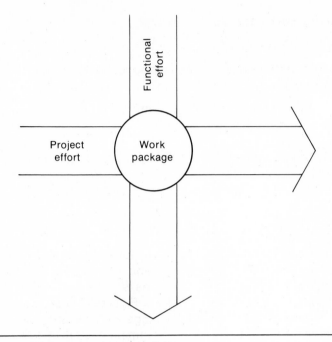

2.6

Confluence of the project and functional effort
around project work package.

the products (hardware, software, services, and other outputs) that
completely define the project. Such a model enables project engineers, project managers, functional managers, and general managers to think of the totality of all products and services comprised in
the project. This model then can be used as the focus around which
the project is managed—reporting progress and status of engineering efforts, resource allocations, cost estimates, procurement actions, etc. More particularly, the development of a work-breakdown
structure with accompanying work packages is necessary to accomplish the following actions in the management of a project:

1   Summarizing all products and services comprised
in the project, including support and other tasks

2   Displaying the interrelationships of the work
packages to each other, to the total project, and to other engineering activities in the organization

3   Establishing the authority-responsibility matrix
organization

4   Estimating project cost

5   Performing risk analysis

6   Scheduling work packages   **43**

7   Developing information for managing the project

8   Providing a basis for controlling the application of resources on the project

9   Providing a reference point for getting people committed to support the project

Each of these actions requires explanation; this explanation is given elsewhere in the text. This chapter is addressed to building the matrix organization and motivating people to support the project.

THE PERSONAL-COMMITMENT FACTOR

The project manager carries out the normal functions of a manager: planning, organizing, motivating, directing, and controlling. The focus around which he carries out these functions is illustrated in Fig. 2.7 in relationship to the project work packages. Figure 2.7 is shown to highlight the importance of the work package with the project management process and the personal-commitment factor. In the matrix organization the management functions come to focus on the project work packages.

The project management process.

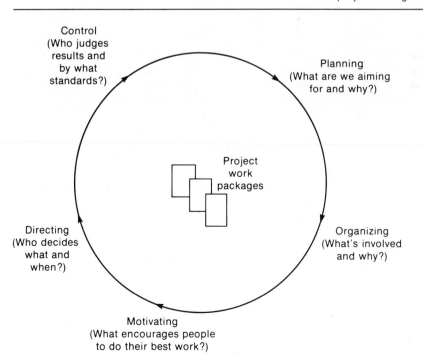

2.7

Control
(Who judges
results and
by what
standards?)

Planning
(What are we aiming
for and why?)

Project
work
packages

Directing
(Who decides
what and
when?)

Organizing
(What's involved
and why?)

Motivating
(What encourages people
to do their best work?)

44   How well people are committed to support the project can make the difference between success and failure of the project. The importance of having individuals committed to support a work package of the project cannot be overemphasized. One engineering manager described the importance of this commitment in the following manner:

> A work package is a commitment made and agreed to by two individuals "eyeball to eyeball" so to speak. It is a personal thing and has nothing to do with charts, graphs, or other inanimate pieces of paper. You don't have to make any commitment you don't agree to—normally. In very unusual circumstances your boss, and only your boss, will make a commitment for you, but this is the exception, not the rule. Only the parties to the commitment may agree to a change or cancellation of the commitment.

Of course, this commitment presupposes that the necessary planning has been done, i.e., the tasks have been adequately defined so that the personal commitment can be made. The same engineering manager noted:

> If you can't commit to what the other person is asking you for, then the two of you had better redefine the task in such a manner that a commitment can be made. "Time and Material" contracts cannot be successful without this agreement. If you don't have a clear understanding of what each party is committed to, the inevitable day will arrive when the "misunderstanding" surfaces and someone's reputation and integrity suffer. These "misunderstandings" are more typically "no understandings," i.e., the problems or definitions of tasks were difficult in the beginning so they were avoided or set aside like they didn't exist. A real commitment between the two parties eliminates the problem and allows a real project management to flourish.

The work-breakdown structure along with the work packages provides a common denominator or framework for monitoring performance, cost, and schedule throughout the life of the engineering project. As an allied benefit the work-breakdown structure is a convenient method of managing by exception, that is, it provides a focus for highlighting critical items or areas of the project that must receive prompt attention by the managers concerned.

So much of getting people committed to a project depends on their understanding of what is expected of them and how they are to work with other people. When engineers are first exposed to management they find that the management process is not nearly as orderly as their engineering discipline. In fact they usually find that management is "messy," particularly in the people relationships; yet so much of the success of an engineering manager depends on how effectively he is able to work with other people—peers, associates, superiors, subordinates, etc. As one recently appointed engineering manager said to one of the authors:

It came as quite a shock to me to realize that my success as a manager was so dependent on what other people could and would do and how effective I was in working with them. It wasn't this way when I was an engineer; then my success depended on how effective I was as an individual working alone. Being a manager is so much different from being a technical person; one must not only be able to see the big picture but see how the parts and people relate and fit into the big picture.

My training as an engineer taught me to look closely at the technical details of the work. This was a form of myopia; my particular engineering discipline helped me to develop a form of tunnel vision. This presented no problem as long as I remained in engineering work.

I was an outstanding engineer—in fact so outstanding that I was promoted to be a manager of engineers. Now my future is in someone else's hands. Now if I don't produce by working through my staff I will be a failure. This is not a comfortable feeling.

One reason engineers find being a manager "messy" is because they usually do not take the time and apply effective techniques for determining the *specificity* of how people should relate to each other as they cooperate in some endeavor. Too often the engineering manager (and other managers as well) tries to understand these relationships by studying traditional organizational charts, job descriptions, policy manuals, and such kindred documentation. These means are inadequate for an understanding of complex organizational interfaces such as found in the scientific and engineering community.

CHARTING THE AUTHORITY-RESPONSIBILITY MATRIX

In the authors' experience the organizational aspects of project management—how people are expected to relate to each other—cause most of the difficulties encountered in adopting project management techniques. Organizing a matrix organization by simply changing the reporting lines on the organization chart is not enough. Further organizational definition is necessary, particularly in developing specific delineations of authority and responsibility.

46   If an engineer were asked to determine the specific interfaces of hardware systems, he would doubtlessly set out very detailed
specifications. These specifications would lay out reliability, quality,
and maintainability criteria. Also, there would be data bases established to provide the justification for setting up tolerances, energy
flow, mechanical intermeshing, and so forth. When the plan for the
integration of hardware was finally laid out, the specifics of the
hardware interfaces would be understood. Yet this same engineer,
when faced with an organizational interfacing situation, would probably not have access to any tools that would help him to better define
the organizational interfaces.

     There are organizational development tools that can help to
give insight into these organizational interfaces. A useful technique is
the linear responsibility chart.

LINEAR RESPONSIBILITY CHARTING

An alternative to the traditional organization chart is found in the
linear responsibility chart (LRC), which goes beyond a simple structural display of the formal alignment of the organization. An LRC portrays the specific relationship of an organizational position and a
task, or work package,[6] to be accomplished. Figure 2.8 shows the essence of an LRC: (1) an organizational position; (2) a work package,
in this case "conduct design review"; and (3) the symbol Δ, which indicates that the director of systems engineering has the primary responsibility for conducting the system design review. When several
work packages are involved, one of the key advantages of using an
LRC to establish organizational relationships becomes apparent, i.e.,
the people are brought into a dialog as to their specific authority and
responsibility to a work package and to other individuals in the organization. Figure 2.9 shows an LRC for an organization involved in
the design, manufacture, and delivery of an industrial system from
the time of receipt of a firm order until the system is accepted by the
customer. In this example, the organizational positions of manager of
programs, information systems, and strategic planning have been
added to the systems engineering position. The LRC depicts how
these four positions relate to the work packages and to each other.

     Project managers (sometimes called "program managers")
report to the manager of programs in this example. The chart also
helps to emphasize how the systems engineering manager has to
work with many other people in the organization.

     Perhaps the greatest value of the LRC is in the process of
getting the people who occupy key positions together to discuss how

---

[6]The reader should understand that a work package can be something other than a piece of hardware. In
Fig. 2.9 the management of the design, manufacture, and delivery of an industrial system is
broken down into work packages.

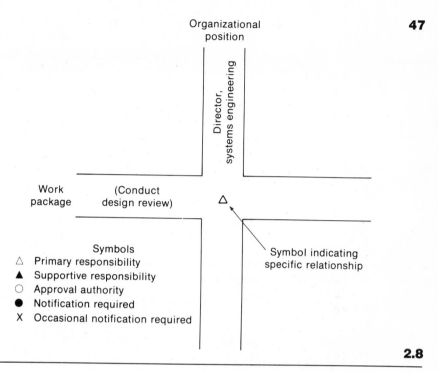

2.8

Model of a simple linear responsibility chart.

each position relates to other positions and to the work packages. In such meetings a dialog provides many opportunities to clarify working relationships, understand the other person's point of view, and in general, educate everyone on how things should be done in the organization.

An LRC can be viewed as an organizational plan; once the organization starts working on a project, the LRC serves as a control mechanism for ensuring that the people work on the right work package. Also, such a chart has a motivational effect on people, since it clarifies the formal work relationships and thus reduces the risk of interpersonal conflict. With the use of such a chart, people stand a better chance of knowing where they stand in the organization and what is expected of them.

DEVELOPING THE LRC

The development of the LRC is inherently a group activity, involving the key people in the organization who have some vested interest in the work that is to be done. If we take the situation that often faces an engineering project manager who is managing a project within an engineering department, we have a good example of the usefulness of

WORK PACKAGE	STRATEGIC PLANNING	INFORMATION SYSTEMS
1 Program definition	●	●
2 Review and design define systems	▲	X
3 Master schedule	X	▲
4 Develop program budget		△
5 Funds allocation		▲
6 Define hardware-software work packages	X	
7 Define outside hardware-software work packages		
8 Schedule work packages		●
9 Schedule other work packages		●
10 Define customer information requirements		●
11 Conduct ongoing design review		X
12 Operate schedule tracking system		△
13 Schedule review		▲
14 Manage customer-initiated changes		▲
15 Manage corporate-initiated changes		▲
16 Develop startup plan		●
17 Startup implementation		●
18 Warranty problems	X	●
19 Customer acceptance		
20 Out-of-warranty problems		

Code:
△ Primary responsibility
▲ Support responsibility
● Notification
○ Approval authority
X Occasional notification

Other:
Contract administration (CA)
Sales
Field sales
Field service
Installation and checkout

2.9

Linear responsibility chart: Design, manufacture, and delivery of an industrial system.

SYSTEMS ENGINEERING	MANAGER OF PROGRAMS (PROGRAM MANAGERS)	OTHER
▲	△	▲ Sales
△	▲ ○	
▲	△	▲ CA ● Field sales
▲	▲ ○	● CA ▲ Sales
	▲○	△ CA ▲ Sales
△	▲	● CA
△	▲	● CA
△	▲ ○	● CA
▲	△	● CA
▲	△	● ▲ CA
△	▲	
		▲ CA
▲	△	▲ CA
▲	△	▲ CA
▲	△	▲ CA
▲	▲ ○	△ Field service ▲ CA ▲ Field sales
▲	△	▲ Field service ▲ Field sales
△	▲	▲ Field service
▲	△	▲ CA
▲	△	▲ Field sales

50   the LRC. In such a situation, the following plan for the development of an LRC should prove useful:

1   Distribute copies of the current traditional organization chart and position descriptions to the key people.

2   Develop and distribute blank copies of the LRC.

3   At the first opportunity, get the people together to discuss:
(*a*)   The advantages and shortcomings of the traditional organization chart.
(*b*)   The concept of a project work-breakdown structure and the resulting work packages.
(*c*)   The nature of the linear responsibility chart, how it is developed, and how it is used.
(*d*)   A way of establishing a symbology to show the work-package–organizational-position relationship. Getting a meeting of the minds on this symbology is very important so that the people will be willing to commit themselves to such a relationship.
(*e*)   The makeup of the actual work-breakdown structure with accompanying work packages.
(*f*)   The fitting of the symbols into the proper relationship in the LRC.

4   Encourage an intensive dialog during the actual making of the LRC. In such a meeting people will tend to be protective of their organizational territory. The LRC by its nature requires a commitment to support and share the allocation of organizational resources applied to work packages. This commitment requires the ability to communicate and decide. The process simply takes time. But it is worth the effort: when the LRC is completed, the people are much more knowledgeable about what is expected of them.

Even though the LRC can be of great assistance in helping to clarify personal and organizational interfaces, the *commitment* of people to support the project is vital. It is in the planning stages of a project that this commitment becomes so important. One manager emphasized this commitment in the context of project planning by noting that "the project plan is simply the summation of all of the individual commitments reached between the project manager and the functional managers. If the project manager . . . does not have these commitments, he is in no position to make a commitment on behalf of the Division. Remember, no one can commit you but yourself. Likewise, if you ever serve as project manager, *you* cannot commit anyone else."

Those engineers who aspire to be project managers should note the importance of this commitment idea. A good way to start is to ask themselves: Am I fully committed to those projects I am currently supporting? Have I provided the kind of personal commitment to the project manager that I would like if I were a project manager? Have I helped to develop an environment in the project work that is conducive to accomplishing the project objectives on time and within budget?

Much of the success of project management depends on how effectively the people work together to accomplish project objectives and gain personal satisfaction. Gaining this satisfaction can be related to how the people—the social groups—feel about the way in which project management is being practiced in the organization. The emotional patterns of the social groups—peoples' perceptions, attitudes, prejudices, assumptions, experiences, and values—determine the organization's "cultural ambience." This ambience in turn influences how people act and react, how they think and feel, and what they say in the organization, all of which ultimately determines what is taken for socially acceptable behavior in the organization.

THE CULTURAL AMBIENCE OF PROJECT MANAGEMENT

The success of project management may be less sensitive to the specifics of the organizational system than it is to the cultural ambience that exists in the organization. From a sociological viewpoint, a culture develops from the social and intellectual patterns within a group of people having a degree of common purpose, goals, language, customs, mores, and traditions. In its organizational context, a cultural ambience for project management deals with the social expression manifested by the participants engaged in managing projects. Within such organizations a culture emerges that reflects certain behavioral patterns characteristic of the members of that organization. Such behavioral characteristics influence the attitudes and the modus operandi of the people. A project team is made up of many participants in different organizational roles—superiors, subordinates, peers, associates—all working together to bring a project to completion. The cultural ambience that ultimately emerges is dependent upon the way participants feel and act within the matrix organizational environment.

An organization's cultural ambience is not just the people involved. It is the people united through policy, plans, procedures, formal and informal organizational relationships for the purposes of accomplishing project objectives through a matrix structure. The in-

52
tended formal relationships of the matrix organization are portrayed in organizational charts, project management manuals, job descriptions, policies, plans, procedures, and such documentation. This documentation is developed to establish the formal way of doing things in the organization.

There are other cultural forces that are not explicitly stated in any formal documentation and that emerge through the informal organization. In the cultural sense, these informal ways of doing things help to develop or inhibit a particular behavior on the part of the people. This behavioral attitude is often demonstrated by the statement: "Well, this is just the way we do things around here."

This cultural ambience reflects the prejudices and modus operandi of key people in the organization, both those who hold formal leadership roles and those who have emerged as informal leaders, able to influence others through their personal charisma, knowledge, expertise, interpersonal skills, social ties, etc.

The cultural ambience for project management is important, because it ultimately determines what is to be accomplished and how it will be done. This ambience affects the manner in which decisions are made and implemented in the matrix organization.

A descriptive summary of the cultural ambience conducive to successful project management reflects factors such as the following:

A consensus approach to decision making through which the project participants actively contribute in defining the question or problem as well as designing courses of action to resolve problems.

An explicit formal model of matrix organizational relations, which delineates the formal authority and responsibility patterns of the project managers, general managers, functional managers, and others working on the project.

An adversary role, which is assumed by any project participant who senses that something is wrong in the management of the project. Such an adversary questions goals, strategies, and objectives and asks the tough questions that have to be asked during the management of a project. A spontaneous adversary role can provide valuable checks and balances to guard against decisions that are unrealistic or overly optimistic. A socially acceptable adversary role facilitates the rigorous and objective development of data bases on which decisions are made.

An actual participation of people in the management of the project. They allocate their time to it and become ego-involved in planning for and controlling the project resources. As active participants in the project deliberations, they are quick to suggest innovative ideas for improving the project or to sound the alarm when things do not seem to be going as they should.

A tendency toward flexible organizational structure and flexible lines of authority and responsibility defining that structure. There is much give-and-take across these lines, with people assuming an organizational role that the situation warrants rather than what the position description says should be done. Authority in such an organizational context gravitates to the person who has the best credentials to make the judgment that is required.

A project manager who is more of a facilitator than an active manager of people. He carries out the management functions of planning, organizing, and controlling by seeing to it that the necessary decisions are made and that strategies are designed for the implementation of decisions in the project. He facilitates an environment for the participants on the project to work together with economic and social satisfaction.

A process to resolve conflicting organizational policies and procedures. Conflict inevitably arises over the allocation of resources among the various projects. This conflict arises out of the nature of the matrix organization, because the participants view problems and opportunities from the perspective of their organizational loyalties. Since there is a process to resolve this conflict, participants know that they will have their "day in court" to negotiate for the scarce resources.

It may well be the cultural ambience of the organization that finally determines whether or not project management will work effectively. An important part of that ambience is the way in which the participants perceive the nature of project management as it is being carried out in the organization.

Project management techniques can be applied to many different types of projects, ranging from a large project such as the Alaska oil pipeline to small research and development projects.

We tend to think of project management in the context of doing something for a customer outside of the company. Project management techniques can be used effectively within a company when the matters to be managed require the use of a focal point to pull things together and to coordinate all the project activities. In this case the customer may be another manager who has a demand for the delivery of a project on time, within budget, and satisfying the operational objective.

ALTERNATIVE USES OF PROJECT MANAGEMENT

In one large United States corporation, the research and development laboratory is organized along project management matrix lines. Each R&D project has a project manager; in some cases a functional

54   or department manager may also serve as a project manager on some development project. These project managers have a project team in the R&D laboratory that provides specialized support to the project. This corporation is organized on a profit-and-loss-center basis; each P&L center buys research and development support from the corporate R&D laboratory. Each P&L center has a P&L center project manager having complete responsibility for the project. Specialized resources of the R&D laboratory are committed to the P&L center project manager by the R&D laboratory project manager. The R&D laboratory manager commits to the scope and objectives of the project, as specified by the P&L center management. The P&L center management is kept informed of the project's progress to the level of detail desired through periodic design and status reviews as the project is carried out.

This type of intracorporate organizational arrangement is used for evaluating emerging strategic projects that require conceiving, prototyping, designing, and building unique, or first-of-a-kind, industrial systems.

Emerging strategic projects are so important to a profit-and-loss center and so complex to manage that the P&L center has to draw on the research and development laboratory resources to support the project. An intracorporate project management system evolved to serve this need; Fig. 2.10 illustrates this arrangement. The

Relationship of intracorporate project managers. Each project manager functions as an integrator-generalist responsible for ensuring project synergy. Each project manager is dependent on the other as they manage their respective project activities.

2.10

Ongoing specialist-to-specialist contact

Project synergy

P & L center functional elements:
  Finance
  Production
  Sales
  Market research
  Strategic planning
  Information systems

P & L center project manager

R & D laboratory project manager

R & D laboratory specialist elements:
  Product design
  Process design
  Manufacturing design
  Equipment design
  Automation systems
  Power electronics
  Control and instrumentation

Project integrity

P&L center appoints a project manager to manage all aspects of the project within the P&L center as well as work with a project manager in the R&D laboratory who manages a project team in the laboratory. Technical guidance is provided to the P&L center project manager by the R&D technical and management personnel; all overall decisions on the project are made by the P&L management team. Engineers, technicians, and draftsmen are assigned to work on the R&D project with the cost being borne by the customer—the P&L manager. The P&L project manager is furnished technical reports along with cost and schedule information in order that project visibility is maintained by such manager.

A pair of important advantages results from this organizational arrangement:

Corporate research and development resources are brought into focus to support the profit-and-loss center need.

The profit-and-loss center project manager facilitates the introduction of technology into strategic projects by virtue of his image as a local sponsor, i.e., a member of the P&L center staff.

Sometimes project management is used to develop new-business opportunities. Osgood and Wetzel have developed the concept of using project teams to identify such an opportunity and turn it into a viable business. According to them "the team would have direct decision-making responsibility for taking a new venture through the initial start-up stages, then gradually transferring this responsibility to administrative managers as the business approaches operating stability."[7]

Project management should be used only when necessary. The matrix organization is a complex way of organizing engineering resources. Each engineering manager should think about the development of criteria for determining when project management techniques are to be used in the organization.

WHEN TO USE PROJECT MANAGEMENT
Project management is not a panacea; it should be used only when its advantages are clear. Usually project management techniques are used in those engineering management situations where management attention is needed to focus attention on, to highlight, or to emphasize the importance of a particular engineering task. Such an activity might be restricted to a particular subfunction of engineering or be related across several functions. One company that

---

[7]William R. Osgood and William E. Wetzel, Jr., "A Systems Approach to Venture Initiation," *Business Horizons,* October 1977, p. 42.

56   introduced project management techniques in its product-line engineering group defined a project in the following manner: "A narrowly defined activity which is planned for a finite duration, with a specific goal to be achieved (such as a new-product introduction). When the project is successfully completed, the project will be terminated."

The use of project management techniques can be considered for a wide variety of engineering management tasks. Criteria to be applied in determining whether or not to organize a project include such factors as (1) magnitude of the effort; (2) unfamiliarity of the undertaking; (3) interrelatedness of the tasks of the effort, and (4) the organizational reputation that is at stake in the undertaking.[8]

An important reason for developing criteria for using project management is to help in understanding some of the systems effects that occur.

THE SYSTEMS EFFECT OF PROJECT MANAGEMENT

When project management is introduced into an organization, a systems effect results. For example, what might appear on the surface to be an organization change when the matrix organization is set up soon becomes something larger. As the formal reporting relationships of the matrix organization are defined, new informal social relationships will emerge. The effective management of a project requires timely and relevant information. Most business systems are set up to provide information following the established organizational lines. When projects are established, the business systems have to realign the information so as to provide the project manager with adequate visibility of the schedule, cost, and performance parameters of the project.

In an industrial organization known to the authors, project managers were being held responsible for the timely accomplishment of project goals. However, the business-systems activity of the firm was not organized to accommodate and portray the information needed to evaluate schedule, cost, and technical performance of the ongoing projects. In self-defense the project managers developed their own crude information systems. Unfortunately, the data these information systems provided were insufficient, were tardy, and had little *predictive* value. As might be expected, these project managers never really knew how things stood. The result: schedule slippages, cost overruns, unreliable technical-performance evaluations, and general customer disenchantment. Fortunately, this problem was quickly turned into an opportunity through the development of an information retrieval system that was tailored to the project manager's needs.

[8]For a further discussion of when to use project management techniques, see David I. Cleland and William R. King, *Systems Analysis and Project Management*, 2d ed., McGraw-Hill, New York, 1975, Chap. 9.

57

The management functions of planning and control of a project require approaches different from those that would be used if project management did not exist. The use of project selection techniques and project control measures are some additional systems effects of using project management. The next chapter will discuss in more detail the life-cycle concept of project management and how management techniques can be developed to accommodate the systems effects of project management.

SUMMARY

Project management is a way of bringing organizational resources into focus to support an ad hoc organizational activity such as an engineering project. Taken in its total context, project management is an activity that affects organizational couplings, information flows, management functions and techniques, and the cultural ambience of the organization itself.

Project management systems is an idea whose time has come. Engineering managers should develop a conceptual knowledge of the project management process and techniques.

DISCUSSION QUESTIONS

1 What do you consider the major drawbacks of the functional organization? Its major strengths?

2 Outline your understanding of the project management organization.

3 Can program management be effected without a matrix organization? Discuss.

4 Select an engineering process with which you are familiar. Identify and describe its work packages.

5 Cultural ambience has been suggested as a key factor in determining the success or failure of a project. Why is this true?

6 Cite examples of situations or operations that would benefit from a project management approach.

7 What is the systems effect of project management?

8 Why is it important to accurately define the authority and responsibility patterns in the matrix organization?

9 The matrix organization is found in environments other than project management. Describe how it might be used in a different context.

10 The project manager and the functional manager have a com-

58   plementary authority relationship. Defend or refute this statement.

11   What are some of the uses to which a work package can be put?

12   Compare the advantages and disadvantages of the linear responsibility chart.

KEY QUESTIONS FOR THE ENGINEERING MANAGER

1   What opportunities exist in this organization for the practice of project management?

2   Is the cultural ambience of this organization such that project management would be successful?

3   What are the key organizational interfaces through which the engineering manager's job is carried out?

4   Is the management principle of unity of command being practiced in this organization? Why or why not?

5   Has the *specific* authority of the project manager vis-à-vis the functional manager been defined?

6   Has a policy document outlining the authority of the project manager and the functional manager been defined in the matrix organization?

7   Do the technical people in the organization understand the idea of the work package in an engineering context?

8   Is the use of a work-breakdown structure understood by key people in the organization?

9   What can an engineering manager do to encourage the professional engineers to become fully committed to support the project managers?

10   One of the greatest values of a linear responsibility chart is the dialog that takes place in completing the chart. Has this aspect of linear responsibility charting been fully put into practice?

11   How would the cultural ambience for project management be described for the organization?

12   Have criteria been developed for the time when project management will be practeiced in the organization?

13   What are some of the systems effects that exist from having established project management in the organization?

BIBLIOGRAPHY

Blanchard, Benjamin S.: *Engineering Organization and Management,* Prentice-Hall, Englewood Cliffs, N.J., 1976.

Burck, Charles G.: "How G.M. Turned Itself Around," *Fortune,* Jan. **59**
16, 1978.

Chironis, Nicholas P.: *Management Guide for Engineers and Technical Administrators,* McGraw-Hill, New York, 1969.

Clayton, Soss: "A Convergent Approach to R&D Planning and Project Selection," *Research Management,* September 1971.

Cleland, David I. and William R. King: *Systems Analysis and Project Management,* 2d ed., McGraw-Hill, New York, 1975.

Crowston, Wallace B.: "Models for Project Management," *Sloan Management Review,* Spring 1971.

Frankwicz, Michael J.: "A Study of Project Management Techniques," *Journal of Systems Management,* October 1973.

Haavind, Robert C., and Richard L. Turmail: *The Successful Engineer Manager: A Practical Guide to Management Skills for Engineers and Scientists,* Hayden, New York, 1971.

"How to Stop the Buck Short of the Top," *Business Week,* Jan. 16, 1978.

Jennett, Eric: "Guidelines for Successful Project Management," *Chemical Engineering,* July 9, 1973.

Joyce, William B.: "Organizations of Unsuccessful R&D Projects," *IEEE Transactions on Engineering Management,* vol. EM-18, May 1971.

Killian, William P.: "Project Management—Future Organizational Concepts," *Marquette Business Review,* vol. 15, no. 2, 1971.

Lawrence, Paul R.: "The Human Side of the Matrix," *Organizational Dynamics,* Summer 1977.

Lodwig, Steven: "Should Any Man Have Two Bosses?" *International Management,* April 1970.

"Management Itself Holds the Key," *Business Week,* Sept. 9, 1972.

Meads, Donald E.: "The Task Force at Work—The New 'Ad-Hocracy,'" *Columbia Journal of World Business,* November-December 1970.

Project Management: Techniques, Applications and Managerial Issues, American Institute of Industrial Engineers, Norcross, Ga., 1976.

Realities in Project Management, Proceedings of the Ninth Annual International Seminar/Symposium, Project Management Institute, Chicago, Oct. 22–26, 1977.

60   Shannon, Robert E.: "Matrix Management Structure," *Industrial Engineering,* March 1972.

Silverman, Melvin: *The Technical Program Manager's Guide to Survival,* Wiley, New York, 1967.

Smith, G. A.: "Program Management—Art or Science?" *Mechanical Engineering,* September 1974.

"Teamwork Through Conflict," *Business Week,* Mar. 20, 1971.

ENGINEERING PROJECT LIFE CYCLE 3

In this chapter we will develop the project (or system) life-cycle idea and discuss its phases. We will also describe some examples where the life-cycle concept has been applied in the engineering context. In our discussion, we will tend to use the terms "project life cycle" and "system life cycle" interchangeably. The life-cycle concept applies to both a project and a system—hence our tendency to interchange the terms.

Any system, whether hardware, software, biological, economic, etc., is dynamic. A system is created, develops, grows, reaches maturity, and may eventually decline and disappear. Product systems—such as a transportation system—go through years of detailed design and development before production. When the transportation system is finally put into operation, many years have elapsed since the emergence of the idea to develop it. A new drug will undergo approximately 10 years of development and testing before it can be sold to the customer. The requirement for registration and testing of the drug under the United States Pure Food and Drug Act has expanded the time required to bring a drug from the basic research stage to the marketplace. Engineering project managers in the construction industry are aware of the additional time required for the completion of construction projects since the passage of land-use laws and the requirement to file environmental impact statements. Virtually every industry has felt the growing impact of legislation affecting such areas as consumer protection, environmental impact, social security, etc., which serves to lengthen the time required to develop, design, produce, and place a system in its operational environment.

Every system has natural phases of development. Even the solar system had a beginning, perhaps a cataclysmic beginning. The seasons of the year, the growth of our children, the coming and going of institutions all serve to remind us how transitory systems are in this world of ours.

62 PHASES OF THE SYSTEM LIFE CYCLE

A system is an assemblage or combination of things—both abstract and concrete—and people for its realization. In this section, we present a generic model of a system life cycle that can be applied to many different situations to be encountered by the engineering manager. Such situations may include:

Moving an engineering organization from one location to another

Reorganizing an engineering department

Managing an engineering project

Managing a research project

Analyzing the requirements for and obtaining capital equipment

Managing a product-improvement engineering effort

Studying a change in the quality and quantity of professional personnel

Managing an engineering change

Developing and building an energy management system

The phases of the life cycle—and what is to happen to the system during its life—depend on the distinctive characteristics of the system. The phases we have identified below serve only to represent a generic approach that can be taken by an engineering manager who plans to use this concept in his activities.[1] Figure 3.1 portrays a schematic model of such a life cycle. Note that the generic model in Fig. 3.1 and Tables 3.1 through 3.5 are intended to provide a broad philosophy of the notion of the life cycle applied to many engineering activities.

THE CONCEPTUAL PHASE
The germ of the idea for a system may evolve from other research, from current organizational problems, or from the observation of organizational interfaces. The conceptual phase is one in which the idea is conceived and given preliminary study and evaluation.

During the conceptual phase, the environment is examined, forecasts are prepared, objectives and alternatives are evaluated, and the first examination of the *objective, cost,* and *schedule* aspects of the system's development is performed. It is also during this phase

[1]In describing the generic approach to a system life cycle, we have drawn heavily from David I. Cleland and William R. King, *Systems Analysis and Project Management,* 2d ed., McGraw-Hill, New York, 1975, as well as from public domain documentation developed by DOD and NASA.

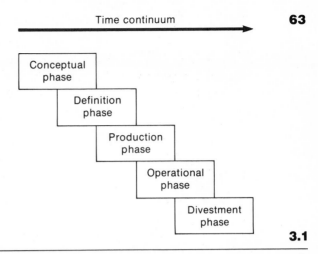

3.1

Generic model of system life cycle.

that basic strategy, organization, and resource requirements are conceived. The fundamental purpose of the conceptual phase is to conduct a feasibility study of the requirements in order to provide a basis for further detailed evaluation. Indeed, the conceptual phase should provide the opportunity to "ask ourselves what questions we should be asking ourselves."[2]

If the conceptual phase is properly approached, it can serve to kill those projects which should not survive. There will typically be a high mortality rate of potential systems during the conceptual phase of the life cycle. This is rightly so, since the study process conducted during this phase should identify projects that have high risk and are technically, environmentally, or economically infeasible or impractical.

Some insight into the value a rigorous application of the conceptual-phase questions can have in sorting out ventures that should be eliminated is given in a recent *Business Week* article. Industry, according to *Business Week,* should learn how to weed out bad ideas early.[3] To that end, the Dexter Corporation has instituted an eight-factor "innovation index" approach to research management that considers such factors as effectiveness of communications, competition, timing, etc., and comes up with an innovation potential for new ideas. Glaser describes the innovation index as a technique to evaluate the innovation of a project.[4] According to Glaser: "If one must err, do it on paper—it's cheaper."[5]

---

[2]We have paraphrased Dr. Albert Einstein's famous remark.
[3]"Vanishing Innovation," *Business Week,* July 3, 1978, p. 46f.
[4]See Chap. 5 for a description of this technique and the role that communication plays in making the technique work.
[5]Milton A. Glaser, "The Innovation Index, " *Chemical Technology,* vol. 6, March 1976, pp. 182–187.

What should be used as a standard yardstick or denominator against which to measure potential new systems in the conceptual phase? We believe there is only one such proper common denominator: *the basic purpose or mission of the organization.* The basic purpose should be described in the master plan for the organization. The length of a system life often extends many years into the future, and the timing of the project with respect to the basic purpose of the organization at some *future* time is a key factor in the evaluation.

In summary, the conceptual phase is important, since it sets the stage for asking a series of questions about the proposed idea vis-à-vis basic organizational objectives. Table 3.1 outlines some of the key issues that have to be dealt with in this phase.

THE DEFINITION PHASE

Once the system has survived the conceptual phase, the analysis of the system progresses to the definition phase. The purpose of the definition phase is to determine, as soon as possible and as accurately as possible, cost, schedule, performance, and resource requirements and whether all elements, work packages, and subsystems will fit together economically and technically.

The definition phase simply tells in more detail what it is we want to do, when we want to do it, how we will accomplish it, and what it will cost. This phase allows the organization to fully conceive

Conceptual phase

3.1	1	Determine existing needs or potential deficiencies of existing systems.
	2	Establish concepts which provide initial guidance to overcome existing or potential deficiencies.
	3	Determine initial technical, environmental, and economic feasibility and practicability of the proposed system.
	4	Determine compatibility of proposed system with the basic purpose of the organization.
	5	Examine alternative ways of accomplishing the system objectives.
	6	Provide initial answers to the questions: (a)　　　What will the system cost? (b)　　　When will it be available? (c)　　　What will it do? (d)　　　How will it be integrated into existing systems?
	7	Identify human and nonhuman resources required to support the system.
	8	Select initial designs which will satisfy the system objectives.
	9	Determine initial system interfaces.
	10	Establish an organizational approach.
	11	Make a specific decision to continue or discontinue the development of the system.

Definition phase

3.2

1	Firm identification of the human and nonhuman resources required
2	Preparation of final performance requirements
3	Preparation of detailed plans required to support the system
4	Determination of realistic cost, schedule, and performance requirements
5	Identification of those areas of the system where high risk and uncertainty exist, and delineation of plans for further exploration of these areas
6	Determination of necessary support subsystems
7	Definition of intersystem and intrasystem interfaces
8	Identification and initial preparation of the document required to support the system, such as policies, procedures, job descriptions, budget and funding papers, letters, memoranda, etc.
9	Establishment of the compatibility of the system with the future basic purpose of the organization
10	Design of prototypes

and define the system before it starts to physically put it into its environment. Simply stated, the definition phase dictates that one stop and take time to look around to see if this is what one really wants before the resources are committed to putting the system into operation and production. If the idea has survived the conceptual phase, a conditional approval for further study and development has already been given. The definition phase provides the opportunity to review and confirm the decision to continue development, create a prototype, and go into production or installation.

Decisions made during the definition phase may also include the cancellation of further work on the system and the redirection of organizational resources elsewhere. The concerns and activities of this phase are described in Table 3.2.

THE PRODUCTION PHASE
After the definition phase of the project has passed and the system has been committed for production, the system progresses to the production phase. In this phase the system takes physical shape and form and is capable of taking its place in the environment.

The purpose of the production phase is to acquire and test the system elements and the total system itself using the standards developed during the preceding phases. The production process involves such things as the actual setting up of the system, the fabrication of hardware, the delineation of authority and responsibility patterns, the construction of facilities, and the finalization of supporting documentation. Table 3.3 details this phase.

During the production phase the inadequacies of the analysis conducted in the two preceding phases will come to light. Cost

Production phase

3.3   1   Updating of detailed plans conceived and defined during the preceding phases, as well as conducting of field tests where applicable

2   Identification and management of the resources required to facilitate the production processes, such as inventory, supplies, labor, funds, etc.

3   Verification of system production specifications

4   Design of final production components

5   Beginning of production, construction, and installation

6   Final preparation and dissemination of policy and procedural documents

7   Performance of final testing to determine adequacy of the system to do the things it is intended to do

8   Development of technical manuals and affiliated documentation describing how the system is intended to operate

9   Development of plans to support the system during its operational phase

overruns, schedule slippages, or system performance deficiencies may appear. System performance inadequacies can be corrected through the judicious use of engineering changes—if such changes are managed properly. Such changes, however, usually impinge on the project's costs and schedule and result in the expenditure of additional resources.

A study conducted by the United States government gives a good indication of the magnitude of cost overruns in the acquisition of large systems.[6] The report uses the following categories to identify cost escalation:

1   Quantity changes—including scope changes

2   Engineering changes—an alteration in the established physical or functional characteristics of a system

3   Support changes—involving spare parts, ancillary equipment, warranty provisions, and government-furnished property and/or equipment

4   Schedule changes—in delivery schedule adjustments, completion date, or some intermediate milestone of development, production, or construction

5   Economic changes—influence of one or more factors in the economy, such as inflation

6   Estimating changes—due to corrections or other changes occurring since the initial or other baseline estimates for program or project costs were made

[6]*Final Status of Major Acquisitions,* Report to Congress by Comptroller General of United States, June 30, 1975.

7   Sundry—changes which do not fall within the above categories, such as environmental costs and relocation assistance for water and highway projects

A cost-growth analysis for 120 projects, other than military weapons systems, is summarized in the accompanying table.[7]

Type of change	Energy Research and Development Administration	Army Corps of Engineers	Department of Transportation	Other civil agencies	Total changes
Quantity	$ 215,610*	$1,860,171	$10,218,288	$ 994.683	$13,288,752
Engineering	501,910	1,832,686	19,870,574	485,006	22,690,176
Support	39,500	0	8,857	0	48,357
Schedule	88,940	11,865	−5,250	77,000	172,555
Economic	670,110	4,353,736	29,445,667	2,163,016	36,632,529
Estimating	209,090	406,742	370,538	199,386	1,185,756
Sundry	229,250	143,431	4,309,375	691,847	5,373,903
Total	$1,954,410	$8,608,631	$64,218,049	$4,610,938	$79,392,028
Number having 100 % or more increase	7	78	15	20	120
Total civil projects	26	170	55	216	467

*Dollar figures in thousands.

A perusal of this data indicates that the principal factor in cost growth is the economic changes, chiefly inflation, which account for 46 percent of the total. Engineering changes account for 28 percent of the total. One can raise questions about the adequacy of the analysis conducted during the conceptual and definition phases. Would a more thorough analysis of these systems in these phases have helped to reduce subsequent cost (and schedule) overruns? We believe it would have helped a great deal.

        The point is this: The conceptual and definition phases of a system life cycle provide an opportunity to analyze and design the project so that the question "Do we know what we are really getting into?" can be answered before a production commitment is made. And that is what these two phases are really all about.

THE OPERATIONAL PHASE

The fundamental role of the manager of a system during the operational phase is to provide the resource support required to ac-

[7]Ibid.

68 complish system objectives. This phase indicates the system has been proven economical, feasible, and practicable and will be used to accomplish the desired ends. In this phase the manager's functions change somewhat. The manager is less concerned with planning and organizing and more concerned with controlling the system's operation along the predetermined lines of performance. His responsibilities for planning and organization are not entirely neglected—there are always elements of these functions remaining—but he places more emphasis on motivating the human element of the system and controlling the utilization of resources of the total system. It is during this phase that the system may lose its identity per se and be assimilated into the "institutional" framework of the organization.

If the system in question is a product to be marketed, the operational stage begins the sales life-cycle portion of the overall life cycle, for it is in this phase that marketing of the product is conducted. Table 3.4 shows the important elements of this phase.

THE DIVESTMENT PHASE

The divestment phase is the one in which the organization gets out of the business that it began with the conceptual phase. Every system—be it a product system, a weapons system, a management system, or whatever—has a finite lifetime. Too often this goes unrecognized, with the result that outdated and unprofitable products are retained, inefficient management systems are used, or inadequate equipment and facilities are put up with. Only by the specific and continuous consideration of the divestment possibilities can the organization realistically hope to avoid these contingencies. Table 3.5 relates to the divestment phase.

Even those engineering projects that have been "failures" can hold something of value in the sense of lessons learned from the experience. The evaluation of failures is usually neglected. However, if it is done properly and without any attempt to submit "report cards" on the people who are involved, the results can be helpful. Otherwise, "unless one learns from history, one is condemned to repeat it."[8]

[8]Hanson W. Baldwin, *Strategy for Tomorrow*, Harper, New York, 1970, p. 10.

Operational phase

3.4	1 Use of the system results by the intended user or customer
	2 Actual integration of the project's product or service into existing organizational system
	3 Evaluation of the technical, social, and economic sufficiency of the project to meet actual operating conditions
	4 Provision of feedback to organizational planners concerned with developing new projects and systems
	5 Evaluation of the adequacy of supporting systems

Divestment phase

1	System phasedown	**3.5**

2	Development of plans transferring responsibility to supporting organizations
3	Transfer of resources to other systems, including transfer of technology
4	Development of "lessons learned from system" for inclusion in qualitative-quantitative data base to cover:

(a)	Assessment of image by the customer
(b)	Major problems encountered and their solution
(c)	Technological advances
(d)	Advancements in knowledge relative to strategic objectives
(e)	New or improved management techniques
(f)	Recommendations for future research and development
(g)	Recommendations for the management of future programs
(h)	Other major lessons learned during the course of the system

An evaluation of the management of projects that fail—*that do not accomplish what was expected*—is essential. Also, an evaluation of the management of successful projects can be useful. Such evaluation should be centered around the harmony of the results with the intended purpose reflected in a higher-level organizational plan. If the organization is successful in a particular business area, there is a strong tendency to continue research and development projects in that area. Drucker alludes to a real strategic danger in this respect when he comments, "It is far more difficult to abandon yesterday's success than it is to reappraise failure."[9]

Taken together, Tables 3.1 through 3.5 provide an outline of the overall system life cycle. Of course, the terminology used in these tables is not applicable to every system that might be under development, since the terminology generally applied to the development of consumer product systems is often different from that applied to other systems. Both in turn are different from that used in the development of a financial system for a business firm. However, whatever the terminology used, the concepts are applicable to all systems.

For instance, the following phases are present in the life of an engineering project: (1) research and development, (2) design, (3) construction and manufacturing, (4) implementation and maintenance, and (5) transfer of technology. Figure 3.2 illustrates the correspondence between the terminology of the system life cycle and the phases identified for an engineering project. Each phase of an engineering project could itself be managed on a life-cycle basis.

It should be noted here that there are not necessarily one-to-

[9]Peter F. Drucker, *Management: Tasks, Responsibilities, Practices,* Harper, New York, 1973, p. 159.

System life cycle	Conceptual	Definition	Production	Operational	Divestment
Engineering project life cycle	R & D	Design	Construction or manufac- turing	Implementation and maintenance	Technology transfer

3.2

Life-cycle phases of an engineering project.

one relationships among the activities of the corresponding phases in the different terminology patterns. As an example, in a major engineering project with one main product—say, the construction of a mass transit system—the sale of the ultimate product should precede as well as be part of the research and development phase. Feasibility studies that are carried out in this phase invariably provide information for a go–no-go decision. If the project survives this phase and moves on to the next, then the potential customer must have been satisfied with the prospects for ultimate accomplishment of the project. In other words, the customer has bought the project.

This concept of a marketing activity at the outset of a project life cycle differs from the traditional view that marketing belongs to the operational phase. Both views are accurate provided one identifies the context in question.

To go one step further, a given undertaking need not formally go through all the life-cycle phases. Indeed, using feedback information from a previous process or system, it may be possible to skip both the R&D and design phases and to commence with the construction phase.

The use of a life-cycle approach is well established in the engineering community. The particular approach used depends on the type of engineering being carried out. Companies in the aerospace industry have developed highly sophisticated life-cycle models to use in managing the design and development process of their products. For example, one company envisions the design and development process of a gas-turbine engine for aircraft applications as consisting of six phases. The phases involve various steps to establish requirements, create and verify designs, qualify the design product, and support production engines during flight operations. The six phases of the design and development process with their purpose and output are shown in Table 3.6.

All the steps in each phase are conducted for a complete engine-development program. If the effort is less than a full develop-

The six phases of the design/development
process*

PHASE	PURPOSE	OUTPUT	**3.6**
1 Conceptual	Understand customer needs Create conceptual designs Select best concept Demonstrate conceptual design	Conceptual design definition Conceptual design hardware Demonstrated conceptual design	
2 Initial design verification	Complete initial product design Verify modified product design concept	Initial product design definition Initial product design hardware Verified initial product design concept	
3 Product design	Modify initial design to incorporate verification test results Verify modified product design	Improved product design definition Improved product design hardware Verified product design concept	
4 Qualification	Develop engine through qualification	Qualified engine	
5 Initial operational	Introduce product into initial operation Identify and fix early problems	Operational aircraft PSP and CIP programs for problem solution	
6 Full deployment	Provide customer with full complement of operational aircraft Provide customer support for life of product	Full operational weapon system deployment Sustaining engineering PSP and CIP programs	

*See "Design/Development Process Guide, Subsystem Guide 3-3," Pratt & Whitney Aircraft Group,
Government Products Division, United Technologies Corporation, p. 6.

ment, only some of the steps would be appropriate. For example, a
"technology contract" would involve only the first portion of the conceptual phase.[10]

LEVELS OF ACTIVITY
The phases discussed in the preceding pages are not mutually
exclusive. Rather, they overlap. In addition, each phase goes through
various levels of activity. There are four discernible groups of
activity. They are initiation, growth, maturity, and decline of the
phase. As indicated in Fig. 3.3, the levels of activity are highest during
the growth and maturity periods.

[10]See "Design/Development Process Guide, Subsystem Guide 3-3," Pratt & Whitney Aircraft Group,
Government Products Division, United Technologies Corporation, for an excellent description of the
design and development process.

72

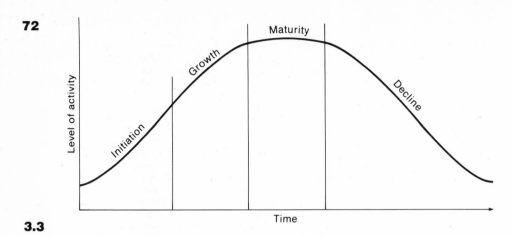

3.3

Levels of activity in a given phase.

Of course, not all projects reach maturity. For example, in the research and development community the success of projects has been poor: about one of twenty programs initiated ever results in significant commercial success.[11]

In those situations of successful products, research has shown an average elapsed time between the initial conception and a positive cash flow averaging 7 to 10 years.[12]

SOME APPLICATIONS OF THE LIFE-CYCLE CONCEPT

To illustrate the ubiquity of the life cycle in the engineering management context, subsequent portions of this chapter will cite some examples where the life-cycle concept is applied. The examples will illustrate the concept in different types of projects imbedded in one or more phases of the generic life cycle.

INDUSTRIAL DEVELOPMENT PROJECT
An oil company contracted for the construction of camp sites in Saudi Arabia. The company that managed and constructed these camp sites used a master life-cycle phasing chart as the basic approach for the planning and control of the construction. Table 3.7 contains the life-cycle chart with its major phases and work packages in each phase.

Of course, each of the work packages has subactivities or subsystem work packages, which in themselves can be dealt with on a life-cycle basis with attendant phases and activities.

[11]P. L. Mullins, "Capital Budgeting for Research and Development," *Management Services,* May-June 1969, pp. 45–50.
[12]Ibid.

Life-cycle phases of camp site
construction

3.7

DEVELOPMENT	DESIGN	ENGINEERING AND CONSTRUCTION	OPERATIONAL AND MAINTENANCE
Conceptual studies	Design Drawings	Estimating Reporting	Logistics Facilities and
Engineering economics	Specifications Estimates	Forecasting Analysis	maintenance Project control
Site planning	Bid packages and	Schedules	Inspection
Environmental impact studies	evaluation Budget analysis	Material control Purchasing	Feeding and housing
Master planning		Expediting Vendor inspection	Communications Community
		Traffic	services
		Subcontracts	Transportation
		Construction planning and methods	Material and equipment protection
		Constructibility reviews	Operator training Waste control and
		Labor relations	disposal
		Quality control	
		Field administration and engineering	
		Safety and first aid	
		Warehousing	
		Engineering: Architectural Civil Electrical Instrumentation Mechanical Piping Sanitary Structural Water supply	

RESEARCH AND DEVELOPMENT PROJECT

An R&D project goes through a life cycle from the idea stage to a prototype model—and, if it is meritable, on to full-scale production. Before a development project is undertaken, a reasonable degree of commercial risk is demonstrated. As the commercial cost-schedule parameters of the project are defined, a profile of the minimum technical parameters that are required will be delineated. At each stage of the R&D pipeline, a series of activities will proceed in parallel with the technical project development program work, such as:

Market feasibility studies

Legal and patent investigation

Financial analysis

74   Promotion engineering

Figure 3.4 shows an expanded research pipeline envisioned by Merrifield.[13]

NEW-PRODUCT DEVELOPMENT

The value of the life-cycle concept in the management of a project is demonstrated in the procedure used by a major industrial firm in its identification, development, and funding for new products and features. At appropriate intervals during the evolution of a process, a review is carried out to assess the status of the project and to recommend whether or not the project should continue.

An "innovation team" gains insight and seeks opportunities for new products and features by continuously monitoring customer needs and wants, socioeconomic and competitive information, interface industries, and new and emerging technologies. Along with others, the innovation team then screens ideas for feasibility and ultimately develops general objectives and quality and costs (preliminary specifications) for the new product. At this point, the product planning council, consisting of engineering and financial managers, reviews the proof of feasibility.

If the evidence is convincing, different functional divisions of the organization then collaborate to search for alternative methods of producing the product within the constraints of the preliminary specifications, as well as examine the product impacts such as marketing prospects. A second product planning council review takes place at this stage. If the outcome is positive, a new-product development team is established.

[13]Bruce D. Merrifield, *Strategic Analysis, Selection, and Management of R&D Projects*, AMA Management Briefing, AMACOM, a division of American Management Associations, New York, 1977.

Stages of the research pipeline. (*From D. Bruce Merrifield, Strategic Analysis, Selection, and Management of R&D Projects, AMACOM, New York, 1977, pp. 28–32.*)

3.4

Conceptual	Definition	Production		Operational	
I	II	III	IV	V	VI
Idea	Feasibility demonstration	Product and process development	Pilot plant prototype	Semi-commercial	Full-scale production

7 to 10 years
(1 out of 20 successes)

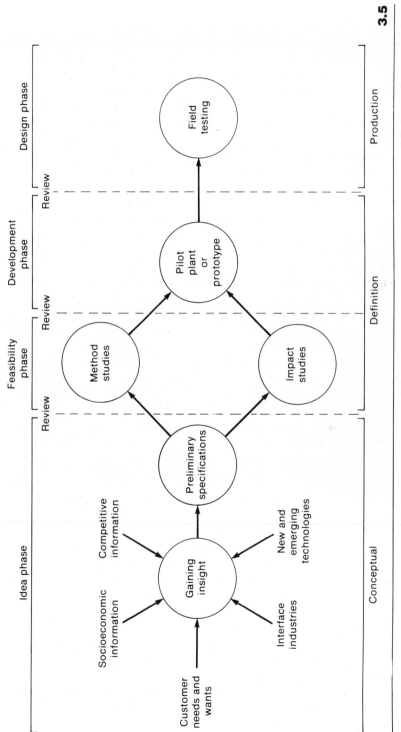

3.5 Life cycle of a new product.

76   The development team carries out pilot plant and prototype studies, and its efforts are subjected to yet another review. Success at this level then leads to the product design phase. A summary of these steps is given in Fig. 3.5, with the terminology of the generic life cycle included to underscore its direct correspondence with the new-product model under discussion.

CONSTRUCTION PROJECT

In Fluor Utah, Inc., a leader in executing engineering, procurement, and construction contracts, projects are divided into identifiable phases. Each phase contains a defined work function with a specific input and goal. As the project moves from phase to phase, the composition and talents of the people assigned to the project change. Figure 3.6 shows how these phases are arranged in a life-cycle relationship.

MANAGEMENT IMPLICATIONS OF THE LIFE-CYCLE CONCEPT

The concept of the life cycle has its greatest impact in contributing to the management of the project. Indeed, the management system for the project can be built around the model of the life cycle of the project. In this section some management implications of the life cycle will be discussed.

    The importance of viewing engineering activities through a system life cycle can be illustrated by taking an engineering project and looking at it from a management perspective. An *engineering project* can be defined as any agreement to deliver an engineering product or service for a stipulated amount of money in a stated time. With this definition, an engineering project can be a subsystem in a larger system (or a work package or work unit) for which some individual is held responsible. The individual so designated may be called a "project manager," "project engineer," "engineering manager," "engineering team manager," "engineer in charge," "work package manager," "subdivision work manager," "engineering task force manager," or other such titles. These titles signify that some individual is held responsible for, and has the authority to manage, the engineering activity in the literal sense of the word. This definition also means that any undertaking to perform some specified engineering work on a specified budget and schedule is a project—unless the undertaking is a subset of a project: a work package, a work unit, or even a job number. Even the subsets of a project will have one individual as a focal point to do the undertaking or to see that the undertaking is carried out on time, within budget, and satisfying the objective of the undertaking.

Phases of project execution **77**
I Planning
II Preliminary engineering
III Review
IV Detail engineering
V Detail engineering and construction overlap
VI Construction
VII Test and commission

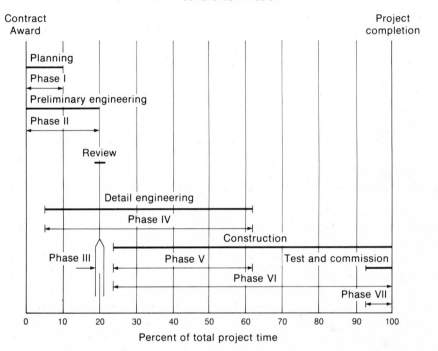

Life cycle of a construction project. (*Adapted from Robert K. Duke, H. Frederick Wohlsen, and Douglas R. Mitchell, "Project Management at Fluor Utah, Inc.,"* Project Management Quarterly, *vol. 8, no. 3, September 1977, p. 37.*)

3.6

An engineering undertaking—a project—can come from a customer or from another part of the organization, or it can originate within an engineering department as an unsolicited proposal for a customer. These engineering projects vary in size, from a few thousand dollars upwards to multimillion-dollar projects. Some future engineering projects are astronomical in size: Consider some of the large engineering projects of the proposed natural pipeline to be constructed from northern Alaska to the continental United States. Engineering projects can range from very large to very small.

What is considered as a small engineering project depends on the viewpoint of the engineering department and the organization.

78   One senior engineering manager in a diversified engineering company in the waste water treatment field defined small engineering projects as those which met most of the following criteria:

> Total duration is approximately 3 to 12 months.
>
> Value is between $5000 and $100,000.
>
> The project team is in daily or at most weekly communication on a personal basis.
>
> Only a few (two to four) cost centers work on the project and they are closely related.
>
> Manual methods of recordkeeping are more effective than automated systems.
>
> The project manager has quick and direct access to those people who have the detail on schedule, cost, and technical content data.
>
> The work-breakdown structure has no more than four tiers.
>
> There are no formal assistant project managers or subproject managers.

In practice, the real differences between small projects and large projects are the amount of detail required for large projects, the more rigorous documentation required, and the large number of work packages. The larger the project, the more complex everything becomes. The principles, concepts, and management processes remain the same.

PLANNING AND THE LIFE CYCLE
Now that we have defined an engineering project, and have concluded that the project has to be managed, what implications does the system life cycle have for such projects? The answer to this question is to be found in one of the first management activities in the life cycle by the project manager, namely, planning the engineering project.

No matter how small a project may be, it needs a plan—a document that lays out a strategy that clearly defines *what* is to be done, *why, by whom, when,* and for *how much.* A plan is needed even if a project is a subset in a larger project.

An engineering project plan is a document that outlines a strategy on how future events in the project will be handled. A plan, which deals with future events, is by its very nature cast in the context of starting something today and finishing it at some time in the future—a life cycle.

The essential elements of an engineering project plan are **79** the following:

A summary of the essentials of the project written in such a manner that anyone can get a grasp of the project. It states briefly what is to be done and how it is to be done, and lists the end-product or service deliverables.

A work-breakdown structure that has enough detail to identify the relationships of work packages, work units, and job numbers.

A citation of tangible milestones that can be counted, measured, and evaluated in such a way that there can be no ambiguity as to whether or not a milestone has been achieved. There should be a correspondence of milestones to project budgets to facilitate adequate monitoring.

An event logic network (an activity network) that shows the sequencing of the elements (work packages, work units, and job numbers) of the project and how they are to be related—which can be done in parallel, series, etc.

Separate budgets and schedules for the project elements and the individual that is to be held responsible for these elements.

An organization interface plan that shows how the project relates with the rest of the world—customer, line-staff organizations, subcontractors, suppliers, and other clientele that have some role in the project.

A plan for how the project is to be reviewed—who reviews the project, when, and for what purpose.

A list of key project personnel and their assignments in the project.

The project plan for a small engineering project need not be elaborate. The project plan that we have outlined above is meant to be only representative. The level of detail and the content of the plan can be expanded as needed from the above "plan for the project plan."

PROJECT CHANGES
Even though a project has been well planned, there will probably be some changes in the project as it is completed. In this section some of the changes affecting a typical project are depicted.

The project changes in its nature during the life cycle. Expenditures, people, level of effort, and organizational involvement all

80   change as the project goes from idea generation through to its operational phase. The uncertainty related to time and cost of completion diminishes as the project nears its completion. Archibald has portrayed this diminishing uncertainty of time and cost in the manner shown in Fig. 3.7. Archibald notes:

> The specified result and the time and cost to achieve it are inseparable. The uncertainty related to each factor is reduced with completion of each succeeding life-cycle phase. Figure [3.7] illustrates the reduction of uncertainty in the ultimate time and cost for a project at the end of each phase. In phase 1, the uncertainty is illustrated by the largest circle. The area of uncertainty is reduced with each succeeding phase until the actual point of completion is reached.[14]

When a decision is made to develop the project, the management function of *organizing* comes into play. The human and nonhuman resources requirements are identified and obtained in the quality and quantity needed to support the project. The human resources used on a system during its life cycle will vary. Different personnel skills are required as the project is initiated, grows, and declines. A single individual may be all that is assigned when the project is in the conceptual stage. After the project is approved for further funding, a team of engineers is assigned to a skeleton organization to begin definition of the project objectives and hardware-software tasks and subtasks. A conceptual framework for the project is finalized; schedules and costs are detailed. Then additional personnel are assigned to carry out the development and design work. Design engineering is completed and actual construction of the product is begun. The project organization continues to exist, but it exists in more of a surveillance mode. A production organization takes over the operation of the project; this organization attains peak strength. The product is delivered to the customers along with field engineering services, installation and checkout, and field support services. Finally, field modification, if required, is carried out. At this time the life cycle of the project has ended; personnel still remaining on the project are returned to their permanent organization, perhaps to be placed on a new project in the organization.

Litton Industries, in their Litton Microwave Cooking Division, provides an example of the role that the organizing function plays in the management of a project. In the engineering department of this division, a functional form of organizational structure has been changed to a task-team structure. For each approved engineering project, a task team is assigned. The task team is composed of a

---

[14]Russell D. Archibald, *Managing High-Technology Programs and Projects*, Wiley, New York, 1976, p. 23.

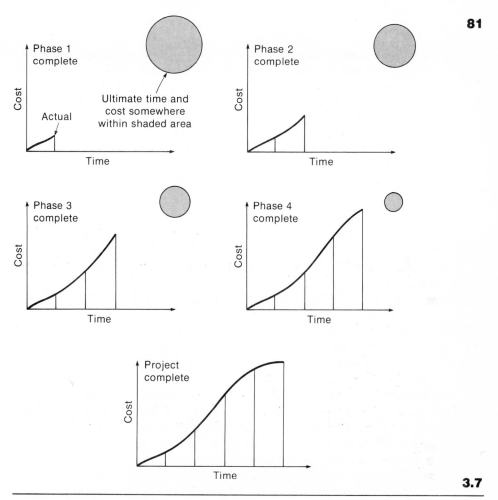

3.7

Relative uncertainty of ultimate time and cost by life-cycle phase. (*From Russell D. Archibald, Managing High-Technology Programs and Projects, Wiley, New York, 1976, p. 23.*)

design engineer, a stylist, technicians, draftspeople, quality and manufacturing engineers, a marketing manager, buyers, and a home economist. As the team progresses through its development cycle, additional people are assigned, such as industrial engineers, cost accountants, quality control personnel, and production managers. An engineering project manager is assigned to manage this interdisciplinary project from the initiation of the development cycle until a steady level of production is reached.[15]

When human resources enter the picture, the *motivating* function of management comes into play. Managers need to become

[15]See William W. George, "Task Teams for Rapid Growth," *Harvard Business Review,* March-April 1977, for further description of how engineering task teams are used at Litton Industries, Inc.

82   concerned about the satisfaction of both the organizational and people needs of the project. The motivation function has application throughout the project's life.

Directing—that management function which is concerned with the decision process—is integral to the development and design of the project. Whether or not to continue development of the project is a key decision that has to be considered at various points in the life cycle of the project. Finally, the *controlling* of the project, i.e., using organizational resources as planned to successfully accomplish project objectives, comes into play. When a manager goes to meetings held for the purpose of reviewing the progress of a project, he may be required to perform the following activities:

> Reschedule the organizational resources being applied to the project.
>
> Approve a change in the technical specifications.
>
> Require that revised cost estimates be established for the development phase of the project.
>
> Approve a change in personnel working on the project.
>
> Recommend to higher-level management whether or not the project should be continued.

Of course, an engineering manager carries out these managerial functions—planning, organizing, motivating, directing, and controlling—in a somewhat continuous fashion throughout the life cycle of the project.

CONFLICT MANAGEMENT AND PROJECT LIFE CYCLE
An important part of the management of a project is dealing with the people who are working on the project, particularly as those people conflict with each other as they work on a project.

The ability to effectively manage conflict is usually given as one of the characteristics of effective project management. Conflicts of one sort or another invariably arise during the life cycle of a project. Several research studies have reported on the nature of conflict in project management. Thamhain and Wilemon have studied conflict management from the perspective of the individual project life-cycle stages.[16] Their study presents the conflict issues associated with the four accepted stages of projects: project formation, buildup, main program phase, and phase-out. Their study also presents some recommendations for reducing the consequences of conflicts.

---

[16]Hans J. Thamhain and David L. Wilemon, "Conflict Management in Project Life Cycles," *Sloan Management Review*, vol. 16, no. 3, Spring 1975.

Thamhain and Wilemon sampled project managers from a **83** variety of technology-oriented companies. Project managers were asked to rank the intensity of conflict experienced for each of seven conflict sources. The seven potential conflict sources were:

1   Conflict over project priorities

2   Conflict over administrative procedures

3   Conflict over technical opinions and performance tradeoffs

4   Conflict over manpower resources

5   Conflict over cost

6   Conflict over schedules

7   Personality conflict

Figure 3.8 presents the mean intensity experienced for each

Mean conflict intensity profile over project life cycle. (*From Hans J. Thamhain and David L. Wilemon, "Conflict Management in Project Life Cycles,"* Sloan Management Review, *Summer 1975.*)

3.8

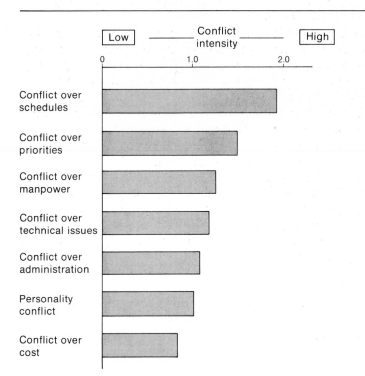

84   of the potential conflict sources over the entire life of the projects. Areas most likely to cause problems for the project manager during the entire life cycle are schedules, project priorities, and manpower resources.

Thamhain and Wilemon suggest some measures to minimize many of the potential problems arising out of the conflict found in the life cycle of projects. Table 3.8 summarizes their specific suggestions.

SIGNIFICANCE OF A LIFE-CYCLE APPROACH

On the surface, the concept of the life cycle that has been put forth in this chapter suggests a checklist to be followed in a mechanical fashion. Certainly no checklist has ever been useful for very long in any situation as dynamic as the management of an engineering project. We do not intend that the system and the questions that complement the model in Tables 3.1 through 3.5 be used as a checklist. Rather, we hope that the engineering manager accepts the notion that engineering activities are generally characteristic of the pattern suggested in these models; the complementary questions associated with these models can become a guide for understanding the totality of the engineering job to be done—and the questions to be asked in the doing of that job.

Answers to the questions outlined in the phases of the life cycle help the engineering manager recognize where a project or task is at any given time. The questions themselves suggest where the project should be at a phase in the life cycle. A good way to view these questions is to look backwards and see the phases the project has been through and then try to determine where it is currently located in its life cycle. When its current position is known, the use of foresight comes into play. Looking ahead provides a good perspective of both the present and the past.

Recognizing the existence of a life cycle in projects can have considerable value for the engineering manager and the people who are bringing products or services through an engineering analysis. To approach engineering management on a life-cycle basis can have value in several ways:

1   It encourages a deliberate engineering management philosophy to be followed. It systematically postulates an approach to managing the effort in advance and reduces the damage that so often results from stopgap, crisis-oriented decisions. The very existence of a life-cycle approach enforces a scientific approach to the planning and control that is to be carried out in the management of the activity.

2   It helps to force a specific analysis and review of

Major conflict sources by
project life-cycle stage

	MAJOR CONFLICT SOURCE AND RECOMMENDATIONS FOR MINIMIZING DYSFUNCTIONAL CONSEQUENCES		**3.8**
PROJECT LIFE-CYCLE PHASE	CONFLICT SOURCE	RECOMMENDATIONS	
Project formation	Priorities	Define plan clearly; joint decision making and/or consultation with affected parties Stress importance of project to goals of the organization	
	Procedures	Develop detailed administrative operating procedures to be followed in conduct of project; secure approval from key administrators; develop statement of understanding or charter	
	Schedules	Develop schedule commitments in advance of actual project commencement Forecast other departmental priorities and possible impact on project	
Buildup phase	Priorities	Provide effective feedback to support areas on forecasted project plans and needs via status review sessions	
	Schedules	Carefully schedule work-breakdown packages (project subunits) in cooperation with functional groups	
	Procedures	Contingency planning on key administrative issues	
Main program	Schedules	Continually monitor work in progress; communicate results to affected parties Forecast potential problems and consider alternatives Identify potential "trouble spots" needing closer surveillance	
	Technical	Resolve technical problems early Communication of schedule and budget restraints to technical personnel Emphasize early technical testing Facilitate early agreement on final designs	
	Manpower	Forecast and communicate manpower requirements early Establish manpower requirements and priorities with functional and staff groups	
Phase-out	Schedules	Monitoring schedules throughout cycle Consider reallocation of available manpower to critical project areas prone to schedule slippages Attain prompt resolution of technical issues which may impact schedules	
	Personality and manpower	Develop plans for reallocation of manpower upon project completion Maintain harmonious working relationships with project team and support groups; try to loosen up "high-stress" environment	

Source: Hans J. Thamhain and David L. Wilemon, "Conflict Management in Project Life Cycles," *Sloan Management Review,* vol. 16, no. 3, Spring 1975.

ongoing projects or tasks. All too often an engineering project is initiated and left on its own. People assigned to the project become dedicated to making the project work; often this dedication becomes so strong that the completion of the project becomes more important than an ongoing evaluation of the work to see whether or not the project should be completed.

3   Answering the questions in the phases of the life cycle is a demanding task. Seldom can one individual answer these questions. A team approach is required, bringing the team together at a regular basis to find the answers—or, perhaps just as important, to recognize that answers to the questions are not yet available.

4   It forces a wider view of the systems nature of the project. An example can be drawn from recent industrial experience to illustrate this point.

During the late 1960s and the 1970s several large United States corporations entered the residential construction business. One such corporation built a plant in the Southeast for the construction of modular houses. This corporation brought its high-technology manufacturing know-how to the operation of this plant. Key innovations drawn from production-line technology were introduced into the fabrication of modular homes. The factory became a good example of production technology. The production engineer who was the original project manager for the design, development, and construction of the plant did a superb job of bringing everything together. The plant was completed on time and within budget. The project manager was appointed as the plant manager of this facility. Production of modular houses started and soon rose to the plant's 40-hour weekly rate potential. At about this time some discomforting problems became evident:

The inventory of modular houses filled up the adjoining warehouses and overflowed onto the outdoor space on the company's property.

At the construction sites local building codes were found to be inconsistent with the configuration of some of the modular houses that had been constructed.

The mounting costs of transporting the modular house units to the construction sites indicated that the plant was not constructed in the right location.

The forecasts for modular housing were allowed to dictate the need for land acquisition—a clear failure to synergize the plant production and land acquisition activities.

The organization that was in the residential construction business was staffed with a mix of managers heavily biased toward production management. Consequently an undue amount of time and resources was spent on the improvement of the modular housing production activity; too little time was spent on the home real estate development and construction business and how that business related to the fabrication of modular housing units.

The reader will probably immediately note that some of these problems were not of an engineering nature and therefore out of the jurisdiction of the engineering manager. However, the systems idea that "everything is related to everything else" has to be considered. If a systems life-cycle profile model had existed for the design and construction of this plant—which it did not—there is a high probability that many of these larger systems issues would have emerged in sufficient time to do something about them.

The main larger systems issue was the whole residential construction business of this company. The plant was just a subsystem. If more careful analysis had been carried out during the development life cycle of this business activity, substantial monies would have been conserved for the corporation.

SUMMARY

In this chapter we have put forth the philosophy of a system life-cycle approach to engineering management and the value of such a philosophy for those people who manage engineering activities. An engineering manager is a person who accomplishes engineering goals through other people. An effective engineering manager influences the course of future events in his organization.

To see the larger whole in any management undertaking is a mind-stretching task—a task left to the manager who facilitates the perception of the larger whole through the use of a model similar to the systems life-cycle concept.

The philosophy of a life cycle in human events is found in many different segments of our society. An approach that uses a philosophy of a life cycle makes good sense. That is why we believe this philosophy should be embraced.

DISCUSSION QUESTIONS

1 What is meant by the statement "Each system has natural phases of development"?

2   Identify and briefly define a model of the phases of the life cycle of an engineering project.

3   The conceptual phase may be the most important phase in a system life cycle. Defend or refute this statement.

4   What are some of the key characteristics of the definition phase?

5   In the production phase of the life cycle, the inadequacies of the analysis conducted in the preceding phases will come to light. Why is this so?

6   What are some of the categories of cost escalation?

7   What conclusions can be reached about a system's economic and technological feasibility if it is in its operational phase?

8   In which phase does a system's sales life cycle begin?

9   An engineering project that is a failure can have some value for the organization. Why?

10   What are some of the management implications of the life-cycle concept?

11   What are the essential elements of an engineering project plan?

12   What are some of the potential conflict sources in the management of a project?

QUESTIONS FOR THE ENGINEERING MANAGER

1   Does the idea of the system life cycle have any particular application in this organization?

2   What are the engineering methods that are currently in use in this organization for evaluating the progress of an engineering project in its life cycle?

3   Have there been any cost overruns in engineering projects in this organization? What actions have I taken to reduce the probability of future overruns?

4   Are there opportunities in this organization to which the system life-cycle concept could be applied?

5   What are the essential elements of an engineering project plan as currently defined in this organization? Do the assigned engineers understand these elements?

6   Do I understand the changes that occur in an engineering project during its life cycle? What degree of control do I have over these changes?

7   How do I manage the assignment of personnel to a project according to the needs of the project's life cycle?

8   Is there any opportunity for a task-team structure to be used in this organization?

9   Have there been any conflicts in the management of engineering projects in this organization?

10   What methods exist for the resolution of conflicts within this or-   **89**
ganization? How successful have these methods been?

11   How effectively are decisions being made in this organization?
What input do the assigned engineers play in the strategic technological decisions of this engineering community?

12   What techniques are currently in use for determining where the
assigned projects are in their life cycles? How effective are these
techniques?

BIBLIOGRAPHY

Abramson, Bertram N., and Robert D. Kennedy: *Managing Small Proj-ects for Fun and Profit,* TRW System Group, 1969.

Archibald, Russell D.: *Managing High-Technology Programs and Projects,* Wiley, New York, 1976.

Baldwin, Hanson W.: *Strategy for Tomorrow,* Harper, New York, 1970.

Bass, Lawrence W.: *Management by Task Forces,* Lomond Books, Mt. Airy, Md., 1975.

Blanchard, Benjamin S.: *Logistics Engineering and Management,* Prentice-Hall, Englewood Cliffs, N.J., 1974.

Cleland, David I., and W. R. King: *Systems Analysis and Project-Management,* 2d ed., McGraw-Hill, New York, 1975.

Davis, D. Larry: *Life Cycle of Human Resources,* Human Resources Institute, University of Utah, Salt Lake City, 1973.

Davis, Stanley M., and Paul R. Lawrence: *Matrix,* Addison-Wesley, Reading, Mass., 1977.

de Kluyver, Cornelius A.: "Innovation and Industrial Product Life Cycles," *California Management Review,* vol. 20, no. 1, Fall 1977.

Drucker, Peter F.: *Management: Tasks, Responsibilities, Practices,* Harper, New York, 1973.

Duke, Robert K., H. Frederick Wohlsen, and Douglas R. Mitchell: "Project Management at Fluor Utah, Inc.," *Project Management Quarterly,* vol. 8, no. 3, September 1977, p. 37.

Final Status of Major Acquisitions, Report to Congress by Comp-troller General of the United States, June 30, 1975.

Forrester, Nathan B.: *The Life Cycle of Economic Development,* Wright-Allen, Cambridge, Mass., 1973.

George, William W.: "Task Teams for Rapid Growth," *Harvard Business Review,* March-April 1977.

90   Glaser, Milton A.: "The Innovation Index," *Chemical Technology,* vol. 6, March 1976.

Greiner, Larry E.: "Evolution and Revolution as Organizations Grow," *Harvard Business Review,* July-August 1972.

Hajek, Victor G.: *Management of Engineering Projects,* McGraw-Hill, New York, 1977.

The Initiation and Implementation of Industrial Projects in Developing Countries: A Systematic Approach, United Nations Industrial Development Organization, New York, 1975.

King, William R.: *Quantitative Analysis for Marketing Management,* McGraw-Hill, New York, 1967.

Lawrence, Paul R., et al.: "The Human Side of the Matrix," *Organizational Dynamics,* American Management Association, New York, Summer 1977.

Levitt, Theodore: "Exploit the Product Life Cycle," *Harvard Business Review,* November-December 1965.

Lock, Dennis L.: *Complete Guide to Project Management,* Cahners Books, Boston, 1968.

Martin, Charles C.: *Project Management—How to Make It Work,* American Management Association, New York, 1976.

Merrifield, D. Bruce: *Strategic Analysis, Selection, and Management of R&D Projects,* AMACOM, New York, 1977.

Mullins, P. L.: "Capital Budgeting for Research and Development," *Management Services,* May-June 1969.

Osgood, William R., and William E. Wetzel, Jr.: "A Systems Approach to Venture Initiation," *Business Horizons,* October 1977.

Thamhain, Hans J., and David L. Wilemon: "Conflict Management in Project Life Cycles," *Sloan Management Review,* vol. 16, no. 3, Spring 1975.

"Vanishing Innovation," *Business Week,* July 3, 1978.

THE HUMAN SUBSYSTEM IN ENGINEERING 4

This chapter presents an overview of the human subsystem in engineering management. The subsystem will be described from the viewpoint of the many roles the engineering manager is required to play as he carries out his managerial responsibilities. One of these roles, that of *motivator,* is particularly significant to the engineering manager, since the effectiveness of motivation in the organization can influence a mission's success or failure. In this chapter we will give particular attention to the manager's role as a motivator.

The reader should gain an appreciation of the "interdisciplinary generalizations"[1] about the human subsystem in modern engineering organizations. Human subsystem involves just about everything associated with the human element. Taken in its extreme meaning, all management matters would fall into the area of the human subsystem, since management means working with other people to accomplish organizational purposes. An understanding of the human subsystem requires some knowledge of sociology, psychology, anthropology, communications theory, semantics, cybernetics, decision theory, logic, philosophy, leadership, and so on! How can the manager hope to develop expertise in all these disciplines in order to be able to manage effectively the human subsystem of an organization? It would be difficult to become an expert in so many disciplines, but the manager can gain an understanding of what these disciplines have to contribute to the management of people. Good management results when managers use their experience, their intuition, and interdisciplinary generalizations to guide them in their relations with other people.

[1]Term credited to F. J. Roethlisberger in *The Elusive Phenomena,* Harvard Graduate School of Business Administration, Boston, 1977.

92   This approach may sound quite unscientific to the engineer. It is, however, the essence of what it takes to succeed in dealing with people: an understanding of the interdisciplinary generalizations about people in their workplace. The application of interdisciplinary generalizations in dealing with people can be approached in a rational manner. Indeed, the development of management thought and theory has been accomplished from such a rational base.

THE RATIONAL NATURE OF MANAGEMENT

Most books on management encourage the manager to develop a systematic approach to the work of management. Such books examine and analyze the managerial functions of planning, organizing, motivation, direction, and control in a systematic, rational fashion. There is much to be said for treating the manager's job in this manner. Much of what the manager does can be learned and is reducible to logical, rational thought processes.

Engineering managers would readily agree there is a need to develop rational thought processes in applying the management functions. But they would also readily admit that many of their challenges cannot always be met by simply applying rational thought. Some of the problems (and opportunities) that come to the manager's attention for resolution may seem to defy any logical solution— particularly where the human subsystem is involved.

These human-subsystem problems are difficult to analyze in a logical manner because the manager himself is a key factor in the situation. Most of what an engineering manager does is to interact with people, both inside and outside the organization. He interacts with a group of "clientele"—a body of people who have a vested interest in the problem (or opportunity) at hand. These clientele are the principal sources of the information the manager needs to make decisions; they are also the group through which the problem has to be worked and the solution implemented. Consequently, the manager's skills in interacting with these clientele are a key factor in how successful he will be in achieving organizational objectives.

It is not enough for the manager to be simply logical and rational in his practice of the management functions. He must also find ways to put himself into the picture of the human subsystem of the organization. To do this requires an ability to "humanize" the managerial role. When the managerial role is so performed, there is a good chance that the rest of the organization will reflect this performance.

The failure to allow for the human element in management was dramatically exemplified by problems at the General Motors Corporation Vega Plant at Lordstown, Ohio. Many of the production line stoppages and other problems at this plant were attributed to a lack of empathy for the production workers and the assembly-line environment in which they worked. Hoping to avoid a similar human **93** conflict, the plant manager of a new Volkswagen plant in New Stanton, Pennsylvania, set up an "unusual training program for assembly line foremen as the . . . plant gears up for production of its Rabbit model next year . . . . The German-based auto builder's approach to worker-management relations was kicked off last week with a three-day seminar for new supervisors on the 'value of the person' in the workplace."[2]

Comparing the experiences of these two situations should help the engineering manager decide what to do about developing his own skills in humanizing his managerial role. Engineering schools are making a contribution, with more schools developing curricula around the theme of "humanizing the engineer."[3]

There is a large body of literature in the behavioral science field devoted to managerial techniques for improving interpersonal relations. This body of literature has an important place in the lexicon of engineering management. It focuses on the engineering manager's ability to work with people—to improve his perceptions of the technologists with whom he works to accomplish organizational ends.

The nature of the interpersonal relationships between the engineering manager and the individuals and groups with whom he works is not so much a matter of organizational structure as it is a matter of *style of interaction*. The manager plays many roles in these interactions. His perceptions of the roles, and the flexibility with which he plays them, are important factors in his success. One must remember that these roles cannot be separated from the rational thought process of managing—planning, organizing, and controlling the activities in the organization. These thought processes are carried out within the context of interpersonal interactions through which the manager accomplishes results by working with other people. As one examines the variety of these interactions, their impact on the nature of the engineering manager's many roles should be clarified.

ROLE PLAYING IN ENGINEERING MANAGEMENT

In this section we examine the roles that an engineering manager plays and how these roles impact on his work as a manager.

A TECHNOLOGIST
To supervise technical people, the manager has to have an understanding of the technology involved. Imberman found that engineers

---

[2]"It's Strictly Carrot, Not Stick, at Rabbit Plant," *The Pittsburgh Press*, Mar. 5, 1978.
[3]See "Engineers Train for the New World of Business," *Business Week*, Jan. 16, 1978.

94   resent supervision by nonengineers. They want managers they can respect for their knowledge, ideas, and expertise.[4]

An engineering organization exists to develop a technology and to facilitate its effective use in some market context. An engineer who becomes the manager of technology will find himself in a conflict situation: *technologist versus manager.* A manager, particularly one in a rapidly changing technical world, faces the difficult challenge of keeping a balance between appraising progress in a technology and facilitating the work of people whose main objective is to maintain a level of the state of the art in that technology.

The engineering manager provides a focus for technical leadership—establishing technical objectives and providing for the application of technical knowledge in the organization. Therefore, he plays a dual role. As a technologist, or a technical strategist, he defines the technical targets, sets the system boundaries, determines the priorities, and develops the guidelines for the measurement of the output. As a critic, he passes judgment on the adequacy of the work of the other technologists.

Engineers recognize that their work must be evaluated. They know that someone—or some group—has to pass judgment on the technical relevance of what they are doing. We often assume that this judgment is done by the peer group. However, someone must organize the peer group to be able to pass judgment on the technical work; it is the manager's job to use the peer group as a vehicle to evaluate the effectiveness with which technical objectives are being attained. Even in this context, the manager's role as a technologist requires that he know enough about the technology involved to be able to ask the right questions—and know if he is getting the right answers.

The role of the technologist-manager as a critic may be illustrated by an example from the academic community. In an engineering school, promotions are based primarily on the judgment of peers. In principle, such judgment is collective and objective, and it is related to the contributions the faculty member makes to the university. These contributions are usually valued highly in the decision as to a faculty member's qualifications for promotion. For example, a key criterion for promotion to the academic grade of full professor is the national-international image the candidate has attained in his field. Evaluations of faculty and of technologists outside the university are sought. Many peer groups are involved in reviewing the fitness of a faculty member. The dean of the engineering school, who has some understanding of the technology involved, serves mainly as a facilitator to organize a process whereby the peer group has an opportunity to function as a group critic. The initiation of new academic programs, the modification of curricula, and the de-

[4]W. Imberman, "What Bugs Engineers?" *Machine Design,* Aug. 24, 1978, pp. 88–92.

velopment of strategies for research are some additional areas **95** where the dean of the school serves as facilitator of peer-group actions.

This example from the academic community illustrates the manager-technologist's responsibility to see that a mechanism exists for eliciting technical leadership from the technologists in the organization to help appraise results. If the engineering manager can organize the technical talent in the organization to evaluate progress, he need not worry about trying to keep up with every new technical development in his field. When he needs a technical evaluation of a problem (or opportunity), he can organize a temporary team of technical experts to advise him. If the manager does this as a matter of course in the organization, it will tend to strengthen the peer group as an evaluation mechanism. Peer groups will take on this role as a part of their ongoing activities. When this happens, the manager can concentrate on the nontechnical aspects of his managerial role.

In effect this peer group evaluation is carried out at Texas Instruments through the manner in which this company manages innovation. A key part of this process is accomplished through the "idea concept" whereby an individual can submit a proposal for funding of a development project. The approval of such a project includes an evaluation by a peer group. After the development project is funded, a team is organized to provide the disciplines required to accomplish the project. The use of such a team at Texas Instruments has helped to foster a cultural ambience where "it's impossible to bury a mistake in this company; the grass roots of the corporation are visible from the top. There's no place to hide in TI—the people work in teams, and that results in a lot of peer pressure and peer recognition."[5]

AN AGENT

A second role that the engineering manager plays is that of an agent—one who represents and acts for another in a transaction. As an agent, the engineering manager "markets" the capabilities and outputs of the technical people to his clientele—customers, higher levels of management, outside clients, and other groups that have a need for the products or services of the organization. Since many of the top-level decisions affect the technical people in the organization, someone has to represent their interests in these decisions. Management meetings provide ample opportunity for the engineering manager to act as an agent of the technical workforce and to negotiate for the best interests of the engineers and scientists. This suggests another role for the manager—the role of negotiator.

[5]"Texas Instruments Shows U.S. Business How to Survive in the 1980s," *Business Week*, September 18, 1978, p. 81. Reprinted by special permission, (c) 1978 by McGraw-Hill, Inc. All rights reserved.

96

A NEGOTIATOR

A negotiator is one who arranges an agreement on an issue. Engineers can see the need for someone who is a competent negotiator for the technical organization and can effectively resolve the conflicts among the various clientele of that organization.

There is inherent conflict in an organization because of the way we divided up the work to be done. Engineering departments usually have many experts with special knowledge that can be used to solve specific technical problems. These experts see the technical world through their specialty. This causes them to develop a form of parochialism that tends to restrict their view of the world: they see only their part of the world and nothing else. An engineering manager has to negotiate the conflict that arises out of this parochialism so that the best overall decision is made for the organization. This negotiation role is also manifested in the engineering manager's responsibilities in serving as a member of a team for labor negotiation and strategies, contract negotiations, and working out the conditions of employment for new employees.

The manager must exercise negotiation skills to protect the interests of his people. Many situations in engineering management involve a give-and-take bargaining, where the engineers expect their manager to look out for their best interests and negotiate the best deal possible for resources to support the organization. Many times the skills of negotiation center around the engineering manager's role as a logistician.

A LOGISTICIAN

A logistician is one who procures, distributes, maintains, and replaces human and nonhuman resources for the organization. The term "logistics" has a military connotation, describing the discipline of providing the supplies to support the fighting forces in their field operations. An engineering manager serves as a logistician for his organization and supplies the resources so that the technologists can do their work. In this role, the technologists expect the manager to be tough, shrewd, and an expert in the political process of organizational in-fighting. Engineers are aware of the resource constraints imposed by the organization's budget. They see the budget-allocation process as a time when many managers in the firm compete for scarce resources. A manager who succeeds in obtaining a good budget for his organization will be viewed as a success by the engineers working for him; they will have the equipment, space, and materials needed to carry out their work. Conversely, if the manager is not successful in competing for these scarce resources through the budgeting process, he may be viewed as a failure irrespective of his technical and managerial skills.

In either case, with an adequate or inadequate budget, the

manager will have to carry out the role of logistician for his people. If **97** the manager is a skillful logistician, this strength can compensate for deficiency in other roles. Engineers can excuse the other inadequacies if their manager obtains the resources they need to carry out their work and to achieve success as individuals as well as contribute to organizational goals.

As a logistician, the manager does a lot of things that no one else wants to do. Perhaps these pedestrian but necessary duties are not properly called logistics duties—some of them may be considered as supply clerk duties. Sometimes they are viewed as trifling and irritating. They involve answering correspondence, smoothing relations with clientele, maintaining social contact with sister organizations, participating in protocol meetings, unsnarling administrative foul-ups, etc. All this busy work has to be done by someone; engineers resent getting involved in such work and typically view it as unproductive or downtime activity, to be avoided at all times. They are grateful to have the manager take care of all this paper shuffling and leave them free to do their technical work.

In the role of organizational logistician, the engineering manager is involved in many activities that have little to do with the technical work being carried out. But if the logistics work is not carried out to the satisfaction of the engineers, the technical work will suffer. Most of this work is short-range in nature, taking care of resource needs for the immediate future. However, there are long-term or strategic logistic needs as well. This suggests the next role of the engineering manager: a strategist for the organization.

A STRATEGIST

A strategist is one skilled in the use of science and art in the development of a sense of future direction for an organization. As a technical strategist the engineering manager is concerned with the selection of engineering objectives and goals and the design of a plan for utilization of organizational resources most economically to accomplish objectives and goals. The senior engineering manager of an organization is expected to work as a member of the top level management team in the design of overall organizational purposes. He represents the engineering community in this respect and is expected to contribute engineering know-how in a synergistic fashion along with marketing, production, and finance strategies to attain the organizational purposes. In the capacity of a technical strategist, the engineering manager is dealing with diverse, fundamental problems, issues, and opportunities. He manages an intellectual process of collaborating with other key people in affirming the basic purpose of the organization. The external engineering world has to be put into perspective, the organization's resources analyzed, the competition assessed, questions concerning societal and environmental con-

98   tributions answered, and finally the purpose for the organization as a whole determined. All this work is carried out so that the engineering effort is consistent with higher organizational objectives. If this role is done well, the manager can face the hardships and uncertainty of the future with confidence and organizational self-reliance. This will be recognized by members of the organization; an attitude of well-being and personal satisfaction will result. Professional people who are given career counseling will often want to know what future the organization has—and how they can fit into that future. These career counseling sessions introduce another key role of the engineering manager, viz., that of *counselor*.

A COUNSELOR

A counselor is a person who participates in an exchange of opinions and ideas and gives advice and guidance. There are many situations in which an engineering manager is called on to serve as a counselor to individuals and to groups of clientele. This counsel covers a wide spectrum ranging from advising top management on program development strategy to the job-performance counseling given to a professional.

A skillful manager will use nondirective counseling to help his professionals make their decisions. In this type of counseling, the manager does not attempt to give advice directly. Instead, he asks probing questions about the decision under study, thus leaving the initiative to the individual. Nondirective counseling is an art. It requires the patience to listen without passing judgment; the individual is led through a series of questions to realize for himself what the manager may already know. When a decision is made, the professional feels that it was his decision, and not one made by the manager.

The authors have worked with the director of a large R&D laboratory who is known internationally for his technical knowledge. This director is more of a technologist than a manager. His skills in the management functions—planning, organization, etc.—are below average; yet his laboratory is successful. Some of the reasons for this success hinge on this individual's skills in counseling his project managers on their research through a nondirective approach in a group situation. When key decisions have to be made—for example, to commit resources to a research project—this director will start a line of questions like: "Have you thought of this?" "What would happen if you did thus and so?" "What are the five or six things you believe to be the most important considerations in this research project?" These kinds of questions and many others are put to the project managers quietly and diplomatically. If there are inadequate answers, the director proceeds to question more intensely in these areas; he never seems to react to an answer except to ask another question. If the research project has not been thoroughly **99** analyzed, this lack of analysis soon comes out. Finally, the project manager says something like: "I would like to study this situation some more and get back in touch with you." The counselor-director responds: "That would be very useful!"

Within his laboratory, this director has an image of a very decisive person, yet his decisions are always worked through his managers along the lines we described. He is much admired and emulated by his people. His ability to do nondirective counseling is by far his most important managerial strength. He provides technical leadership by drawing on the talents of his senior technical people and thus multiplies the total technical leadership available to the organization.

Even though the engineering manager serves in the capacity of a nondirective counselor, he must not forget his role as a disciplinarian in the organization.

A DISCIPLINARIAN

A disciplinarian is one who enforces organizational policies and procedures and keeps the organization on the most promising path toward objectives. This is a role that draws on the manager's teaching and taskmaster skills. It is also a role where the manager has to make tough decisions about an individual's potential growth in the scheme of things. A disciplinarian's role is not an easy one—and it is a role that managers like to forget, for some of the decisions that have to be made are unpopular. Fairness, equity, and objectivity are all considerations that are important to keep in mind as the manager carries out the role of a disciplinarian.

In carrying out this role, the manager puts a lot of emphasis on the control function of management, i.e., seeing that things are happening according to the established plan. If a study of actual organizational progress shows a departure from the progress that was planned, then corrective action has to be instituted. It is important that this corrective action be instituted quickly so that organizational wasted effort is reduced and so that the professionals perceive the ambience of the organization as motivational in nature.

A MOTIVATOR

In the capacity of a motivator, the engineering manager is responsible for providing an environment in which people attain social, psychological, and economic satisfaction in their work. Engineers are used to dealing with problems and opportunities that are typically well-defined; the forces bearing on these problems and opportunities can be counted, measured, and evaluated. When the engineer comes into contact with the managerial aspects of his job, there is a tendency to look for those forces that can be measured. Many of the

100 forces of management, particularly the motivation of people, may at times be so nebulous as to deny any real definition and evaluation.

An important aspect of the behavior of people centers around the motivation patterns that are found in the human subsystem. An engineering manager must work through and for others as he performs his managerial role. Consequently, he requires an understanding of the forces that motivate the individual in the workplace.

To motivate people is to stimulate them to action. People are motivated by a wide variety of stimuli. Sociologists, psychologists, psychiatrists, and behavioral scientists are some of the professionals who have studied the theory and practice of motivation. It would be impossible to summarize in one chapter all their findings about the factors and forces that motivate people. About all we can do is make some observations about the factors that seem to have an influence in motivating an individual to action—to do something in support of organizational objectives.

To motivate an engineer, the manager should try to understand what engineers look for and aspire to in their work. Raudsepp comments on this as follows:

> First, engineers want to feel that their work is interesting; they want a feeling of personal professional growth.... Secondly, an engineer wants a salary at least equal that of other engineers on his level.... Third, engineers want status that is more than symbolic; they do not value grandiose titles and regard them as placating devices to deflect their attention from more concrete symbols of status and higher salaries. Fourth, they want effective communication with other components of the organization; they want to feel that top management knows what they are doing; they want to be kept informed about company policies, plans, and operations outside their own work areas. Fifth, engineers want effective managers who can become involved in the engineers' projects.... Sixth, engineers want modern, up-to-date facilities and ample working space.[6]

As one might expect, engineers are motivated by factors directly related to their job situation.

JOB MOTIVATIONAL FACTORS

Over many years of dealing with engineering organizations in a workshop situation, the authors have conducted a survey using the questionnaire shown in Table 4.1. This questionnaire is given to profes-

---

[6]Paraphrased from E. Raudsepp, "Motivating the Engineer: The Direct Approach Is Best," *Machine Design*, Nov. 24, 1977, pp. 78–80.

Job motivational factors

Please indicate the five items from the list which you believe are the most important in motivating you to do your best work.   **4.1**

_____	1	Steady employment
_____	2	Respect for me as a person
_____	3	Adequate rest periods or coffee breaks
_____	4	Good pay
_____	5	Good physical working conditions
_____	6	Chance to turn out quality work
_____	7	Getting along well with others on the job
_____	8	Having a local house organ, employee paper, bulletin
_____	9	Chance for promotion
_____	10	Opportunity to do interesting work
_____	11	Pensions and other security benefits
_____	12	Having employee services such as office, recreational, and social activities
_____	13	Not having to work too hard
_____	14	Knowing what is going on in the organization
_____	15	Feeling my job is important
_____	16	Having an employee council
_____	17	Having a written description of the duties in my job
_____	18	Being told by my boss when I do a good job
_____	19	Getting a performance rating, so I know how I stand
_____	20	Attending staff meetings
_____	21	Agreement with agency's objectives
_____	22	Large amount of freedom on the job
_____	23	Opportunity for self-development and improvement
_____	24	Chance to work not under direct or close supervision
_____	25	Having an efficient supervisor
_____	26	Fair vacation arrangements
_____	27	Unique contributions
_____	28	Recognition of peers
_____	29	Personal satisfaction

Source: Adapted from David I. Cleland and William R. King, *Management: A Systems Approach,* McGraw-Hill, New York, 1972, p. 370.

sionals in engineering organizations—engineers, scientists, and engineering-scientific managers—as a way of getting them to think about the conditions of motivation in their organization. After the respondents fill out these questionnaires, distribution of the responses is then tabulated for the participants as a group and posted for all to see—to use as a focus for a discussion on the subject "What motivates people to do their best work?" A lively discussion then follows, as the participants are given a good idea of what motivates them and their peer group. A pattern of responses has emerged from these people, indicating the items listed in alphabetical order in Table 4.2 as being the most important in motivating them to do their best work.

From these results, one might make the following observation: Good pay is an important factor in motivation—but so are many other factors. It is important to note that many of these factors deal with the workers' environment and the interpersonal relationships

Job motivational factors–
most frequent response

4.2	Chance for promotion
	Chance to turn out quality work
	Feeling my job is important
	Getting along well with others on the job
	Good pay
	Large amount of freedom on the job
	Opportunity for self-development and improvement
	Opportunity to do interesting work
	Personal satisfaction
	Recognition of peers
	Respect for me as a person

between the engineering manager and the peer group. It is also obvious that the manager has a significant impact on how well the members of his organization are motivated to do their best work.

When an engineering manager studies Table 4.2, he might ask himself the question: "Given that these are the factors that motivate an individual to do his best work, what have I done to create an environment in the organization to highlight these factors?" If the manager wants to go further, he can administer this questionnaire to his people to get an insight into how they feel about the organizational climate for motivation. In administering the questionnaire to his engineers, the manager should be careful to do the following:

Get everybody together.

Explain the purpose of the questionnaire.

Have the people fill out the questionnaire.

Preserve the anonymity of their responses.

Immediately tabulate the results.

Discuss the results with the participants.

Be prepared to be "put on the spot."

Take action about the suggestions that are bound to come out of the discussion.

Provide the people with feedback on what is being done.

If the manager is willing to do this, he should gain benefits for his organization by way of a better climate for motivating people. However, the manager must be careful, because if the use of this questionnaire is not handled properly, it could create a negative impression. For example, if the manager administers the questionnaire, shares the results with his people, and then does nothing about **103** their suggestions, the climate for motivation could be impaired.

There are many different factors that motivate people to do their best work. A project engineer may be motivated more by having a "large amount of freedom on the job" than by "adequate rest period or coffee breaks," since he can probably take a coffee break anytime it is convenient for him to do so. By contrast, a worker who fills a key position on a production line may feel that having an "adequate rest period or coffee break" is more important to him as a motivator. This suggests that individuals are unique in themselves and how they view their place of work and what motivates them. Individuals are conditioned by many factors in their lifetime, including their cultural background, education, and previous experience, to name a few. These factors play a role in the way an individual perceives his environment and how he is motivated. Jucius[7] has identified several intangible factors that an individual brings to his job that influence his behavior. These intangible factors include:

Psychological attitudes

Physiological condition

Political beliefs

Moral standards

Professional standards

Prejudices and habits

The above factors are very difficult to measure; an understanding of their existence can give insight into how they may influence an individual's on-the-job behavior. A manager should be aware of their effect on the individual's job performance and should try to understand why an individual acts as he does. A person's attitude about his job and the effectiveness with which he performs his duties may be influenced by early experiences, apart from the current job situation. Each of us is a victim of everything that has gone before us in our lifetime. All these previous experiences and attitudes combine with our current job situation to make our motivation and behavior complex and not necessarily rational.

UNDERSTANDING HUMAN BEHAVIOR

Which of the roles described above are the most important for the manager? Of course, this depends on the situation at hand; the manager "plays many parts." His perceptions of these many parts and of the time to play each part are important factors in his success. These

[7]Michael J. Jucius, *Personnel Management,* Irwin, Homewood, Ill., 1963, p. 86.

104   roles do not exist independently of each other or of the rational thought processes of the manager. Managers vary in their ability to play these roles; strength in one role can help assure success for a manager. Managers who are interested in a personal development program for themselves should strive to improve their capabilities in all roles.

The real impact these roles have on the manager is the realization that all of them have a strong person-to-person characteristic. As the manager plays the many parts, his interpersonal skills determine how well he is able to represent the engineering community in these various roles.

There are no peopleless organizations. This statement may seem redundant and trite, yet as one observes the behavior of some engineering managers trying to manage technical people, one is struck by their insensitivity to the human subsystem in the organization. Why is this so? For many reasons, no doubt. We believe that a large part of this problem stems from a lack of exposure for the engineering manager to the subject of the behavior of people.

An important generalization about the behavior of people deals with their needs and how these needs affect their motivation. Maslow,[8] an early writer in the field of motivation, gave us a good insight into the subject of needs and motivation.

A CONCEPT OF MOTIVATION
Maslow put forth the idea of a "hierarchy of human needs" that placed the various needs of individuals in relationship with one another. This hierarchy of needs was developed to describe certain levels of human needs that motivate individuals in general. Maslow's concept is best portrayed in Fig. 4.1.

The relationships of the needs illustrate the idea that certain basic physiological requirements (thirst, hunger, sex, sleep, etc.) have to be satisfied before an individual can be expected to seek satisfaction of higher-level needs. As higher-level needs are satisfied, the individual shifts his drive for satisfaction to another category. Put another way: A satisfied need is no longer a motivator of action.

Modern management theory has accepted Maslow's hierarchy-of-needs concept as a reasonably good description of reality. The higher-level needs in this hierarchy have to do with the mind and the spirit rather than the body. One management author considers these higher-level needs "nebulous in nature." According to Davis:

> Dissatisfied workers usually attribute their dissatisfaction to something more tangible, such as wages. Many wage dis-

[8]A. H. Maslow, "A Theory of Human Motivation," *Psychological Review*, vol. 50, 1943, pp. 370–396, and *Motivation and Personality*, Harper, New York, 1954.

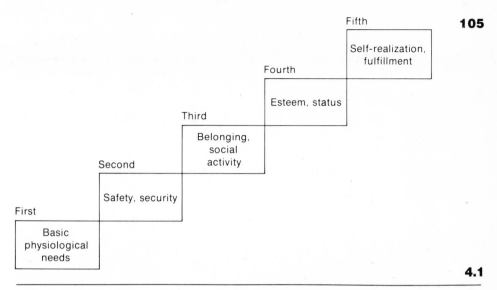

Fifth

Self-realization, fulfillment

Fourth

Esteem, status

Third

Belonging, social activity

Second

Safety, security

First

Basic physiological needs

4.1

Order of priority of human needs. (*Adapted from Keith Davis,* Human Relations at Work, *3d ed., New York, 1967.*)

putes are consequently not really wage disputes, and meeting the wage request does not remove the basic dissatisfaction which existed . . . . [Such] needs have the following characteristics:

1   They are strongly conditioned by experience.

2   They vary in type and intensity among people.

3   They change within any individual.

4   They work in groups, rather than alone.

5   They are often hidden from conscious recognition.

6   They are nebulous feelings instead of tangible physical needs.

7   They influence behavior. It is said that we are logical only to the extent our feelings let us be.[9]

Experienced managers will attest to the real-world implications of human needs and motivation in their organizations. For such managers the use of participatory techniques to help satisfy the professional's needs and to bring about a higher degree of motivation on the part of the professional is essential.

[9]Keith Davis, *Human Relations at Work,* 3d ed., McGraw-Hill, New York, 1967, p. 333.

106   PARTICIPATION IN MANAGEMENT

Participation is defined by Keith Davis as "an individual's mental and emotional involvement in a group situation that encourages him to contribute to group goals and to share responsibility for them."[10]

Three important ideas are inherent in this definition: first, an ego involvement is necessary on the part of the individual; second, participation stimulates the individual to make a contribution to the organization; and third, participation encourages people to accept responsibility for an activity.[11] Participation can help the people in an organization develop a sense of belonging. It can raise productivity, improve morale, and provide an outlet for the creativity and innovation skills of people.

The desire of people to participate in their organization's activities is "rooted deep in the culture of free men around the world, and . . . . is probably a basic drive in men."[12] Comparative studies in England and the United States suggest that participation is a basic human drive rather than a cultural acquisition.[13] The Japanese approach to decision making puts emphasis on getting the participation of the people who are to be affected by the decision. The use of project teams, task forces, and similar approaches to specific organizational problems is an effective way of getting people to participate.

Litton Industries, Inc., has used task teams operating within a functional structure to achieve and manage a very rapid growth. These task teams foster a high degree of team participation, which is considered to be more important than individual contribution. The results speak for themselves in the Microwave Division: "The division's sales grew from $13 million to $130 million in five years, its profit increases have averaged 75% per annum, it enjoys market leadership in every product category, with market shares ranging from 25% to 75%, and its unit production has increased from 125 units per day to more than 3,000 per day."[14]

Participation is not all "roses." There are practical limits to what it can do for the manager. If people do get involved, they will expect to be kept informed on what is happening in the organization, particularly on those matters in which they have participated. Thus the manager will have to be careful to keep the communication channels in the organization open to provide relevant feedback.

When project teams are used in the management of an engineering project, a high degree of participation of the professional

---

[10]Ibid.
[11]Ibid.
[12]Ibid.
[13]See N. R. F. Maier and L. R. Hoffman, "Group Decision in England and United States," *Personnel Psychology*, vol. 15, Spring 1962, p. 86.
[14]Reprinted by permission of the Harvard Business Review. Excerpt from "Task Teams for Rapid Growth," by William W. George (March–April 1977). Copyright © 1977 by the President and Fellows of Harvard College; all rights reserved.

people occurs. In such a situation, most important decisions require **107** the participation of all team members. Unilateral decision making rarely happens. If the manager is used to a clear line of authority to make unilateral decisions, the participation of team members in the decision process can be frustrating. The result, however, is worth the effort, as George notes: "The strength for this decision-making process is, of course, the high caliber of the decisions themselves and the high degree of individual commitment toward carrying out the decisions in an effective manner. Both are direct results of broad participation in the decision-making process."[15]

Participation should be meaningful so that the people see their contribution assisting in the accomplishment of organizational goals and objectives. Engineers participate in the engineering organization's work through providing specialized skills when and where required. Participation is a form of personal involvement over and above the providing of specialized skills when and where needed. Participation means that the individual gets involved in an activity to the extent of sharing the responsibility for seeing that something is accomplished. Authority for making a decision about what is to be done still rests with the responsible manager, who makes that decision after considering the viewpoint of his team members. Participation does not mean that the manager gives up his legal right to make decisions on behalf of the organization; it means that he provides an opportunity for the people in the organization to contribute advice and counsel as to what that decision should be. A couple of examples known to the authors will illustrate this point.

The president of a manufacturing company became interested in instituting project management in the engineering department in his organization. He set up a team of engineers representing the various disciplines assigned to the engineering department. They were asked to select a leader and were given the objective: "Determine if the company should institute project management." The team was provided financial resources to cover its operating expenses, such as travel, clerical help, etc. It studied this problem for several months, interviewed many of the company's engineers and engineering managers, visited other companies that were using project management, and developed a bibliography on project management. Periodic briefings were held for the president and his staff. Over a period of several months, many questions were studied: What can project management do for the company? What problems might project management create? What strategy should be adopted to implement the organizational changes? Finally, recommendations were made as to how the company could reorganize its engineering department along a project-management approach.

---

[15]Ibid., p. 80.

108   The team continued in existence during the transition to a project-management approach, to act as a sounding board on the organizational change being brought about and to suggest fine tuning of the strategy being implemented. When the change was completed, the team was disbanded.

In another situation, the manager of a manufacturing development laboratory in a large R&D organization needed to invest substantial monies in additional equipment to enter a new market area. A team of engineers was organized to study the organizational needs, determine what was available in new equipment, and make recommendations to the R&D laboratory director on what equipment to procure. The team's recommendations were accepted by the manager and the equipment was subsequently obtained.

Using teams of professionals to participate in making decisions helps to satisfy the needs of the members and the organization. A valuable system of checks and balances comes into play as the members participate in building the data bases necessary to support the decision. By participating, the team members gain insight into the work of management as well as a sense of contributing to organizational objectives.

An important aspect of participative management deals with the manager himself. He, too, is subject to all the forces of human behavior that affect his organization, many of which have been presented in the previous pages. His decisions, which affect the entire organization, are not solely economic, and technical in nature but are conditioned by his prejudices, ethics, politics, attitudes, morals, etc. He may decide issues on ethical or moral bases rather than on the basis of what makes the best engineering sense for the organization. He may make a decision based on what he personally wants to do rather than what is the most logical to support the organizational goals. Christensen, Andrews, and Bower have given us some insight into such a situation:

> We must acknowledge at this point that there is no way to divorce the decision determining the most sensible economic strategy for a company from the personal values of those who make the choice. Executives in charge of company destinies do not look exclusively at what a company might do and can do. In apparent disregard of the second of these considerations, they sometimes seem heavily influenced by what they personally want to do.[16]

This is not a complimentary thing to say about managers— but they are human, too, and are only as rational and logical as their feelings permit them to be!

[16]C. Roland Christensen, Kenneth R. Andrews, and Joseph L. Bower, *Business Policy: Text and Cases,* 4th ed., Irwin, Homewood, Ill., p. 448. © 1978 by Richard D. Irwin, Inc.

Another facet of human behavior that can facilitate or **109** negate the management of an organization is the attitude held by managers about people in general. In the next section, we explore the role that attitudes play in the management of people.

ATTITUDES TOWARD PEOPLE

An attitude is a state of mind or feeling with regard to some matter. The attitude that a manager has about the nature and behavior of people will be reflected in how successful he is in leading people toward the organizational goals. In 1960, Douglas McGregor challenged some of the traditional attitudes about managing people in his book, *The Human Side of Enterprise.*[17] In this book, McGregor put forth the idea of "theory *X*" and "theory *Y*" as representative of the general attitudes held by managers about the role of people in organizational life. According to McGregor, behind every managerial decision or action are assumptions about human nature and human behavior. For theory *X* these assumptions are:

1   The average human being has an inherent dislike of work and will avoid it if he can.

2   Because of this human characteristic of dislike of work, most people must be coerced, controlled, directed, threatened with punishment to get them to put forth adequate effort toward the achievement of organizational objectives.

3   The average human being prefers to be directed, wishes to avoid responsibility, has relatively little ambition, wants security above all.[18]

Theory *Y* assumptions about human behavior are in dramatic contrast to these:

1   The expenditure of physical and mental effort in work is as natural as play or rest. The average human being does not inherently dislike work. Depending upon controllable conditions, work may be a source of satisfaction (and will be voluntarily performed) or a source of punishment (and will be avoided if possible).

2   External control and the threat of punishment are not the only means for bringing about effort toward organizational objectives. Man will exercise self-discretion and self-control in the service of objectives to which he is committed.

---

[17]Douglas McGregor, *The Human Side of Enterprise,* McGraw-Hill, New York, 1960.
[18]Ibid., pp. 33–34.

3   Commitment to objectives is a function of the rewards associated with their achievement. The most significant of such rewards, e.g., the satisfaction of ego and self-actualization needs, can be direct products of effort directed toward organizational objectives.

4   The average human being learns, under proper conditions, not only to accept but to seek responsibility. Avoidance of responsibility, lack of ambition, and emphasis on security are generally consequences of experiences, not inherent human characteristics.

5   The capacity to exercise a relatively high degree of imagination, ingenuity, and creativity in the solution of organizational problems is widely, not narrowly, distributed in the population.

6   Under the conditions of modern industrial life, the intellectual potentialities of the average human being are only partially utilized.[19]

McGregor's viewpoint points out the inadequacies of an approach to dealing with people that limits the cooperation (and the participation) in the organization. The theory *Y* approach to management strongly implies that the manager's attitude toward the people in the organization limits or encourages those people to contribute to an organizational goal. If a manager is perceived as believing that "people are no good" and have to be managed on a theory *X* basis, the people will respond accordingly. Conversely, if he is perceived as being sensitive to the needs of people and as recognizing that their work can be a source of satisfaction to them, that they usually like to accept responsibility, and that they are willing to participate in the affairs of the organization, they will respond with enthusiasm and creativity. One might note then that the state of human relations in the organization is in great part dependent on the quality of leadership that the manager provides.

LEADERSHIP
Leadership is the capacity to be a leader—the ability to lead. A leader is one who leads others along a way.

Leadership is so interwoven with the human element of organizations that it is difficult, if not impossible, to distinguish where leadership and human behavior stop and start. Leadership is a responsibility of all managers. A leader's role is to get things done through other people; he can do this only by influencing the behavior

---

[19]Ibid., pp. 47–48.

of the people who have a responsibility to do something for the or- **111**
ganization.

Leadership in the political, business, military, church, and other environments has been written about extensively in literature. To consider all the viewpoints about leadership is far beyond the scope of this book. We will examine very briefly some of the approaches that have been used to investigate leadership. The engineering manager can then consider his own philosophy about his role as a leader as he does further reading in the field of leadership.

Prior to the 1930s, leadership was viewed as something that an individual possessed—certain traits and characteristics of successful leaders. Early scholars in the field identified such traits as integrity, decisiveness, intelligence, energy, teaching skills, etc. Many lists of these traits were developed in research studies in the field. For the most part, these traits were believed to be inherited rather than acquired. The list of traits that successful leaders had grew and grew. Finally, over 100 of these traits or characteristics were identified. One of the basic difficulties in using the trait approach is that "seldom, if ever, do any two lists agree on the essential traits and characteristics of effective leadership."[20]

More recent views of leadership have taken a new direction, looking not only to the individual's traits but to the behavior of the leader and those people who are being led. McGregor[21] summarizes some of the generalizations from recent research on leadership, which portray "leadership as a relationship." According to McGregor:

There are at least four major variables now known to be involved in leadership: (1) the characteristics of the leader; (2) the attitudes, needs, and other personal characteristics of the followers; (3) characteristics of the organization, such as its purpose, its structure, the nature of the tasks to be performed; and (4) the social, economic, and political milieu. The personal characteristics required for effective performance as a leader vary, depending on the other factors.[22]

McGregor's viewpoint is important because "it means that leadership is not a property of the individual, but a complex relationship among these variables."[23]

Other approaches have explained leadership as a *type* of behavior: e.g., (1) dictatorial, (2) autocratic, (3) democratic, and (4) laissez-faire. The operating styles of these types of leaders consist of

---

[20]S. G. Huneryager and I. L. Heckmann, *Human Relations in Management,* 2d ed., South-Western, Cincinnati, 1967, p. 243.
[21]McGregor, op. cit., pp. 179–189.
[22]Ibid., p. 182.
[23]Ibid.

112   getting work done through fear (the dictator), centralizing decision making in the leader (the autocrat), decentralizing decision making (democratic leadership), and allowing the group to establish its own goals and make its own decisions (laissez-faire leadership).

Still other approaches deal with leadership as being specific to the particular situation in which it occurs. Therefore, a leader has to take cognizance of the groups of people (superiors, subordinates, peers, etc.) to which he is related as well as the organizational structure, resources, goals, time variables, etc. This suggests the real complexities of trying to understand what leadership is all about. A basic conclusion can be drawn: A successful leader must be adaptive and flexible—always aware of the needs and motivations of those whom he tries to lead toward fulfillment of organizational goals. He must be aware of how he performs as a leader and how his behavior affects the performance of those on whom he depends. As a leader tries to change the behavior of his people, a change in his behavior may also be needed.

Research into the matter of leadership continues, since we really know little of what it is and why it works sometimes and fails at other times. The engineering manager who strives to improve his leadership abilities is encouraged to do reading in the area. The references at the end of the chapter can be helpful in this respect.

SUMMARY

The engineer who is a purist might say that the approach we have suggested in this chapter is unscientific. We might agree. But can people be managed scientifically? Are people always rational? Is the management of people more a science than an art? Questions like these have been debated by management theorists and practitioners for a long time. Let's take the easy way out and say that the management of people includes elements of both science and art. It is the artful aspect that we were primarily concerned with in this chapter.

We described the many roles that a manager plays as he maintains a multiplicity of interfaces with his clientele. His success in accurately and effectively maintaining these roles is reflected in the degree to which the organization—and the people—accomplish respective objectives.

If one treats the management of people as primarily an artful activity, then one can make observations rather than establish scientific fact. One can study the behavior of people in a particular situation and make general observations about the nature of their behavior. Change the environment a little and their behavior may change slightly or significantly. This does not alter the importance of observing behavior and drawing observations from that behavior. **113** These observations can be integrated into the manager's strategy and modus operandi for managing people.

The viewpoint put forth in this chapter is a practical view of some of the factors that motivate professional personnel. Motivation is related to the needs of people; the engineering manager must understand the nature of these needs if he is to deal successfully with the challenge of motivating professional people.

We ended the chapter on the subject of leadership. In the next chapter, we will examine an important role of the leader—that of an effective communicator to his clientele.

DISCUSSION QUESTIONS

1 What is meant by the phrase "interdisciplinary generalizations"?

2 In what manner might the work of management be described as rational in nature?

3 How might the human subsystem of an organization be defined?

4 What is meant by "humanizing" the engineer?

5 The authors take the position that the engineering manager plays many roles in discharging his responsibilities. What is meant by this statement?

6 Identify and describe the typical roles of an engineer as defined in this chapter. Should any additional roles be added?

7 If an engineering manager is not strong in an aspect of technology in his organization, what should he do to overcome this weakness?

8 What is the most important role that an engineering manager plays? Defend your choice.

9 What is meant by "nondirective counseling"? When should such counseling be used by the manager?

10 What do engineers look for in their work?

11 What are some of the key motivational factors affecting job performance in an engineering community?

12 Distinguish between a theory X and theory Y engineering manager.

KEY QUESTIONS FOR THE ENGINEERING MANAGER

1 Have I ever designed a "conceptual model" of the human subsystem in this organization?

114

2   What professional reading have I done which gives me some background in the interdisciplinary nature of engineering management?

3   What action have I taken to humanize my engineering manager's role? How do the engineers assigned to this organization perceive my role in this respect?

4   What do I perceive as my many parts in managing this organization?

5   Have I organized peer groups in this engineering organization to help me evaluate the adequacy of technology being provided?

6   Can technological mistakes be buried in this organization? What can I do to prevent this sort of thing from happening?

7   How successful has my counseling been in this organization? Do the professionals in the organization feel comfortable in coming to me with their problems—and suggestions?

8   What factors motivate the engineers in this organization? Would it be possible to determine what key factors motivate these engineers?

9   Does Maslow's concept of hierarchy of needs give me any insight into the behavior of people in this organization?

10   What means have I provided to encourage participation in this engineering community? Has there been meaningful participation? Why or why not?

11   Are all the decisions made in this organization made on a rational basis? Why or why not?

12   Do the engineers perceive me as a theory X or theory Y manager? What influence does my leadership style have on this perception?

BIBLIOGRAPHY

Albrook, Robert C.: "Participative Management: Time for a Second Look," *Fortune,* May 1967.

Christensen, C. Roland, Kenneth R. Andrews, and Joseph L. Bower: *Business Policy: Text and Cases,* 4th ed., Irwin, Homewood, Ill., 1978.

Cleland, David I., and William R. King: *Management: A Systems Approach,* McGraw-Hill, New York, 1972.

Cyert, Richard M., and James E. March: *Theory of the Firm,* Prentice-Hall, Englewood Cliffs, N.J., 1963.

Davis, Keith: *Human Relations at Work,* McGraw-Hill, New York, 1967.

Dowling, William (ed.): *Effective Management and the Behavioral Sciences,* American Management Association, New York, 1978.

Dubin, Robert: *Human Relations in Administration,* Prentice-Hall, **115** Englewood Cliffs, N.J., 1961.

"Engineers Train for the New World of Business," *Business Week,* Jan. 16, 1978.

Galbraith, J.: *The New Industrial State,* Houghton Mifflin, Boston, 1967.

George, William W.: "Task Teams for Rapid Growth," *Harvard Business Review,* March-April 1977.

Hodgetts, Richard M.: "Leadership Techniques in the Project Organization," *Academy of Management Journal,* June 1968.

Huneryager, S. G., and I. L. Heckmann: *Human Relations in Management,* 2d ed., South-Western, Cincinnati, 1967.

"It's Strictly Carrot, Not Stick, at Rabbit Plant," *The Pittsburgh Press,* Mar. 5, 1978.

Jucius, Michael J.: *Personnel Management,* Irwin, Homewood, Ill., 1963.

McClelland, David C., and David H. Burnham: "Power Is the Great Motivator," *Harvard Business Review,* March-April 1976.

McGregor, Douglas: *The Human Side of Enterprise,* McGraw-Hill, New York, 1960.

Maier, N. R. F., and L. R. Hoffman: "Group Decision in England and United States," *Personnel Psychology,* vol. 15, Spring 1962.

Maslow, A. H.: *Motivation and Personality,* Harper, New York, 1954.

————: "A Theory of Human Motivation," *Psychological Review,* vol. 50, 1943.

Roethlisberger, F. J.: *The Elusive Phenomena,* Harvard Graduate School of Business Administration, Boston, 1977.

————:*Management and Morale,* Harper, New York, 1941

Sobel, Robert: *The Entrepreneurs,* Waybright and Talley, New York, 1974.

COMMUNICATION IN ENGINEERING MANAGEMENT 5

A few years ago alumni of Purdue University were asked what had been most conspicuously lacking in their education. These alumni, predominantly engineers, declared overwhelmingly: "Communication."[1]

An engineer may be judged more by the way he is able to express his ideas than by his degree of technical knowledge. Everyone with whom the engineer comes in contact is able—or at least thinks he is able—to judge the way the engineer is able to speak and write.

The possession of technical knowledge is the reason engineers are hired; when they are hired there is an assumption that they can communicate this knowledge in their place of work with their work associates—superiors, subordinates, and peers. An engineer does not generally have a good image as a communicator—one who conveys information about some matter. He may be effective in communicating with fellow engineers, but when he comes in contact with other people with whom he must work, communication can become a problem. For example, an engineer who is a member of a project team may work with people from other disciplines, such as marketing, finance, accounting, information systems, or production. If he is not able to communicate what needs to be said about engineering to these people, misunderstanding can occur.

Early in his career an engineer is chiefly concerned with communications of a technical nature. Later on in the engineer's career, nontechnical, management communications become more important. Typically, engineers come to a decision point between 3 and 7 years after graduation. At that point, they choose either to stay in their engineering specialty or to move into engineering management. The communications needs of the two paths are different. Those remaining in engineering specialties write technical reports and

[1]Cited in Robert Haakenson, "Improving Oral Communication," *Effective Communications for Engineers*, McGraw-Hill, New York, 1974, p. 48.

118   present the results of their work to their associates, whose interests are similar to their own. Those in the engineering management path communicate in a broader spectrum. In addition to reviewing technical reports and presenting their views to their peers and superiors, they also deal frequently with subordinates and superiors on a face-to-face basis in discussing nontechnical matters such as hiring and firing of employees, justifying resource-allocation decisions, and motivating people to perform effectively in their place of work. Both the engineer and the manager use communications skills to sell themselves, their decisions, their ideas, and their technical judgments.

Peter Drucker, an eminent management scholar and consultant, has stated that the ability to communicate effectively heads the list of criteria for success: "As soon as you move one step from the bottom, your effectiveness depends on your ability to reach others through the spoken or written word. . . . In the very large organizations. . . this ability to express oneself is perhaps the most important of all the skills a man can possess."[2]

Probably no good engineer has ever been fired because he was not a good communicator of his ideas; but there are not many good engineering managers who are not effective communicators! Many times engineers have missed out on promotions and other rewards because their ability to communicate has been poor. Their technical knowledge may be adequate; but their inability to communicate that knowledge puts real limitations on their opportunities to progress into the managerial ranks.

A survey of United States corporations by the *Harvard Business Review* gives some insight into the ability to communicate as a factor in promotability.[3] A portion of this study consisted of a listing of personal attributes and their importance in promotion. "Technical skill based on experience" placed fourth from the bottom in the list of twenty-two attributes. Other attributes, such as ambition, maturity, capacity for hard work, ability to make sound decisions, getting things done with and through people, flexibility and confidence, all ranked high. "Ability to communicate" was at the top of the list.

Other studies have reached the same conclusion. Engineers who have become managers realize the importance of communications as they carry out the many roles required of them in filling a management position: technologist, agent, negotiator, logistician, strategist, counselor, disciplinarian, motivator, and so forth.

In this chapter an overall perspective of the role of communications in engineering management will be presented. This overall perspective should give the reader insight into the nature of communications and how his ability to communicate can be improved.

[2]Peter F. Drucker, "How to Be an Employee," *Fortune*, May 1952, p. 126.
[3]"What Helps or Harms Productivity," *Harvard Business Review*, January 1964.

THE NATURE OF COMMUNICATIONS   119

Communication is the process by which information is exchanged between individuals through a common system of symbols, signs, or behavior. Humans carry on communication with each other by three principal means: (1) by an actual physical touch—such as a tap on the shoulder, a pat on the back, or the ritualistic expression of the handshake; (2) by visible movements of some portions of their bodies—such as the pointing of a finger, the wink of an eye, or a smile, nod, or grimace; (3) by the use of symbols, spoken or portrayed, which have some meaning based on experience.

Most present-day failures in communication can be traced to misunderstandings of the symbols that play an important part in the process of interhuman communication. Such misunderstandings come about largely because of our inadequacies in creating, transmitting, and receiving these symbols, both written and spoken. Engineers will readily recognize the value that symbols have in communicating phenomena in the engineering disciplines. Unfortunately the symbols that are used in communication in the management discipline are not so precise in their meaning.

Words and combinations of words give us more difficulty in communicating than other kinds of visual symbols. Take the word "manage" as an example. *The American Heritage Dictionary of the English Language*[4] defines "manage" as follows:

1. To direct or control the use of; handle, wield, or use (a tool, machine, or weapon). 2. To exert control over; make submissive to one's authority, discipline, or persuasion. 3. To direct or administer (the affairs of an organization, estate, household, or business). 4. To contrive or arrange; succeed in doing or accomplishing, especially with difficulty: *I'll manage to come on Friday—intr.* 1. To direct, supervise, or carry on business affairs; perform the duties of a manager. 2. To carry on; get along: *I don't know how they manage without him.*

The term "management" is defined as:

1. The act, manner, or practice of managing, handling, or controlling something. 2. The person or persons who manage a business establishment, organization, or institution. 3. Skill in managing; executive ability.

The real meaning of words—or symbols—depends primarily on how the reader or listener perceives them. A word such as "management" would have one meaning for a corporate executive and

120 another meaning for a trade union executive. The meaning would depend on the image that the word holds for the individual. Such meaning is something we have inside ourselves. Words and phrases that an individual uses may not evoke the same image in someone else's mind; knowing this, one should be as specific as possible in using words. Assuming that everyone knows what you are talking about is usually a poor assumption. One is reminded of the meaning of a word as made famous by Lewis Carroll: "'When *I* use a word,' Humpty Dumpty said in rather a scornful tone, 'it means just what I choose it to mean—neither more nor less.'"[5]

Elsewhere in this book we have emphasized the use of models in engineering management. Models are also useful in depicting the communication process.

A COMMUNICATIONS MODEL

A model of communications has been developed by theorists in the field. A widely known model is described by Shannon and Weaver. The basic ingredients of this model include a source, an encoder, a message, a channel, a decoder, a receiver, feedback, and noise. Other models of the communication process have been described in much the same way.[6] Gibson, Ivancevich, and Donnelly have developed a model of this process. (See Fig. 5-1.) Each of the elements in the process of communication can be described as follows:

SOURCE Communication originates somewhere. In an organizational sense, the source is some member of the organization who has information to communicate to someone else in the organizational system. Superiors, subordinates, peers, and associates are all sources from which a communication process can originate.

ENCODER An encoding process takes place when the originator's ideas are translated into a systematic set of symbols in a language that expresses what the source wishes to express. Encoding provides a structure in which the ideas, purposes, intentions, etc., can be expressed as a coherent message.

MESSAGE The message is "the actual physical product of the source encoder."[7] The purpose of the source is expressed in the form of a message. Basically, this is what the source hopes to communicate effectively to some receiver. The form that the message takes depends to a great extent on the channel that is used.

[5]Lewis Carroll, *Alice's Adventures in Wonderland and Through the Looking Glass*, Macmillan, New York, 1966, 4th printing.
[6]Claude Shannon and Warren Weaver, *The Mathematical Theory of Communication*, University of Illinois Press, Urbana, 1948, and Wilbur Schramm, "How Communication Works," *The Process and Effects of Mass Communication*, University of Illinois, Urbana, 1953.
[7]David K. Berlo, *The Process of Communication*, Holt, New York, 1960, p. 54.

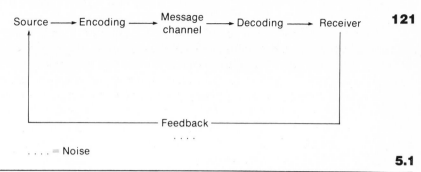

. . . . = Noise

5.1

A communication model. (*From James L. Gib-son, John M. Ivancevich, and James H. Don-nelly, Jr., Organizations, Structure, Processes, Behavior, Business Publications, Dallas, Texas, 1973, p. 166.*)

CHANNEL   Channel refers to the medium through which the message is carried from the source to the receiver. In an organizational sense this could be a face-to-face communication, a written message, a telephone call, word of mouth, informal communications, group meetings, closed television circuits, and many others.

DECODER-RECEIVER   The decoder-receiver provides for the process of communication to be completed. The message has to be decoded in terms of some meaning to the recipient. Each recipient interprets the message, usually in light of his own previous experience, value system, frame of reference, etc. The closer the decoded message is to the intent desired by the sender, the more effective the communication will be.

FEEDBACK   Feedback gives insight into the degree of fidelity that the message has. Message fidelity is rarely perfect. The provision, however, for feedback in the communication process is desirable. The feedback gives the source the opportunity to determine whether the message has been received and whether the message has produced the intended purpose.

NOISE   Noise is anything causing a breakdown, interference, or distraction within the communication process. Shannon and Weaver are credited with the first identification of this notion of noise.[8] These individuals worked for the Bell Telephone Laboratory and were simply describing an electronic communication. Behavioral scientists have picked up this idea and have found it useful in describing the process of human communication. Noise is basically those factors which dis-

---

[8]Shannon and Weaver, op. cit., p. 5.

122　　tort the signal. In a behavioral sense, noise includes those factors in each element of the communication that can reduce message fidelity, such as human distractions, misunderstandings, or misinterpretations by both receivers and senders, semantic difficulties, time sensitivity, status within the people, preconceived value judgments on the part of the receiver as well as the sender, etc.

　　The discussion of the model of communication has been very brief. It is intended only to provide an idea of the model of communications that is generic in all fields of communication. It should be useful in terms of helping the engineer to understand why some of the messages that he sends are not received as he had hoped they would be. The model shows the elements that are involved in the communication process. For example, the care that is taken in encoding a message is important. Also, recognizing the opportunity for noise, the sender should think about the receiver and how the message may be distorted by the receiver as he considers that message because of his status, prejudices, value systems, perceptions, etc. This suggests that when one is sending a message, he should place himself in the position of the receiver, and prepare and encode the message accordingly. It is often said that one should use the language of those to whom one is speaking; this is what the communications model process brings out. The manager should understand and should appreciate the process of decoding as he does the encoding of the message for transmittal.

　　Communication specialists have concluded that one of the most important factors in a breakdown in communications occurs in the variation that takes place in the encoding and decoding processes. When the encoding and decoding processes are similar, then communication is usually the most effective. When these processes become heterogeneous, then a breakdown in communications tends to occur. Weiner refers to this problem as "entropy" or the tendency for human processes to break down rather than come together.[9]

　　We offer communication as an important aspect of an engineering manager's job. It is so inherent to managing people that it is impossible to manage effectively without giving consideration to the effectiveness with which communication is carried out within the context of the management functions. Communication is the capstone of the management functions; without effective communication, the management functions will not be carried out. For example, in planning, the manager has to work with his people in selecting organizational objectives. After these objectives have been selected, they must be communicated clearly to those people who are responsible for attaining the objectives. The objectives must be com-

[9]Norbert Weiner, *Cybernetics: On Control and Communication in the Animal and the Machine*, Wiley, New York, 1948.

municated initially and reinforced from time to time as organizational **123** effort is carried out.

One engineering manager known to the authors starts his monthly management review meetings with the question: "Do we all understand what this department's objectives are?" Usually some member of his management team will raise a question about an aspect of the department's objectives. This engineering manager will take time to have a group discussion and clarify departmental objectives. Occasionally he will ask some of his managers to take the lead in a discussion about department objectives. Opportunities come out of the discussions to reinforce department objectives. He is practicing good management in such situations. He is communicating with his people, a fact which is reflected in the effectiveness with which the organization's objectives are being attained.

CHARACTERISTICS OF COMMUNICATION

The nature of communication can be further understood by considering some of its characteristics.

When one communicates, there is usually a purpose for so doing. Someone is to be told something or to be convinced to do something or not to do anything. Another reason might be to obtain information from an individual or to solicit his opinion on an issue. Another purpose might be to just pass the time of day. In an organization, various forms of documentation—letters, memoranda, plans, schedules, specifications, blueprints, and so forth—all serve as purposeful means for communicating ideas, concepts, policies, and regulations.

To communicate we need a sender and a receiver acting cooperatively with a purpose in mind. Bilateral communication goes like this: "You have told me to do something. I understand what you want, and accept your idea; I will work to carry out your instructions." A legal contract is an example of a bilateral communication. In a contract the agreeing parties read the terms and conditions of the contract; when such terms and conditions, prices, schedules, etc., are understood and accepted, the parties to the agreement sign the contract. Communication has taken place.

Meetings are an interesting study in communication. When a manager calls the members of his organization together to discuss some problem or opportunity, the stage has been set for communication to follow. The manner in which the meeting is carried on influences the degree of communication that has taken place. A skillful manager will be careful to plan for the meeting and develop a style for carrying out the meeting that helps to facilitate communication among the participants.

A manager who hopes to bring about communication in his organization should consider the analogy of the transmitter and

124 receiver used earlier and seek to provide an environment wherein transmitting and receiving are accomplished. If it occurs, then communication has taken place.

COMMUNICATION IS FORMAL AND INFORMAL Every organization has its formal structure—portrayed by the organization chart—through which formal communications are directed. Sometimes formal communications are addressed in general fashion to all members of the organization. Formal communications can be written or verbal; nonverbal communication can also be considered as formal in nature, e.g., the hand signals given by a police officer directing traffic are formal. Informal communication may be viewed as the communication that is carried out within the informal organization. Whenever people are brought together in a workplace, a formal organization system is created.

Even though a formal organization system exists, some elaboration will occur—out of the formal organization, a second organizational system, an informal structure, will emerge. Membership in this informal organization is dependent on common ties, such as friendship, kinship, social status, etc. The need for an informal as well as a formal organization lies both in the psychological and social needs of human beings and their desire to accomplish personal and organizational objectives. People join informal groups at their place of work for social contact, companionship, emotional support, and such things of value coming out of a particular community to which an individual belongs.[10]

What can the engineering manager do about the communications within the informal organization? His interpersonal style of management will probably have some effect on these informal communications. We offer a few suggestions in this matter:

> Accept the notion of informal communication in the organization and what such communication can do and not do for enhancing organizational effectiveness.

> Find ways of getting feedback from the informal organization. Identify the "informal leaders" and spend time listening to what they have to say.

> Use these informal leaders as a source for testing technical approaches, ideas, strategies, administrative actions, reorganizations, and such other things whose acceptance by the members of the organization is required for success.

> Recognize that much of the cultural ambience of the organization is reflected in the attitudes and behavior of people in

[10]See Ross A. Webber, *Management*, Irwin, Homewood, Ill., 1975, Chap. 21, for a thorough discussion of groups and informal organization.

the informal organization. Insofar as possible, work with the **125** informal organization in support of organizational purposes.

COMMUNICATION IS VERBAL AND NONVERBAL   As we consider nonverbal communication, a whole series of physical gestures come to mind: facial expressions, nodding, hand and body movements, eye movements, etc. Those of us who have visited the modern version of a burlesque theater would surely recognize the value of nonverbal communications in certain situations.

It would be difficult to portray nonverbal communication as formal or informal, since it depends on the context in which it is carried out. Consider, for example, what occurred in a small airport in Ohio when one of the authors was writing this material on communication. The author was waiting, along with many other people, for a scheduled flight back to Pittsburgh, Pennsylvania. The flight was delayed by an hour and a half. All the people were sitting in the gate area reading, talking, etc. A ticket agent came down the corridor and took his position behind the check-in counter. Immediately, a long line of people formed in front of the ticket counter. Nothing was said by the ticket agent. His presence at the check-in counter brought about a form of nonverbal communication and people reacted.

Another situation occurred while one of the authors was consulting for the chief executive officer of a large United States corporation. This consultation was done during the time that pantsuits were starting to be the fashion in women's apparel. The secretary to the president of this corporation had been contacted by several senior executive secretaries in the corporation about the acceptability of wearing pantsuits to work. This secretary drafted a memorandum for the president to sign suggesting that pantsuits were acceptable in the company. The president—a subtle manager—decided not to issue any formal statement on the matter. Instead, he asked his secretary to start wearing a pantsuit to work and see what would happen. Within a week the wearing of pantsuits became commonplace in this corporation. Effective nonverbal communication? We think so!

One researcher has provided us with some insight into the importance of nonverbal communication in social interaction: "Psychologist Albert Mehrabian has done extensive lab measurements on what happens when one person talks to another. He finds that only 7% of a message's effect is carried by words, while 93% of the total impact reaches the 'listener' through nonverbal means—facial expression, vocal intonation. Feelings are communicated mainly by nonverbal behavior."[11]

It seems to be true that certain movements of the body

[11]Cited in Layne A. Longfellow, "Body Talk: The Game of Feeling and Expression," *Psychology Today*, October 1970. Reprinted from *Psychology Today* magazine, © 1970 by the Ziff Davis Publishing Co.

126 during a communication process have meaning. Once these movements are learned, understood, and transferred, an honest and effective level of communication has taken place.

COMMUNICATIONS REQUIRE REINFORCEMENT Communications are perishable and must be repeated if a particular message is expected to get through and remain. For example, an organizational policy—a guide to the way people are expected to think in the organization—should be reinforced from time to time if it is expected to be effective. A policy concerning the objectives of the organization can be forsaken unless the organizational members are reminded periodically of the reason for which the organization is in existence.

 To be effective, communication requires practice—and more practice. The process of communicating which involves transmitting and receiving information, requires practice on the part of those who transmit and those who receive.

COMMUNICATION IS MULTIDIMENSIONAL Communicating is a multidimensional process. This is almost a trite saying, yet many engineering managers attempt to communicate only downward with their engineering staff and fail to communicate laterally with others who are assisting them in their work. Such managers see communications as a one-way process of transmitting what they want the subordinates to know. Communication channels in an engineering department are complex, however; downward when the engineering manager issues instructions to his staff and upward when the members of the engineering staff report on their progress on a project. Individual engineering project team members communicate on a lateral basis when they have shared interests in a project. If an engineering project is highly technical in nature, it may be necessary for an individual team member to talk with the vice president of engineering on some technical matter involving the project. This may be done in the formal sense, as during a project review meeting, or informally during a social activity. Figure 5-2 illustrates the multidimensional nature of communication.

 The engineering management functions of planning, organizing, motivating, directing, and controlling are only carried out through the ability of a manager to communicate within his clientele group. These managerial functions are established in order to exchange information, ideas, and work relationships among members of the team, whether that team be an engineering project, a department, or the entire engineering organization. For example, when a plan is prepared for an organization, a great deal of communication has to go on among those people who prepared the plan, but the plan itself, after it is in existence, is not effective until it has

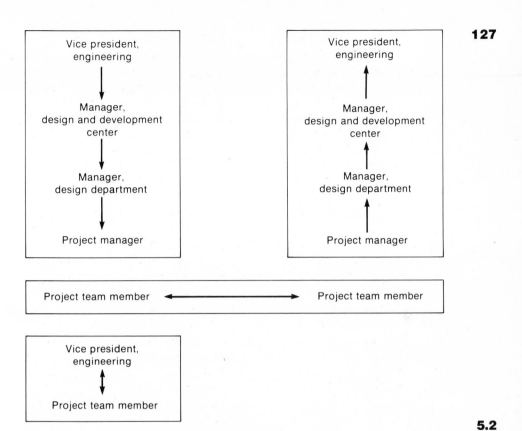

5.2

been communicated—understood and accepted by those that have the responsibility for implementing it.

COMMUNICATION INCLUDES LISTENING   Good listening is an artful skill that some people have and others lack. To become a good listener, a person must work at developing this skill through studying good listening practices and applying them to his own conscious program of self-development.

Listening, according to Weismantel, involves both listening for facts and listening for feeling.[12] Listening for facts is important as an educational role. A manager who gives instructions to his subordinates on how to do a job provides an opportunity for learning. A briefing given on a proposed engineering design of a new product involves communication of facts to the listeners. Listening for feeling involves hearing what the speaker said in its proper context, as

[12]Guy E. Weismantel, "Management—By Listening," *Effective Communication for Engineers,* McGraw-Hill, New York, 1974, p. 32.

128   well as what was not said. If the listener does not respond to what a person has said, then communication may have broken down. A gesture or other form of nonverbal communication can involve an emotional factor, unspoken but very real, which results in another communication failure.

How and when a person remains silent can indicate his feelings about a subject. An individual who is generally talkative about new ideas usually shows eagerness or criticism and will discuss the ideas for their good and bad points. If during a conference this individual says nothing, his silence is meaningful and should be "listened to" for feelings.

A manager who is skilled at listening can learn many things about what is going on in his organization. Employees appreciate a manager who gives them time and is skillful at listening to them. The manager who is skilled at listening has an important technique at his disposal for sizing up the personalities of his people. Employees will take note of how good a listener the manager is, drawing conclusions from what he says as well as his manner, facial expression, gestures, etc.

Keith Davis, in his book *Human Relations at Work,*[13] presents ten commandments for good listening:

1	Stop talking!
2	Put the talker at ease.
3	Show him that you want to listen.
4	Remove distractions.
5	Empathize with him.
6	Be patient.
7	Hold your temper.
8	Go easy on arguments and criticism.
9	Ask questions.
10	Stop talking!

It is important that the engineering manager enter into the communication process with a positive attitude and a desire to listen.

The manager (or the engineer) who feels the need to evaluate his own listening ability or initiate a program for his subordinates should obtain professional assistance. There is a lot of literature dealing with listening. Colleges, universities, professional associations, and consulting firms offer programs that give the individual a

---

[13]Keith Davis, *Human Relations at Work,* 3d ed., McGraw-Hill, New York, 1967, p. 333.

chance to enhance his listening ability. The important thing is that **129**
the individual recognize when improvement in listening is needed by
someone, and then seek help.

To really listen, one must use all the senses. True listening is
more than hearing. A smile, a frown, or silence can communicate a
message without a word being spoken.

The importance of listening is underscored graphically by a
Sperry advertisement that appeared in the *Wall Street Journal.* The ad
read, in part, "It's about time we learned how to listen.... [There's]
the problem of people not knowing how to listen. Most of us spend
about half our waking hours listening. Yet research studies show that
we retain only 25% of what we listen to.... Because listening is the
one communication skill we're never really taught..."[14]

ON-THE-JOB COMMUNICATIONS

Communicating with employees about their job responsibilities is an
important form of communications within an organization. This involves the transmittal to them of work-related information concerning what their job is in such a way that the employees understand
their responsibilities and the expectations that the manager has for
them in the performance of that job. Usually, this is done in the form
of a written job description; however, a written job description is a
form of unilateral communication unless the manager has discussed
with the employee what is expected in work performance, including
an explanation of methods, standards, and policy guidelines. This
discussion has to be carried out in such a way that the subordinates
and the manager understand the job requirements and their reciprocal authority and responsibility. Communication is part of the management of human resources, conveying management's idea of what
is expected of the work of the subordinate. Reviewing with the employee on a periodic basis his position description and the requirements of his job serves as an excellent feedback to inform the employees on how well the work is being done, how the work might be
done better, and what areas of improvement are necessary. This feedback is necessary and is particularly useful for salary appraisal or
promotion merit evaluation.

Dealing with people in organizations is all too frequently accomplished only through written communications. The very size of
many organizations today precludes a direct verbal communication
relationship on the part of the manager with all employees. At some
level in the organization, however, every employee has a manager
with whom he should have candid, ongoing, face-to-face communications. Employees need information properly communicated to

---

[14]*The Wall Street Journal,* Sept. 11, 1979, p. 11.

130   them in order to be able to work efficiently. Effective communication is the primary instrument in the development of human resources. If employees are not sure of their responsibility and authority, the most sophisticated of management systems will not have any real effectiveness. For most managers, the greatest opportunity for increasing the effectiveness of work lies with the process of communications between themselves and their employees.

One small company, concerned about the adequacy of communication with its employees, instituted an "employees' lunch program" to improve dialog between management and hourly rated employees. Under this program two members of management would take several employees to lunch outside the plant two or three times each week. At these luncheons anything could be discussed in a wide range of topics from the "strategic direction of the company," to some specific aspect of working environment in the plant. The managers, who acted as hosts for the lunch, tried to do much more listening than talking. The luncheons provided an opportunity to answer questions, discuss company policy, listen to employee suggestions for improving working conditions, solicit suggestions, and in general receive feedback on how things were going as perceived by the employees. If questions were asked which could not be answered by the employees, an answer would be given in the next issue of the company internal newsletter.

These luncheons proved valuable to management. After each luncheon the host managers were required to submit a summary of the matters discussed to the general manager through the director of human resources. The submission of this summary form enabled the director of human resources to follow up and manage the employee luncheon program. Were the results worth the cost? Most definitely, claims management, for out of these luncheons came information and suggestions for quality improvement, productivity increases, identification of potential grievances, and other matters which improved the effectiveness of the work force. Moreover, the employees appreciated having the opportunity of getting a broader picture of the company's problems and opportunities. Higher morale resulted from these lunches and the employees developed a stronger sense of belonging to the company.

There was an indirect benefit to the managers who were afforded an opportunity to see the company through the eyes of the employees. Many of the managers were isolated from the workers' environment—indeed seeing it principally through the eyes of several layers of subordinate managers. By participating in the employees' lunch program, these managers were able to keep in touch with the real world of the worker.

The company was quite successful in terms of profitability. It

131

set the pace for product development in its industry. Managerial, professional, and hourly rated employee turnover was very low. Although the working force was unionized, several years had elapsed since any serious grievance had been filed. There has not been a strike in this company for over 10 years.

Clearly, this was a well-managed company that was producing outstanding results. The degree to which its success can be attributed to the employees' lunch program is open to speculation. However, these lunches contributed something worthwhile to communication in the organization.

Communication can serve several purposes. It may be a means of giving information, of expressing ideas or feelings, or of influencing others. A mere comment about a situation or an event intended to have no real influence on behavior may sometimes result in a major response. Senior executives are sometimes surprised that a purely casual remark on their part made during a visit to a subordinate element of the organization may result in the expenditure of a substantial amount of energy to do something because it was interpreted as a request for action. The reverse may also be true. A major attempt to influence behavior by communication may have only very minor effects.

AUTHORITY AND COMMUNICATION

A manager's authority may be more dependent on his ability to persuade people (superiors, subordinates, peers, associates) to his viewpoint than on the position he occupies in the organizational hierarchy. McGregor makes some interesting points about the manager's ability to persuade others through the communication process:

A related aspect of what is often a rather naive faith in the power of communication as a method of persuasion is closely connected with management views of "rational-emotional" man: "Tell them the reasons and they will do what we want." Management tends to assume quite naturally, but often incorrectly, that its decisions and policies are rational and objective. It further assumes that "proper communication" will transform its audience into rational recipients. Thus the primary requirement for obtaining compliance with management desires becomes that of making sure that the reasons for a given action are understood. Of course this is a necessary requirement but it is by no means sufficient. Considerably more important are the characteristics of the system to which the communication input is made.[15]

---

[15]Douglas McGregor, *The Professional Manager*, McGraw-Hill, New York, 1967, p. 151.

132   McGregor identifies one characteristic in an effective managerial team as open communications through sound decisions and their successful implementation. Feelings and emotions, according to McGregor, are relevant facts. They are inextricably interwoven with intellectual facts and logical arguments and a rational-emotional man. He further states:

> The process of managing conflict by working through differences—in contrast to denying or suppressing them—is complex and difficult. There is no simple technique like the use of authority or the vote for accomplishing it. The leader cannot manage conflict, but he and the group together can do it. They must start by creating certain underlying conditions, modifying certain group variables, which as we have seen are hard to change, but once established can be maintained.
>
>   The first of these necessary conditions is authentic communication among the members and between them and the group leader. This means that each member (including the leader) be genuinely free to express his feelings as well as his intellectual ideas openly in the group. It means also that such expressions will be sympathetically explored by the group until they are fully understood (although not necessarily agreed with). It means that other members (or the leader) will not intentionally ride roughshod over any individual feelings or express snap judgments publicly or privately about his competence or his value to the group because his expressions may seem deviant, unclear or even irrational. It means that every member of the group (including the leader) will understand what is involved in genuine listening and will have acquired some skill in this difficult art.[16]

Managers sometimes assert and really believe that communications are authentic and open within the groups they lead; however, this is not usually the case because of the pressures typical of hierarchical organizations, the difficulty in objectivity, the emotional aspects of the individual, and the power relationships within a group. In such situations, McGregor notes, "open communication would be foolhardy. . . . Concentrated effort of a skilled kind, extending over months or even years, is necessary to bring about a high degree of authenticity in communications within a managerial team."[17]

    Authentic communications cannot be produced by fiat or even by a plea from a leader. They depend upon other conditions, one of the most important of which is a kind of mutual trust and support within the organization.

[16]Ibid., pp. 191–192.
[17]Ibid., p. 192.

Innovational potential

5.1

FACTOR	INDEX WEIGHTING
1 Effectiveness of communications	20
2 Scientific and technical competence	20
3 Presence of a champion	15
4 Recognition of market opportunities	15
5 Recognition of technical opportunities	10
6 Degree of top management interest	10
7 Competitive factors	5
8 Timing	5

Source: Milton A. Glaser, "The Innovation Index," *Chemical Technology,* vol. 6, March 1976. Reprinted with permission. Copyright 1976 American Chemical Society.

INNOVATION AND COMMUNICATIONS

Engineering managers are at the forefront of innovation in an organization. The importance of communication in the innovation process within organizations is illustrated by the experiences of the Dexter Corporation.

The Dexter Corporation uses an innovation index to evaluate the innovation of a project.[18] Each weighting factor shown in Table 5.1 is multiplied by the "degree of perfection" each element has in a specific project. A total for all elements is computed; the total for each project is compared with totals for other projects in reaching a decision on funding of the projects.

One of the most important factors is "effectiveness of communication." Communication is divided into three subcategories:

Technical to marketing

Technical to customer

Technical to technical

TECHNICAL-TO-MARKETING INTERFACE   This interface is considered to be the most important communication line in the integration of a project. According to Glaser: "Innovation is impossible if technical people continually ignore advice and suggestions of the marketing people. Innovation is equally impossible when marketing people are authoritarian, do not involve the project technologists directly with prospective customers, or if they continually disparage the progress or competence of the technical group."[19]

[18]Milton A. Glaser, "The Innovation Index," *Chemical Technology,* vol. 6, March 1976, pp. 182–187. Reprinted with permission. Copyright 1976 American Chemical Society.
[19]Ibid., p. 183.

134   Failure can often be ascribed to the fact that the technical and marketing people do not know each other. Communication is impossible under such circumstances. Even when the parties know each other, communication does not come about unless there is respect and trust. Each marketeer and technologist must have empathy—be capable of seeking the other person's viewpoint.

When there is a substantial hierarchical difference between the technologist and the marketeer, a meaningful discussion of an issue will be difficult if the senior person cannot objectively discuss the issue with the junior individual. At this interface, the senior people should be of approximately the same executive rank. If there is not a reasonable equality in rank, or if the senior people have difficulty in communicating with each other, then the informal organization may be used to help bring about the desired communication. Communication in the informal organization can be an effective way of getting the word to those who can do something about the situation. Nurture these communications channels!

TECHNICAL-TO-CUSTOMER INTERFACE   This is another important communications link, particularly for the technical community to appraise a customer's technical needs. Marketeers may be reluctant to permit too much contact between the vendor and customer technical persons. There are, however, many situations in which a project has been delayed because of a failure of the vendor and the customer technical people to discuss and come to a common understanding of the specifications of a project. There is no doubt that cross-discipline communication is difficult at best. If the communication between two parties is filtered through a third party, then the difficulties can become more troublesome.

Motivation of the technical people can be helped by giving them a key role in working with their counterparts in the customer's organization. If the technical person who is responsible for managing the in-house technical portion of a project is told or made to believe that he should not talk with his counterpart in the customer's organization, he may be resentful, and his productivity may suffer. Glaser notes: "A common reason for low chemist morale and low chemist success in innovating is that they do not have enough opportunity to visit customers, hear their needs first-hand, and participate as closely as possible in customer evaluations and reappraisals. He would incidentally get the psychological lift that such contact brings."[20]

TECHNICAL-TO-TECHNICAL INTERFACE   In this interface the scientific and technical competence of the professional can be affected. Professionals whose knowledge and skills require updating will be

[20]Ibid., p. 184.

particularly helped if there is an ongoing dialog with their col-   **135**
leagues who have more competence in the technical area. Technical-
to-technical communications include exposure to colleagues in the
organization, in the industry, at professional meetings, and during
continuing education programs. A key element of this interface is the
thoroughness with which the engineering manager supports the professional development of the people assigned to him. If the professional has access to a good technical library, his ability to communicate with his technical community can be enhanced.

COMMUNICATIONS AND PRODUCTIVITY

Productivity improvement is a responsibility of every engineering
manager. White-collar productivity is estimated to be about half of
that of blue-collar workers; consequently, the near-term potential in
the white-collar area appears to be immense. Though opinions differ
on the strategies that can be developed to improve blue-collar productivity, all concerned agree on one point—the absolute need for,
and tremendous gains from, effective two-way communication
throughout all levels of an organization.[21] According to one company
president:

A good communications program must include regular
supervisor-employee meetings that provide for feedback—and credible response to that feedback. Videotape
communications can be an effective supplement to those
meetings. Supervisor-employe meetings and videotapes
should be supplemented by roundtable discussions and
talks by the General Manager—or Plant Manager—to all
employes concerning the local plant situation—followed by
an adequate question-and-answer period.

You may wonder how this amount of time away
from productive work can be justified for all employes? If,
for example, you take a half-hour of productive time every
week to communicate better with all your employes, a 1 1/4
percent improvement in productivity would be needed during the remaining 39 1/2 hours to justify this investment.

If you improve communications—and develop a
better understanding of the needs of your business—and its
relationship to the employes' job security—the productivity
gains can be an order of magnitude greater than that: Some
plants have achieved 20 percent or higher productivity improvements using these basic techniques.[22]

There are, of course, many ways to improve productivity:

[21]Paraphrased from "Productivity and Strategy," speech to the Management Council by Thomas J. Mur-
rin, President, Public Systems Company, Westinghouse Electric Corporation, May 17, 1979, used by permission.
[22]Ibid.

136   Adding more efficient equipment, reorganization of resources, effective management of human resources, and so on. Designing and developing an effective two-way communications program is a good way to start. For example, Corning Glass installed escalators instead of elevators in a new corporate engineering building to make face-to-face communication easier. The installation of the escalators was one of several improvements undertaken to improve productivity of scientists and engineers.[23]

COMMUNICATING TO GROUPS

An engineering manager spends most of his time in some aspect of communications: writing, reading, listening, or speaking. Effective speaking to groups calls for a rigorous adherence to three cardinal rules:

1.   Tell them what you are going to tell them!

2.   Tell them!

3.   Tell them what you told them!

To follow these rules a simple organization of the material to be given to the group is needed along this schema:

Central idea _____

Main points _____

No more than five or six central ideas should be presented to a group at any one time. Why only five or six? The human mind can only accommodate a few ideas at one time. The speaker should identify and develop general ideas that are to be communicated to the group. If more ideas surrounding a situation are recognized, a good assessment of the situation is needed, because chances are that some overlap exists.

In one of Plato's dialogs, Socrates comments that a speech is like a living being with a head, body, and feet. A talk is like this: it includes an introduction (tell them what you are going to tell them), a body (tell them), and a conclusion (tell them what you told them). After a preview in the introduction, an elaboration in the body, and a recapitulation at the end, what further needs to be said?

The central idea to be conveyed to a group may come easily, but the development of the main points to illustrate and reinforce the central idea is a stimulating task. There are many ways to improve one's skill in speaking before groups. Enrollment in public speaking classes at a local university is one way. Joining the local toastmaster club can be helpful. Above all, one should seek opportunities to

[23]"How Three Companies Increased their Productivity," *Fortune*, March 10, 1980.

speak before groups and practice the essentials of good public **137** speaking.

COMMUNICATING IN MEETINGS

Many (if not most) people think a meeting is a waste of time; and many meetings are a waste of time. Meetings are usually called for the purpose of:

1   Telling—passing on information such as explaining the reorganization of the engineering department.

2   Selling—winning the group's acceptance of the leader's proposal. Example: Selling a merit-promotion evaluation plan.

3   Solving—working with the group to come up with a solution to a problem. Example: Meeting with a project team in a design review to select product design.

4   Educating—instruction to update the knowledge and skills of the group members, such as an in-plant workshop on computer-aided design.

An engineering executive spends something in excess of 50 percent of his time in meetings of some sort. These meetings are held for a purpose—and the purpose can only be accomplished if effective communications among the participants are carried out. If meetings are looked on as an opportunity to communicate with the participants on some subject, then those who call a meeting will take care to prepare a plan for how the meeting is to be conducted. There are a few key things to keep in mind in preparing a plan for a meeting.

To start, remember that the conduct of a meeting—which includes even a conference between a superior and subordinate—is a managerial activity and can be looked at from the standpoint of the management functions of planning, organizing, and controlling. Using these management functions as a guide, we can suggest a strategy that can be used to prepare for and conduct a meeting.

PLANNING A MEETING Planning for a meeting involves making decisions on the agenda, time, place, and purpose for getting everyone together. In planning for the meeting careful thought should be given to the expected outcome of getting the people together. Information that is needed to help the participants bring about the outcome should be assembled and made available for study before the meeting time. A key question to consider when planning a meeting is simply this: Is this meeting really necessary? In this context Jay has noted: "The most important question you should ask is: 'What is this meeting intended to achieve?' You can ask it in different

138   ways—'What would be the likely consequences of not holding it?' 'When it is over, how shall I judge whether it was a success or a failure?' '[24]

Often the need for a meeting can be eliminated by judicious analysis of a problem before calling people together to seek some resolution. A practical process that can be followed that will help reduce the number of meetings is to prepare a brief, informal memorandum to yourself that addresses these questions:

1   What is the issue (specific problem or opportunity) for which the meeting should be held? Many times meetings are held without any definition of what the output of the meeting will be.

2   What are the facts? Problems or opportunities don't exist in an information vacuum. There are some facts that bear on the situation; that cause the problem or suggest the opportunity.

3   What are the potential alternatives or solutions—and the associated costs and benefits that relate to these alternatives? Even a cursory thinking about alternatives can prove useful in deciding whether or not a meeting should be held.

4   What specific recommendations can be proposed to deal with the problem or opportunity *at this time,* and could be suggested to the meeting participants?

5   What will happen if the meeting is not held?

Answering all these questions should help one find a solution and possibly eliminate the need for the meeting. Even a subordinate who wishes to meet with his boss will find that trying to answer these questions will reduce the frequency with which he needs to consult with his boss.

Part of the planning for a meeting is a *plan for the organization* of the meeting.

ORGANIZING A MEETING   An important part of preparing for a meeting is to organize the resources—a suitable location, identifying and notifying the participants, and designating a chairman. The chairman is responsible for organizing the running of the meeting, seeking points of agreement, limiting debate, and encouraging everyone to give a point of view. An effective chairman should periodically

---

[24]Reprinted by permission of the *Harvard Business Review,* Excerpt from "How to Run a Meeting" by Antony Jay (March-April 1976) p. 47. Copyright © 1976 by the President and Fellows of Harvard College; all rights reserved.

summarize points of agreement—or disagreement, as the case may **139**
be. When agreement is reached and action is required, an individual
should be designated to carry out whatever action is required, with
follow-up if necessary.

Meetings are a group activity. Douglas McGregor has
described "in everyday common-sense some of the things that are
characteristic of a well-functioning, efficient group."[25] According to
McGregor:

1   The "atmosphere," which can be sensed in a few
minutes of observation, tends to be informal, comfortable,
relaxed. There are no obvious tensions. It is a working atmosphere in which people are involved and interested.
There are no signs of boredom.

2   There is a lot of discussion in which virtually everyone participates, but it remains pertinent to the task of
the group. If the discussion gets off the subject, someone
will bring it back in short order.

3   The task or the objective of the group is well understood and accepted by the members. There will have
been free discussion of the objective at some point until it
was formulated in such a way that the members of the group
could commit themselves to it.

4   The members listen to each other! The discussion does not have the quality of jumping from one idea to
another unrelated one. Every idea is given a hearing. People
do not appear to be afraid of being foolish by putting forth a
creative thought even if it seems fairly extreme.

5   There is disagreement. The group is comfortable
with this and shows no signs of having to avoid conflict or to
keep everything on a plane of sweetness and light. Disagreements are not suppressed or overridden by premature group
action. The reasons are carefully examined, and the group
seeks to resolve them rather than to dominate the dissenter.

On the other hand, there is no "tyranny of the minority." Individuals who disagree do not appear to be trying
to dominate the group or to express hostility. Their disagreement is an expression of a genuine difference of
opinion, and they expect a hearing in order that a solution
may be found.

Sometimes there are basic disagreements which
cannot be resolved. The group finds it possible to live with
them, accepting them but not permitting them to block its
efforts. Under some conditions, action will be deferred to

[25]Douglas McGregor, *The Human Side of Enterprise*, Mcgraw-Hill, New York, 1970, p. 232.

permit further study of an issue between the members. On other occasions, where the disagreement cannot be resolved and action is necessary, it will be taken but with open caution and recognition that the action may be subject to later reconsideration.

6   Most decisions are reached by a kind of consensus in which it is clear that everybody is in general agreement and willing to go along. However, there is little tendency for individuals who oppose the action to keep their opposition private and thus let an apparent consensus mask real disagreement. Formal voting is at a minimum; the group does not accept a simple majority as a proper basis for action.

7   Criticism is frequent, frank, and relatively comfortable. There is little evidence of personal attack, either openly or in a hidden fashion. The criticism has a constructive flavor in that it is oriented toward removing an obstacle that faces the group and prevents it from getting the job done.

8   People are free in expressing their feelings as well as their ideas both on the problem and the group's operation. There is little pussyfooting, there are few "hidden agendas." Everybody appears to know quite well how everybody else feels about any matter under discussion.

9   When action is taken, clear assignments are made and accepted.

10   The chairman of the group does not dominate it, nor on the contrary, does the group defer unduly to him. In fact, as one observes the activity, it is clear that the leadership shifts from time to time, depending on the circumstances. Different members, because of their knowledge or experience, are in a position at various times to act as "resources" for the group. The members utilize them in this fashion and they occupy leadership roles while they are thus being used.

There is little evidence of a struggle for power as the group operates. The issue is not who controls but how to get the job done.

11   The group is self-conscious about its own operations. Frequently, it will stop to examine how well it is doing or what may be interfering with its operation. The problem may be a matter of procedure, or it may be an individual whose behavior is interfering with the accomplishment of the group's objectives. Whatever it is, it gets open discussion until a solution is found.[26]

[26]Ibid., pp. 232–235.

McGregor then takes us to a description of the observable **141** characteristics of a poor group—one that is ineffective in accomplishing its purpose:

1   The "atmosphere" is likely to reflect either indifference and boredom (people whispering to each other and carrying on side conversations, individuals who are obviously not involved, etc.) or tension (undercurrents of hostility and antagonism, stiffness and undue formality, etc.). The group is clearly not challenged by its task or genuinely involved in it.

2   A few people tend to dominate the discussion. Often their contributions are way off the point. Little is done by anyone to keep the group clearly on the track.

3   From the things which are said, it is difficult to understand what the group task is or what its objectives are. These may have been stated by the chairman initially, but there is no evidence that the group either understands or accepts a common objective. On the contrary, it is usually evident that different people have different, private, and personal objectives which they are attempting to achieve in the group, and that these are often in conflict with each other and with the group's task.

4   People do not really listen to each other. Ideas are ignored and overridden. The discussion jumps around with little coherence and no sense of movement along a track. One gets the impression that there is much talking for effect—people make speeches which are obviously intended to impress someone else rather than being relevant to the task at hand.
        Conversation with members after the meeting will reveal that they have failed to express ideas or feelings which they may have had for fear that they would be criticized or regarded as silly. Some members feel that the leader or the other members are constantly making judgments of them in terms of evaluations of the contributions they make, and so they are extremely careful about what they say.

5   Disagreements are generally not dealt with effectively by the group. They may be completely suppressed by a leader who fears conflict. On the other hand, they may result in open warfare, the consequence of which is domination by one subgroup over another. They may be "resolved" by a vote in which a very small majority wins the day, and a large minority remains completely unconvinced.
        There may be a "tyranny of the minority" in which an individual or a small subgroup is so aggressive that the majority accedes to their wishes in order to preserve the

peace or to get on with the task. In general only the more aggressive members get their ideas considered because the less aggressive people tend either to keep quiet altogether or to give up after short, ineffectual attempts to be heard.

6   Actions are often taken prematurely before the real issues are either examined or resolved. There will be much grousing after the meeting by people who disliked the decision but failed to speak up about it in the meeting itself. A simple majority is considered sufficient for action, and the minority is expected to go along. Most of the time, however, the minority remains resentful and uncommitted to the decision.

7   Action decisions tend to be unclear—no one really knows who is going to do what. Even when assignments of responsibility are made, there is often considerable doubt as to whether they will be carried out.

8   The leadership remains clearly with the committee chairman. He may be weak or strong, but he sits always "at the head of the table."

9   Criticism may be present, but it is embarrassing and tension-producing. It often appears to involve personal hostility, and the members are uncomfortable with this and unable to cope with it. Criticism of ideas tends to be destructive. Sometimes every idea proposed will be "clobbered" by someone else. Then, no one is willing to stick his neck out.

10   Personal feelings are hidden rather than being out in the open. The general attitude of the group is that these are inappropriate for discussion and would be too explosive if brought out on the table.

11   The group tends to avoid any discussion of its own "maintenance." There is often much discussion after the meeting of what was wrong and why, but these matters are seldom brought up and considered within the meeting itself where they might be resolved.[27]

McGregor's description of the characteristics of effective and ineffective groups points out vividly the role of communications in such groups, and how the operation of the group can affect its efficiency.

CONTROLLING A MEETING   To control means to verify if everything occurs in conformity with the plan adopted. Control in a meeting is that management function carried out to make sure the meeting ac-

[27]Ibid., pp. 236–238.

complishes its purpose. Meetings can be enormously useful for com-   **143**
municating information to people who have a legitimate need for
such information. Control of a meeting starts with the plan for the
purpose of the meeting, how the meeting will be conducted, and such
related matters. The following elements should be emphasized in
controlling a meeting:

1   Establish firm time limits and adhere to those
limits.

2   Start off with a statement of the objective of the
meeting. Reinforce the purpose of the meeting as required.
Prepare an agenda—and use it!

3   Limit the discussion to the purpose of the meeting; try to avoid having anyone dominate the discussions.

4   Summarize progress or lack of progress appropri-
ately during the conduct of the meeting.

5   Encourage—and   control—disagreements.   A
good debate can help to get at the issues in the subject material.

6   Take time during the meeting to assess how well
things are going and what might be done to improve the effectiveness of the discussion.

The best advice of all for a meeting chairman is to study
McGregor's description of a well-functioning group and a poor group
and seek to manage meetings accordingly. A checklist for the conduct of a meeting can be a useful device to save time for everyone
who is associated with the meeting. Table 5.2 is such a checklist.

SUMMARY

An engineering manager has the key responsibility for providing a
cultural ambience in his organization that facilitates efficient and effective communications. Most of the manager's time is spent in some
form of communication with his clientele to satisfy their needs for information on the organization and what it is providing for them.
Effective communication is two-way—going from a transmitter to a receiver with enough feedback to be sure the right message gets across. Effective communication can be managed,
planned, organized, and controlled to accomplish what was intended.

5-2

Before your next meeting, ask yourself: Is this meeting necessary? Do you have a good reason for holding it? Is the purpose:

1 To receive reports from all or most of the participants?

2 To reach a group judgment as the basis for a decision?

3 To define, analyze or solve a problem?

4 To gain acceptance for an idea, program or decision?

5 To achieve a training objective?

6 To reconcile conflicting views?

7 To provide essential information for work guidance or for combating rumors about an impending change?

8 To assure equal understanding on a specific matter by all present?

9 To obtain immediate reactions when speedy response is important?

10 To take up a matter that has been stalled in ordinary channels?

Once you decide to call a meeting, give some thought to strategy. Review these points:

1 Have you made your own survey of the expectations of the participants—how each will relate himself to the goals of the meeting?

2 What problems, points of view and barriers may be brought out in the light of participants' experience with similar problems?

In planning the agenda, consider these factors:

1 As a matter of building participation, have you involved meeting participants in developing the agenda?

2 Is the agenda limited to the number of topics that can be given adequate attention?

3 Have you limited the duration of the meeting—if it is routine—to not more than an hour or an hour and a half?

As a final check on arrangements, take another look at these details:

1 Does each participant know what's expected of him?

2 Have you identified any problems for which you should be prepared in the meeting?

3 Have you prepared a checklist to use as a "meeting leader's guide"?

4 Are the room arrangements adequate?

Even when you have a good reason for calling the meeting, consider reasons for not holding the meeting. Consider these questions:

1 Have you estimated the cost of the meeting, and compared this with the probable gains?

2 Can the business of the meeting be handled more efficiently through (a) personal supervisory action, (b) written communication, (c) individual telephone calls?

3 Can you be ready for the meeting? Will all the key participants be available? Is the timing right?

If the light is still green, make sure you justify the participation of each person—bearing in mind that the ability to communicate effectively diminishes with an increase in the number of participants.

Ask yourself about each participant:

1 Is he to carry out a decision that will be reached at the meeting?

2 Will he contribute unique information?

3 Will his approval be needed?

4 Is he the man with normal functional responsibility for the matter to be discussed?

5 Will he add to the meeting? Is he an idea man, an elder statesman who lends stability, a person who may provoke discussion, one who balances controversial points of view?

Four guideposts for any meeting

1 *Hold a meeting for a definite purpose*—and only when it is likely to achieve results.

2 *Be thoroughly prepared before the meeting.* This includes the agenda, preliminary discussions of meeting topics, and materials to be used in the meeting.

3 *Arrange for lively participation.* Have a balanced mixture of discussion and visual presentations, whenever appropriate.

4 *Bring the meeting to a definite conclusion.* If it is a problem solving or action meeting, make sure that all are quite clear on the decisions reached. Arrange for followup on decisions and actions. If it is an information meeting, summarize or recap the points made.

Start on time—adjourn on time

Source: By John F. Whitcomb—Condensed in AMA's *SupervisoryManagement*, November 1967, from *Steel* vol. 160, no. 19. © 1967 by the Penton Publishing Co.

DISCUSSION QUESTIONS

1 Why is the ability to communicate such an important skill for the engineering manager? For the engineer?

2 According to Peter Drucker, the ability to communicate heads the list of criteria for success. Do you believe this statement of Drucker's is relevant for the individual who is solely interested in a technical career?

3 How is "communication" defined? What are the methods by which one is able to communicate?

4 What are the basic ingredients of Shannon's model of communication? Can this model be related to a real-world situation?

5 Where is communication most likely to break down? How can the chances of a breakdown be reduced?

6 What are the characteristics of communication?

7 What are some of the nonverbal elements of communications? Can nonverbal communication be effective in conveying a message to someone?

8 Why is listening considered a key element of communication?

9 If one wishes to improve one's ability to listen, what self-improvement practices should be put in effect?

10 What is meant by "on-the-job communications"? Why is this part of communications so important to the engineer and the engineering manager?

11 What are some basic rules for planning a meeting?

12 What are some of the characteristics of *effective* and *ineffective* groups?

KEY QUESTIONS FOR THE ENGINEERING MANAGER

1 What importance does the engineering manager place on communication in this organization?

2 Do the engineers in this organization generally have a good image as effective communicators? Why or why not?

3 How do the engineers perceive the words "manage" and "manager" in this organization?

4 What have I done in the past year to improve communications in this organization? How effective have these actions been?

5 What organizational or personal failures have occurred in the past year in this engineering community that can be attributed to poor communications?

6 Are the communications efforts in this organization clearly purposeful in nature?

7 Are the meetings held in this organization representative of good communications? Why or why not?

146   8   Do I hold meetings in this organization that might be perceived as unnecessary by the attendees?

9   Have I used the informal communication channels as a technique to get messages to the engineers? How effective have these efforts been?

10   What recent key messages in the management functions (planning, organizing, motivating, directing, and controlling) have I recently transmitted to the engineers? How successful have these efforts been?

11   Have I seen any situations where my authority in the organization has been impaired because of my lack of communication skills?

12   Are the meetings in this organization most typical of McGregor's *effective* or *ineffective* group? How do I perceive these meetings vis-à-vis the perceptions of the engineers?

BIBLIOGRAPHY

Berlo, David K.: *The Process of Communication,* Holt, New York, 1960.

"Body Language—Insight into Engineering Teams," *Automation,* November 1975.

Cherry, Colin: *On Human Communications,* Science Editors, New York, 1961.

Davis, Keith: *Human Relations in Business,* McGraw-Hill, New York, 1957.

Drucker, Peter F.: "How to Be an Employee," *Fortune,* May, 1952.

Ericson, R. F.: "Organizational Cybernetics and Human Values," *Academy of Management Journal,* 1970.

Gibson, James L., John M. Ivancevich, and James H. Donnelly, Jr.: *Organizations, Structure, Processes, Behavior,* Business Publications, Dallas, Tex., 1973.

Glaser, Milton A.: "The Innovation Index," *Chemical Technology,* vol. 6, March 1976.

Haakenson, Robert: "Improving Oral Communication," *Effective Communications for Engineers,* McGraw-Hill, New York, 1974.

Heckman, W. K.: "Managing for Action: How to Communicate with Meaning," *Pipeline & Gas Journal,* July 1977.

"How Successful Executives Handle People," *Harvard Business Review,* January-February 1953.

Jay, Antony: "How to Run a Meeting," *Harvard Business Review,* **147** March-April 1976.

Leavitt, Harold J., and R. A. H. Mueller: "Some Effects of Feedback on Communications," *Human Relations,* vol. 4, 1951.

Lee J.: "Communication, the Key to Successful Design," *Engineering,* February 1976.

Leonard, V.: "Better Communications Unsnarl Production Tie-Ups," *Production Engineering,* January 1977.

Longfellow, Layne A.: "Body Talk: The Game of Feeling and Expression," *Psychology Today,* October 1970.

McGregor, Douglas: *The Human Side of Enterprise,* McGraw-Hill, New York, 1970.

————: *The Professional Manager,* McGraw-Hill, New York, 1967.

Mok, P. P., and D. Brant: "How to Come Across, Why and What to Do About It," *Chemical Technology,* December 1977.

Raudsepp, E.: "Communicating Is More Than Just Talking," *Machine Design,* Nov. 13, 1975.

Schramm, Wilbur: "How Communication Works," *The Process and Effects of Mass Communication,* University of Illinois, Urbana, 1953.

Shannon, Claude, and Warren Weaver: *The Mathematical Theory of Communication,* University of Illinois, Urbana, 1948.

Webber, Ross A.: *Management,* Irwin, Homewood, III, 1975.

Weiner, Norbert: *Cybernetics: On Control and Communication in the Animal and the Machine,* Wiley, New York, 1948.

Weismantel, Guy E.: "Management—By Listening," *Effective Communication for Engineers,* McGraw-Hill, New York, 1974.

Westley, Bruce, and Malcolm Maclean, Jr.: "A Conceptual Model for Communications Research," *Journalism Quarterly,* Fall 1957.

"What Helps or Harms Productivity," *Harvard Business Review,* January 1964.

Whittaker, P. N.: "Better Communications: A Creative New Tool of Management," *Microwave Journal,* January 1978.

Wickesburg, A. K.: "Communications Networks in the Business Organization Structure," *Academy of Management Journal,* 1968.

MANAGING CHANGE IN ENGINEERING ORGANIZATIONS 6

In some organizations change is brought about only through death or retirement. To change means to do something different. Many changes have occurred in the engineer's way of life in the past two decades. In the future, more dramatic changes appear likely. In modern technical organizations, we might adopt the slogan, "Subject to change without notice." The reasons for change are varied; perhaps the most basic of all is the continuing increase in the world's population. With the increase in population densities and the emergence of underdeveloped nations into world trade, the world is becoming smaller. The continuing technical and management challenges posed by foreign competition, both in the United States domestic and international markets, is a constant reminder of the change facing us.

The engineering manager who knows how to deal with change through the application of a management process will be better able to prepare his organization for its future. This chapter will present some general ideas about the nature of change and suggest ways that the engineering manager can deal with change.

Many examples of change could be given to illustrate how unfixed is the technological world in which we live. It is not our purpose to belabor an already overdone subject by citing anticipated changes in tomorrow's world. We accept change as normal; the prescription we offer deals with the gaining of an understanding of how to anticipate and deal with change as it comes about.

Humanity has always been concerned with change and the impact change would have on particular cultures. In past periods the changes were serial—appearing in successive parts. Today one might hypothesize that our changes tend to be parallel and systems-related—appearing in several simultaneously successive subsystems. Those energy-related changes that will affect our future tech-

150   nological application will truly be of systems nature in which "everything is related to everything else."

Changes affecting engineering organizations come internally as well as externally. Changes that are externally generated are often more visible than changes that come within the organization itself. When a competitor brings out a new product, a change has come about—and such change has visibility. Dramatic changes such as those brought about by the oil embargo of 1974 are easy to recognize. Fortunately, changes of this magnitude are not too common.

In the larger systems environment, economic, social, technological, political, and legal forces and events are at play. These forces and events contain harbingers of future events. The organization's strategic planning office has a responsibility to track these forces to see how they might affect the firm's future. Engineering managers can be tied into the distribution of information about changes in these areas; the tracking of technological changes and an assessment of these changes are a responsibility that the engineering manager must assume if he is to avoid being surprised by changes in his environment. Although the engineering manager is concerned with many changes that can affect his work, he should be concerned primarily with technological change.

Technological change includes those changes which transform resources into a new product or service. In the traditional sense, technology means machines, processes, or techniques, both in the fabrication of the product or service and/or embodied in the product or service itself.

Our interest lies more in the impact of technological changes on the management of the organization, particularly on the structure of the organization and on the behavior of the people in the organization. Some writers have described the role of technology as a key determinant of organizational structure.[1] They observe that firms with a simple and stable technology should adopt a traditional bureaucratic organization; firms with a complex and dynamic technology ought to tend to a more open, participative organizational structure. Other writers have made the point that the use of project-type organizations explained in Chap. 2 is a way of developing a more open, participative form of organization.

The pursuit of new technology by the firm involves a decision to adapt an organizational structure to fit that technology. The introduction of a new technology in the marketplace by a competitor or by the firm itself usually has far-reaching effects on the organization. The degree and extent of structural and behavioral changes depend on the magnitude of the technological change.

[1]Joan Woodward, *Industrial Organization,* Oxford University, New York, 1967, and Frank J. Jasinski, "Adapting Organization to New Technology," *Harvard Business Review,* January-February 1959, pp. 79–86.

Sometimes the change is traumatic—for example, the development **151** of the transistor caused deep and basic changes in the electronic industry. The automation of a previously human-paced manufacturing process causes change as well.

Engineering managers are responsible for keeping their general managers appraised of the technological problems and opportunities emerging in the marketplace. General managers can be so immersed in the totality of the organizational operations that they may not be able to perceive emerging technological ideas that can hold promise for a new product or service. Even if such an opportunity is noted, it may not be exploited. Stories of missed opportunity abound. None is more dramatic than the classical case of xerographic copying, where the managements of such corporations as Addressograph-Multigraph, Lockheed, Eastman Kodak, and IBM failed to see the future demand for xerographic copying. Other companies turned down the opportunity to back this process. Finally, the Haloid Company went it alone and developed a technological idea into what has become the billion-dollar Xerox Corporation. The xerographic process was "by no means the only example of a situation in which general managements with impressive technological experience have been oblivious to the potential of major new technological opportunities."[2]

General managers depend on their engineering managers to develop a technical sense of direction for the organization. Technology managers who accept this responsibility should develop rigorous approaches to understanding and communicating to general managers the direction the firm's technology is taking. To do this, a strategy for managing change must be adopted and communicated to the people managing the technology part of the organization. An important part of this strategy is the creation of a cultural ambience within the organization to define and promote the process of organizational change.

Change comes about through a process of *invention, innovation, imitation, reaction* or *obsolescence.* These terms are used in describing the overall organizational approaches in dealing with changing environments in the organization's future.

INVENTION

Invention is the act or process of originating a new device or process from study and experimentation. The commonplace products of today were ideas in the minds of some who came earlier. In some cases the invention was more an adaptation of an existing tech-

---

[2]Reprinted by permission of the *Harvard Business Review.* Excerpt from "How to Spot A Technological Winner" by George R. White and Margaret B. W. Graham (March-April 1978), Copyright © 1978 by the President and Fellows of Harvard College; all rights reserved.

152   nology into a contemporary form, as occurred in the automobile or "horseless carriage." Sometimes an invention is more an adaptation of an existing technology from the environment. Some inventions by human beings are primarily an application of a "technology" practiced by animals, as in the case of the glider taken from the observation of birds in flight. The honeycomb construction used in the airdynamic surface of an airplane looks similar to the construction used in the honeycomb in a beehive. A study of the configuration of a hornet's nest gives insight into the design of insulation in a building. The funnel-shaped fishing net of the larva of the caddis fly and the "communication system" used by honey bees serve to remind us that many modern inventions are more the borrowing of ideas from the biological environment that surrounds us than original thoughts. Some other practices of the animal and insect world which have been adopted by human beings:[3]

> Use of columns to support horizontal combs in hornet's nest
>
> Use of "spindle and shuttle" in weaving nests by weaver ants
>
> Air conditioning in termite dwellings
>
> Construction of an arch by ants (macrotermes natalensis)
>
> Use of glue to hold nest together by stickleback fish
>
> Use of fermentation of compost pile to provide incubator for eggs of Australian brush turkey
>
> Construction of beaver dams
>
> Design of bubble nest of paradise fish

We do not mean to imply that all inventions had their forerunners in prototypes already developed in nature, but an impressive number of modern inventions have been preceded by an earlier application in nature. Certainly the transistor, the computer, and such inventions have no counterpart in nature. On the other hand, are there any similarities between the navigation system of the modern airplane and the "navigation system" of the migrating birds?

There is one parallel: each invention of human beings has emerged to serve a need; the practices of the biological world also serve needs of particular species. In the case of human beings, the need may have existed since antiquity before it emerged in some social context. The simple invention of the cotton gin had enormous implications for the whole economy of the southern states of the United States. Bessemer's steel-making process provided the world

---

[3]Some of these inventions suggested by Karl Von Frisch in *Animal Architecture*, Harcourt Brace Jovanovich, New York, 1974.

with an important construction material. The invention of gunpowder **153**
contributed to the end of the feudal system of the medieval world.
The invention of the transistor dramatically changed the technology
of communication and information processing. Radar has made
modern air travel safe and space exploration possible. Modern word
processors show promise of significantly improving the productivity
of administrative functions. The computer has helped the engineer
do more thorough design and development work.

And so it goes! Each invention seems to solve some
problem. But then the oncoming problems seem more complex than
the ones that have been solved.

Some inventions have required technical or scientific
knowledge. Others have seemed so simple—at least in hindsight.
Movable type was a simple idea but a great improvement over the
carving of a block for each page when printing a book. The vulcaniza-
tion of rubber was the result of random experiments by
Charles Goodyear. The wheel is so basic and simple that we take it
for granted. Yet the great Indian civilizations of the Aztecs, Mayans,
Toltecs, and Incas all did without the wheel.

To the engineer an invention is extremely important to his
career. The rare engineer who receives a patent for his invention joins
those individuals who have helped to make a better world through
improving the state of the art in the discipline in which they work.

The subject of invention is so broad that all the authors hope
to do is capture the reader's attention and encourage him to give the
subject some thought. A study of the history of inventions is a fas-
cinating subject that should interest the engineer and the engineering manager.[4]

INNOVATION

In an innovative organization, the managers of the organization actively encourage people to be creative and to innovate ways to further
the organization's objectives. An organization that innovates improved products, services, and methods of arriving at those products
and services has a cultural ambience that encourages people to seek
better ways of doing things. In such organizations there is an explicit
belief that the organization's future depends more on the motivation
of the people to prepare for the future than on simple response to future needs. Strategic planning, competitive analysis, and an ongoing
assessment of organizational strengths and weaknesses are a way of
life. People are expected to initiate improved methods and organiza-

---

[4]See Edward De Bono (ed.), *An Illustrated History of Inventions From the Wheel to the Computer*, Holt, New
York, 1974. See also Karl von Frisch, *Animal Architecture*, Harcourt Brace Jovanovich, New York, 1974.
We have drawn from these books in our brief discussion of invention.

154   tional processes. The cultural ambience expects change to lead the way to improved organizational effectiveness. Individuals who take the lead in initiating change are typically viewed as the elite of the organization.

To initiate through the innovation process an improved product, service, or method requires that people in the organization are committed to the practice of innovation.

Innovation is the process of putting things together in a manner that has not been done before. It is the successful integration of related subprocesses: the development of an idea into a useful device and the development of a new market. As the forces of development, production, and marketing are integrated successfully, technological change and perhaps social change have come about.

Schoen notes that innovation is more often the product of recognition and adaptation than of truly original invention.[5] Morton holds that innovation is not a single function or a random act but rather a total process with specialized, interconnected parts all responding in some coordinated way to overall systems goals.[6] Writing in another publication, Morton sees innovation as a renewal process:

> Innovation means renewal. It means the improvement of the old and the development of new capabilities of people and their organizations. Innovation is not the anarchistic destruction of the old—rather, it is the adaptive change and improvement of existing systems. Technological innovation means growth in the power of man's technology to provide new and improved products and services. It is essential to the birth and growth of industrial enterprises and to the economic, social, and political health of nations.[7]

In this chapter, we are mostly concerned with technological innovation, yet it must be recognized that a much wider view of innovation is held. Haggerty captured this wider view when he described innovation as a process that can occur in the creation, making, or marketing side of an organization.[8] In the broadest sense, then, innovation is any act that has the potential to favorably affect the future of the organization through a product, service, process, idea, or whatever. In our concern with the management of technology, we place a premium on that individual who can initiate an innovation.

[5]Reprinted by permission of the *Harvard Business Review*. Quoted from "Managing Technological Innovation" by Donald R. Schoen (May–June 1969), p. 156. Copyright © 1969 by the President and Fellows of Harvard College; all rights reserved.

[6]Jack A. Morton, "A Systems Approach to the Innovation Process," *Business Horizons*, Summer, 1967.

[7]Jack A. Morton, *Organizing for Innovation*, McGraw-Hill, New York, 1971, p. 1

[8]Patrick E. Haggerty, "Strategies, Tactics, and Research," *Research Management*, vol. 9, no. 3, 1966, pp. 141–159.

In today's competitive international markets there is really **155** little choice but to build an organizational strategy around the need for innovative technologies. Dr. Frank Press, scientific advisor to President Carter, summed up the need for innovative strategies in American industry when he remarked: "The harsh truth is that we are now very much locked into a dynamic system of global economic growth, and it is one based largely on technological change and innovation."[9]

Most innovation grows out of some need. The innovation developed in the United States space program grew out of a national need expressed in President John Kennedy's statement of a national goal: "To land a man on the moon and return him safely to earth within this decade." Some innovation comes from a knowledge of a technology followed by a search for its application. For example, knowledge of the process of nitrogen fixation in certain plants is well known. The challenge to agricultural researchers today is to find other plants with an agricultural potential which will fix nitrogen and reduce the need for commercial fertilizers.[10]

Knight describes innovation as existing in three categories: *routine,* which is programmed by management through defined policies and procedures; *distress* innovation, which is undertaken by a firm in trouble in some product, service, or process; and *opportunity* innovation, which is practiced by those firms with sufficient resources to pursue a program of research.[11]

If we view these thoughts in the context of who have been the innovators in the past, some insight can be gleaned from a report based on National Science Foundation data. This report shows that independent inventors, including inventor-entrepreneurs and small technically oriented companies are producing a remarkable percentage of the important inventions and innovations of this century—according to Patrick E. Haggerty, "a much larger percentage than their relative investment in these activities would suggest."[12]

THE EFFECT OF BUREAUCRACY

Such insight into the ability of large organizations to innovate raises questions about the suitability of the bureaucratic form of organization as the mechanism through which to foster innovation. If small organizations seem better able to innovate, why is this so? What is

[9]Walter Sullivan, "A Call for Industrial Research as a Key to Growth," *The New York Times,* February 1978, sec. II, p. 15, column 1. © 1978 by the New York Times Company. Reprinted by permission.
[10]Ibid.
[11]For further discussion see K. E. Knight, "A Descriptive Model of the Intra-Firm Innovation Process," *Journal of Business,* vol. 40, no. 4, 1967, pp. 478–496.
[12]Patrick E. Haggerty, "Long-Term Viability—The Business Problem," *IEEE Student Journal,* September 1968, p. 25.

156 there about large technical organizations—bureaucracies—that contributes to an environment where innovation may be stifled? Some possible reasons:

> The large organization, because of its size, is managed in a formal manner, depending on a system of policies and procedures to ensure uniformity of action.

> The control of large bureaucracy requires the existence of strategies and plans which set up performance standards that can be measured. Top-level managers can then institute any necessary corrective action to help to attain these performance standards. Unfortunately, these objectives often become self-limiting; subordinate managers cannot change the performance standards except through a lengthy appeal process.

> As the company gets bigger and extends its markets, it gets more complex, thousands of people get involved, and "administrators" tend to take over the business—the innovator finds himself immersed in a bureaucratic maze of systems and procedures. The rewards go to the successful administrators, those who are able to turn a profit or forestall a loss. The innovator develops an acute feeling of frustration, a feeling that things are not as they should be for the entrepreneur, and either gives up or seeks opportunities elsewhere.

> The growth of the large organization brings about a high degree of organization, decentralization into successively smaller echelons—into groups, divisions, departments, branches, sections—and the total job is divided into units that permit a manager to "get his arms around" his organization. This technique makes sense for good administration. The company has been broken into units with a manageable span of control. But the decentralization process has compounded the problems of coordination, information flow, territoriality, "not invented here" syndrome, and such things that can discourage the small-group synergism that facilitates innovation.

> In the large company, the system of rewards typically centers around current performance, often expressed in profit or loss. The administrator devotes most of his time to current matters. Organizational resources are first applied primarily to promoting efficiency in the organization and innovation comes in too late for its share of resources. The pragmatic manager who must choose between an uncertain, risky innovation and a hard, tangible current operating problem or opportunity will be sorely tempted to choose the latter and postpone the former.

ORGANIZATIONAL IMPLICATIONS   **157**

Innovation requires some form of organization albeit an informal one. Evidence suggests that innovation tends to be interdisciplinary in nature; this suggests the need for managers to recognize the shortcomings of the bureaucracy to provide the environment supportive of innovation. Some form of team or task force is required to provide the professional people an organization upon which to base their work. The challenge, according to Hammerton, "is to create within a large organization the type of environment offered by a small company. The positive aspects of the latter are close cooperation among all participants, ease of identification with the production activity, absence of make-work tasks, minimum managerial restriction, little need to sell and resell tasks and projects to upper levels of management."[13]

The project team helps to foster an environment that facilitates establishing and maintaining good communications and professional respect. Anshen, commenting on centralizing the long-range planning (including technological planning) function at a single location in the organization, notes: "In these structures, radical change is painful. It is viewed as disruptive and costly by managers committed to the present and appraised by their administration of the present. In their constrained sighting they are right. They resist change because it is in their perceived economic interest to resist and because change threatens their status and their intellectual and emotional commitments."[14]

According to Anshen, this may help to explain why so many of the major innovations—new ideas—come from companies other than the corporate giants. In speaking of the project team, which is free of this resistance to change, Anshen notes: "They (the project clusters) offer the important advantages of tailor-made design to fit unique tasks, flexible resource commitments, defined termination points, and an absence of enduring commitment that encourages resistance to radical innovation."[15]

The project team established to manage emerging technological ideas offers unique characteristics over the more traditional bureaucratic organization. Flexibility in adjusting to new tasks, easy management across organizational boundaries, specific performance measurement, and a motivation affect through the operation of a peer-group relationship are evident. Regarding the matter of organizational flexibility, George Lesch, president of Colgate-Palmolive, noted:

[13]© 1970 by the Regents of the University of California. Reprinted from *California Management Review*, Volume XIII, Number 2, pp. 57 by permission of the Regents.
[14]Reprinted by permission of the *Harvard Business Review*. Excerpt from "The Management of Ideas," by Melvin Anshen (July-August 1969, p. 103). Copyright © 1969 by the President and Fellows of Harvard College; all rights reserved.
[15]Ibid., p. 103.

158

I credit the task-force approach with contributing to our flexibility—and ultimately to our success. *It is an excellent way to bring about innovative change.* Task forces bring together the people whose expertise is needed to gather the necessary information. It gives a variety of different thoughts an airing. And it includes the people who can see that the decisions are carried out. It is the best way to come up with new approaches and to implement them rapidly.[16]

At Texas Instruments, a unique approach to organizing for innovation is given credit for contributing to the growth of the company. Texas Instruments is a $2-billion company with approximately 68,000 workers in forty-five plants spread throughout eighteen countries. Its expectation for growth is 15 percent annually in a worldwide business keyed to rapid technology change and declining unit costs. To meet this challenge Texas Instruments has successfully evolved one of the most formal planning systems in existence. Innovation is emphasized as the lifeblood of the company, and the key to Texas Instrument's success has been its novel, highly complex, system to stimulate and manage innovation. One of the keys to innovation at Texas Instruments is a program dubbed IDEA, started back in 1973. In this program Texas Instruments splits up $1 million annually among forty IDEA representatives, usually senior technical staffers, who have the authority to pass out grants of up to $25,000 to employees with ideas for a product or a process improvement. If one's idea is turned down by an IDEA representative, he can take his idea to another.

In the IDEA program within Texas Instruments, an individual who has an innovative contribution can be provided an opportunity to defend this to a board; if it is approved, that individual becomes an "IDEA manager" in the development of the project. IDEA at Texas Instruments is given credit for "starting an amazing number of innovative products, particularly in the consumer area. The $19.95 digital watch that tore apart the market in 1976 got its start as an IDEA program in the Semiconductor Group of Texas Instruments."[17] In the case of the digital watch, an individual was able to demonstrate for less than $25,000 that this new type of watch could be manufactured. The individual who came up with the idea worked with a team, and in less than a year the team was able to announce the first "all electronic, analog watch that uses a new type of liquid crystal display to show the traditional hands."[18]

In another example, an individual within Texas Instruments

---

[16]Henry O. Golightly, "What Makes a Company Successful?" *Business Horizons*, June 1971, p. 114. (Italics added.)
[17]"TI's Magic in Managing Innovation," *Business Week*, September 18, 1978, p. 84. Reprinted by special permission, © 1978 by McGraw-Hill, Inc. All rights reserved
[18]Ibid.

along with a project team developed a low-cost speech synthesizer **159** built on one tiny chip, with voice quality equal to that of the telephone system. The product as developed by this team was a breakthrough in a talking-learning aid to teach spelling. The subsequent product, called "Speak and Spell," is the first of what is expected to be a flood of products over the next years using the revolutionary chip.

While one wonders what effect TI's innovation strategy really had on improving productivity, some hint of this is gained in the realization that Texas Instruments "has accomplished what one hundred or more U.S. companies tried and failed to do in recent years: it has built a flourishing profitable business in consumer electronics. In doing so, TI has beaten the Japanese at their own game and probably is now No. 1 in worldwide revenues in both calculators and digital watches. Some experts believe that Texas Instruments has more potential in consumer electronics than any other company in the world."[19]

Project teams are also used in planning within TI. For example, a team is used to prepare the annual marketing plan. This team is led by the marketing product manager and includes the sales vice president, advertising, sales promotion, sales training, and public relations managers, and, as required, representatives from financial planning, distribution and transportation, customer service, design engineering, and general accounting. The planning begins with a two-day marketing seminar where key issues are identified. Two months later the team is required to complete its work on the overall plan, along with specific sales plans for the 5-year period in the future. It is reviewed by top management; once the plan has been approved, the task team participates in the implementation phase, each team member being responsible for implementation in his respective area. Since the same people are responsible for planning and implementation, the transition from planning to implementation is smooth.[20]

In 1971, the Microwave Cooking Division of Litton Industries, Inc., reorganized its engineering department from a functional structure to a task-team structure. For each engineering project, a task team was assigned, headed by a project manager. Members of the team included a design engineer, drafting people, a stylist, technicians, quality and manufacturing engineers, a marketing manager, buyers, and a home economist. Each team was assigned specific responsibilities. As a project progressed through the development cycle, other personnel would be added, such as engineers, quality control personnel, cost accountants, and production managers. The engineering project manager assigned at the beginning of a develop-

[19]"Texas Instruments Shows U.S. Business How to Survive in the 1980's," *Business Week*, September 18, 1978, p. 67.
[20]Ibid.

160   ment cycle was responsible for managing the team until the product went into production and a steady level of production was reached. At that time the task team was released, and the team members went on to work on other tasks. In some cases, however, the project team might stay together for additional time in order to work on product improvement, cost reduction, and such ad hoc activities.

The task team at Litton becomes the central organizational mechanism by which the development activity is reviewed. In this review members of top management and key department heads meet with the task team to assess progress against project goals, to discuss the problems encountered, and, in general, to get a good assessment of how things are going on the project. At these review meetings decisions can be made as to whether the project should proceed and what changes in direction or objectives on the project should be taken. In the Microwave Cooking Division of Litton Industries, Inc., today all engineering projects are managed on this basis. A ninety-person engineering department is organized into fourteen task teams and can handle approximately forty projects at any one time.[21]

Some of the cultural changes that have come about within the Microwave Cooking Division of Litton Industries are summarized by George as follows:

> The openness of the organization is perhaps the single most feature of the company to outsiders and newcomers. This openness is demonstrated in a variety of ways: (a) A receptiveness to new ideas and newcomers, (b) an open sharing of information, generally restricted in other organizations, and (c) a willingness of individuals to share their professional problems, challenges and frustrations with others at all levels in the organization.[22]

Other beneficial cultural changes are noted. For example, managers within the company are taking a broader view of their jobs. There is a change taking place in the interactions between the departments and functions. The task-team approach seems to result in less time being spent on interpersonal conflicts and a greater focus on goals and objectives in the organization, and functions at the lower level in the organization are being more effectively integrated.[23]

Innovation depends on someone being creative! Creativity is simply the ability or power to create something; the creative process is characterized by originality, expressiveness, and imagination.

---

[21]Reprinted by permission of the *Harvard Business Review*. Excerpt from "Task Teams for Rapid Growth," by William W. George (March–April 1977). Copyright © 1977 by the President and Fellows of Harvard College; all rights reserved.
[22]Ibid.
[23]Ibid.

Creative behavior in the engineering organization is manifested in **161** new products, machines, services, processes, or technology. Creativity takes an idea and develops the time and place for the idea to come into its own.

THE ROLE OF IDEAS

The development of an idea requires a commitment on the part of people to support the idea. In this respect, Schoen notes: "If there is any single point on which all who have written about the innovation process agree, it is the dependence of innovation on individuals committed to ideas."[24]

To facilitate a flow of ideas in the organization, the approach should be adaptive and flexible enough to encourage and facilitate the emergence and development of ideas from any source. The generation of good ideas is not the sole province of those managers who occupy the top positions in the organizational hierarchy.

Since new products or services do indeed emerge as ideas in an organization, these ideas need to be managed in their life cycle from initial concept through production and distribution in the marketplace. Emerging ideas are harbingers of future change. A system that permits the management of these ideas facilitates the formulation of strategy in the organization. As emerging ideas are identified those which show promise can be segregated out for further study and analysis. Thus, the idea is taken as the point of departure for developing an organization to manage innovation.

Having a good idea is not enough. The idea has to be taken, organized, and integrated into the strategy of the organization. This process of integration cuts through all systems of the organization: technical, social, market, financial, plant, equipment, and so on.

Innovation takes a creative idea and integrates the idea into the systems fabric of the organization. The innovator says: "That's a good idea. How can that idea be integrated into our organizational system to improve our performance? What organizational changes are needed to nurture this idea?"

Innovation—taking a creative idea and applying it to the solution of a problem or the exploitation of an opportunity—is a rare and demanding process for the innovator. In this respect Machiavelli noted: "It must be remembered that there is nothing more difficult to plan, more doubtful of success, nor more dangerous to manage than the creation of a new system. For the initiator has the enmity of all who would profit by the preservation of the old institutions and merely lukewarm defenders in those who should gain by the new ones."[25]

---

[24]Schoen, op. cit., p. 160.
[25]Machiavelli, *The Prince*.

162   The history of innovation and invention is replete with resistance by people who did not welcome the innovation and saw it as a threat to their way of life. Innovations, whether leading to an invention or something like an improved organizational process, "affect the organizational structure and human behavior, resulting in turn in planned and unplanned organizational and social change."[26]

The key to growth in a high-technology firm depends on the firm's ability to innovate technology and successfully introduce that technology to the marketplace. To be effectively introduced, an innovation must be perceived as compatible with both individual and organizational needs. An individual will view an innovation with the question "What's in it for me?" If the individual feels that he will benefit from the innovation, then the organizational acceptance will be facilitated.

This, of course, presupposes that the organization's senior managers have reached a consensus on a statement of organizational objectives and goals and the organizational strategies by which these objectives will be attained. This consensus, reflected in an outline of strategic needs, states the result that the change should bring about. According to Howe et al.:

> Most research suggests that the acceptance of an innovation within the organization is directly related to the knowledge and skills of internal personnel who are in the best position to introduce innovation in that they: (1) know the system, (2) understand and speak the language of the organization, (3) understand the norms of the organization, (4) identify with the organization's needs and goals, and (5) are familiar with the potential users.[27]

The innovative process has its beginning in the emergence of an idea for doing something differently. Innovation, simply put, is the process of working through creative insights and ideas to create something that does not currently exist. Once technological innovation has come about, then the challenge is in having the innovation accepted by those people who have to live with it.

IMITATION

An innovation developed by one company is soon adopted by another. Such is the way of technological competition. A great deal of product imitation of one manufacturer's product by another has al-

---

[26]Ross A. Webber, *Management*, Richard D. Irwin, Inc., Homewood, Ill., 1975, p. 653. © 1975 by Richard D. Irwin, Inc.

[27]Reprinted from Roger J. Howe, M. G. Mindell, Donna L. Simmons, "Introducing Innovation Through OD," *Management Review*, February 1978 (New York: AMACOM, a division of American Management Association, 1978), p. 53.

ways existed. What has not been so apparent is the opportunity of an **163** organization to imitate another's product as a deliberate technological strategy.

Levitt views imitation as a special kind of innovation. According to Levitt:

A simple look around us will, I think, quickly show that imitation is not only more abundant than innovation, but actually a much more prevalent road to business growth and profits. IBM got into computers as an imitator; Texas Instruments, into transistors as an imitator; Holiday Inns, into motels as an imitator; RCA, into television as an imitator; Litton, into savings and loans as an imitator; and Playboy into both its major fields [publishing and entertainment] as an imitator.[28]

The organization that imitates another organization's products and services may be just as successful as—or more successful than—an organization that is innovative. An organization may deliberately pursue a policy of imitation, allowing someone else to carry the research and development costs. Sometimes the initiation process leads an organization to extraordinary behavior in order to be able to discover and imitate a competitor's product. An example of such behavior is reflected in a *Fortune* magazine article, where Bylinsky speaks of the effects of the Japanese to obtain competitive intelligence from the semiconductor companies in an area near San Francisco known as Silicon Valley. According to Bylinsky, it was a:

Japanese effort to pry the secrets....of....high technology businesses out of American entrepreneurs. Japanese agents are aggressively gathering information by both overt and covert means, and they are eagerly buying samples of innovative tools and instruments and sending them back to Japan.[29]

While this action by the Japanese raises questions of business ethics (by United States standards), the advantage such imitation gives is made clear considering the goal of the Japanese, viz., "to leapfrog the United States in semiconductor technology and thus gain dominance of the world's computer and electronics markets."[30] The advantage such imitative strategy gives is apparent: "What normally might take them [the Japanese] a year to discover at home with a tremendous amount of costly research and development can often be obtained in a single conversation in Silicon Valley."[31]

[28]Theodore Levitt, "Innovative Imitation," *Harvard Business Review*, September–October 1966, p. 63. Reprinted by permission. Copyright © 1966 by the President and Fellows of Harvard College; all rights reserved.
[29]Gene Bylinsky, "The Japanese Spies in Silicon Valley," *Fortune*, February 27, 1978, pp. 74–75.
[30]Ibid., p. 75.
[31]Ibid.

We will not discuss the ethical factors in this imitative behavior of the Japanese. We cite the example simply to illustrate the nature of their strategy of imitation to gain dominance of the world's computer and electronics market.

If imitation is the keystone of an organization's technological strategy, then some means is required for surveillance of the competitor.

COMPETITIVE SURVEILLANCE

Competitors are an influential force in changing the opportunities available to an engineering organization. Competitors' strategies and actions can threaten as well as enhance the engineering performance. In this section we will outline the major elements of a strategy for evaluating competitors as a way of keeping abreast of technological changes facing the engineering organization.[32] These elements will suggest key questions to be answered and identify legitimate sources of competitive information.

Perhaps the typical engineering manager has not thought of assessing competitive changes as a part of his responsibility, rather relying on some other organization in the firm, such as marketing research, to accomplish this important assessment. Someone in the firm should assume the responsibility for designing and operating a competitive business intelligence system for the firm. If such a system is functioning, provisions should be made for the engineering department to contribute to the technological evaluation of competitors. An engineering manager working through his engineering department should be responsible for evaluating the significance of the competitor's technology and the use of that technology in the marketplace. No other department in the firm is competent to do this evaluation.

WHERE TO GET COMPETITIVE INFORMATION

The engineering manager and the marketing manager working together can develop the strategy, policy, and procedures for the collection, dissemination, and evaluation of competitive information. The marketing manager can provide most of the general competitor information such as annual reports, sales brochures, Dun & Bradstreet reports, etc., that would be of value to an engineering manager in discerning where the competitor stands in his understanding of the technology involved. For example, an annual report often provides insight into new-product development work and direction.

Information on where the competitor stands in his understanding and application of technology can be gleaned from a variety of other sources, such as:

---

[32]Material in this section is paraphrased from David I. Cleland and William R. King, "Competitive Business Intelligence System," *Business Horizons*, December 1975.

Patents filed with U.S. Patent Office and foreign govern-   **165**
ments

Licensing practices

Investment in research and development as a percentage of
sales

Papers presented at technical meetings

Technical sales brochures

Purchase, disassembly, and evaluation of a competitor's
product

Articles appearing in technical industry publications

Conversations with suppliers who also supply competitors

Studying speeches of key technical executives of the competitor

Industry surveys

Size and modernity of plant and facilities

Information on competitors will always be sketchy; some of
the information will be from reliable sources, some may come
from gossip or from questionable sources. In analyzing such information, assumptions will have to be made about the credibility and the
source. The analyst has to make provision for modification, simulation, rejection, and confirmation as the analysis process is carried
out.

The analysis process has to focus on the strategic implications of the information, i. e., what does the information tell you about
a competitor's probable technological strategies? How capable is he
in conceiving and designing new products, and manufacturing processes? Does he innovate or tend to be a follower of someone else's
innovation? What are the technical skills of his lead engineers? Are
there any single innovators who stand out as entrepreneurs for the
firm?

These and many other questions help to give insight into the
competitor's strategies. The selection of a set of strategies that seem
to be appropriate for the competitor is a keen judgment process for
which there are no specific formulas. Nothing can take the place of
an ongoing dialog on these questions by technical people who are
competent to develop hypotheses on what the competition might do.

Assessing the competitor's overall performance can be accomplished using information from many sources, such as:

Annual reports

Financial statements—using financial ratio analysis

166

Business periodicals

Brokerage firms

Investment bankers

Reports filed with government agencies, e.g.,

Security Exchange Commission

Sales brochures

Trade gossip

The marketing department should provide the engineering manager with analysis of the competitor's overall marketing performance. When the engineering manager receives this analysis, he should factor it into the technological evaluation he is carrying out. As the engineering manager comes across information that might be relevant to other departments of the firm, he should forward that information on to the marketing department for distribution.

For a competitive evaluation to be effective, there are several key questions that should be asked about the competitor:

What are the competitor's technological strengths and weaknesses?

What technological strategies are being followed?

How successful has performance been?

What self-image does the competitor hold?

What image does the competitor's technical ability have in the marketplace?

When these questions are answered, a model of the competitor's technological image can be derived. Such an image is useful in drawing general conclusions about the competitor's overall technological strategy.

There are many other considerations to follow in trying to assess a competitor's strategy. The Bibliography at the end of this chapter should be useful for the engineering manager who wants to do more research on the subject. The important point to be made is simply this: A study of competitors can help provide insight into how to understand and deal with the inevitable change in modern engineering organizations. The example in the next section can serve to illustrate the importance of doing competitive analysis.

REACTION

Innovation and imitation are aggressive ways of dealing with change. Reaction is a passive form of adaptation to change. Reaction occurs when an organization pursues a phlegmatic form of imitation; usually **167** this imitation is passive, pursued when shifts in technology and markets have already occurred. The reactive organization is often playing catch-up ball. Others have innovated or imitated; change is forced on the reactive organization by the circumstances of the marketplace. Imitation is attempted, but often the reaction is desperate and not thoroughly evaluated before the change is set in motion. When this happens, the change that comes about is not what was desired.

An example of such unanticipated results—in a nonengineering context—was realized by the managers of the W.T. Grant Company, when they entered the credit card business to bolster furniture and large household appliances sales. Store managers "were ordered to push the credit card campaign above everything else. . . . On New York's insistence, only cursory credit checks were conducted. . . . By last year [1975] Grant's credit-card receivables totaled $500 million, and half of that was deemed uncollectible."[33] Much of the blame for W.T. Grant Company's bankruptcy can be placed on their reactive behavior to other large retailers' success in the marketing of furniture and large appliances. Unfortunately, their catch-up ball strategy did not work out at all.

Often the need for some reaction in the engineering community is brought about by some far-reaching global change. This global change then places demands for change in the engineering of a product. We might explore this idea by looking at the global changes that hit the business community when the Arab oil embargo of 1974 was instituted. In the case of General Motors, the need for change came about from the federal government's demand for better gas mileage in autos and the oil embargo itself. At that time GM had the worst average gas mileage among United States automakers. As buyers turned away from "gas guzzlers" in the year following the oil embargo, GM's market share plummeted to its lowest point since 1952. GM's effective response to the need to change its product line resulted in a reengineering of its product line. The outcome: Within three years the average mileage of GM cars was the best of the big three automakers.

The focus of GM's strategy was a "downsizing" strategy. At the heart of this strategy was a redesign of all of GM's regular car lines so that all the cars would be in a more fuel-efficient category. A reorganization within GM was undertaken to provide an organizational focus to engineer its cars in a new way, using new design techniques at a time when the margins for error and correction would be more stringent than usual. "Particularly under these circumstances, GM could no longer afford the old problem (by no means unique to GM) of what Estes calls 'N-I-H, not invented here,' a kind of disease

[33]Rush Loving, Jr., "W.T. Grant's Last Days—As Seen from Store 1192," *Fortune*, April 1976, p. 110.

168   engineers have. An engineer suffering from N-I-H resists new ideas that originate outside his bailiwick."[34]

GM adopted the *project center* concept as a way of organizing engineering resources to redesign automobile components. Project centers, composed of design teams, work on plans and engineering problems common to all divisions, such as frames, electrical systems, steering gear, etc.[35] The real point of this GM example is to develop the idea of how a change in the global environment of which GM is a part reverberated through the marketplace and eventually impacted on how GM is organized. One can easily see how the individual engineering manager is affected, finding himself now working for two bosses.[36] His supervision takes a little different nature than previously. His day-to-day working relationships with his peers are somewhat different. According to Burck:

"Many of GM's engineers feel the project center innovation has actually helped enhance the division's individuality of freeing some of them to work on divisional projects."[37]

OBSOLESCENCE

Another form of adaptation to change is to allow oneself, or one's organization, to become a victim of obsolescence. *Obsolescence* is defined as the process of passing out of use or usefulness. An obsolete technology is one no longer used or useful because of an outmoded state-of-the-art practice. Many engineering managers today are running a frightful race between obsolescence and retirement. Those who see the significance of change going on around them and learn to develop adaptive mechanisms to cope with change may win this race along with their organizations. Others who see tomorrow's technology as only a minute change from today's technology may lose this race and take their organization along with them.

Harold G. Kaufman, of the Polytechnic Institute of New York, recently presented research findings concerning the obsolescence of engineers. Using a systems model that had been supported by his research of practicing engineers, Kaufman's work clarifies four factors that contribute to professional obsolescence:

1   Rapid environmental change, especially the information explosion in science and engineering

2   The organizational climate and reward system in which an engineer works

[34]Charles G. Burck, "How GM Turned Itself Around," *Fortune*, January 16, 1978, p. 96
[35]Ibid.
[36]See Chap. 2 for further discussion of the project-center concept in the context of the matrix organization at General Motors.
[37]Charles G. Burck, op. cit.

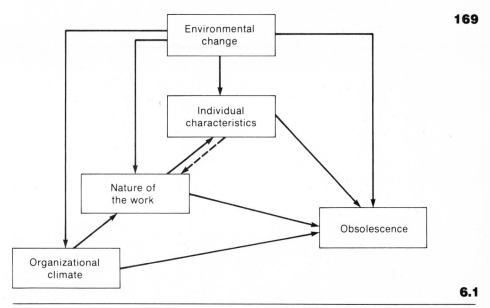

6.1

(From Harold G. Kaufman, "(Continuing Education) = (.05) (Career Management)," New Engineer, August/September, 1978, p. 31.

3   Characteristics of the work to which engineers are
assigned, particularly the degree of engineering knowledge
necessary to do that work

4   Individual characteristics of the engineer, such as
the need for growth and challenge, self-image, and intelligence

Kaufman's model, shown in Fig. 6.1, reflects the organizational climate affecting obsolescence directly, by encouraging or failing to encourage updating, as well as indirectly, by affecting the nature of the work carried out by the engineers, which in turn has a
direct impact on their obsolescence. The model also predicts that individual characteristics of engineers may affect the type of work they
select or are delegated, and vice versa.

What is the relative importance of each of these factors in
explaining obsolescence? To answer that, Kaufman designed a way
of measuring obsolescence and a way of measuring factors through a
"self-assessment questionnaire" that focuses on behaviors that are
thought to be symptomatic. He devised a questionnaire measuring
the three factors of organizational climate, individual characteristics,
and the engineer's work. Results from the survey group that was used
to validate the obsolescence questionnaire suggested that organizational climate was a less important contributor to obsolescence than

170   individual characteristics or the nature of the work being done. Kaufman comments: "A company facing obsolescence on the part of its engineering workforce had better understand the cause. If it tries to attack the problem by forcing special courses upon its employees, or by changing its organizational structure, and the real problem lies in the nature of the tasks the employees are to do, it could waste a lot of money."[38] Kaufman cautions however that sometimes the company's obsolescence problems may be organizational; effort will be wasted trying to redesign the tasks in such cases.[39]

To forgo obsolescence one must be prepared to change. It is not unusual for an engineering manager who has tried to introduce a change in technology into his organization to find the change resisted. It is particularly frustrating when he sees the need for a change but is not able to bring about any change because of the resistance of those people affected by the change. The manager becomes frustrated and often rationalizes a small change or no change at all.

RESISTANCE TO CHANGE

Engineering managers who face resistance to change in their organizations will find that resistance is not directed to technological change. Engineers readily recognize the need for changes in technology. Their resistance is of a different sort—a resistance to the social change that follows technological change. The idea of the social aspect of a technical change was stated back in 1969 by Paul R. Lawrence. Writing in the *Harvard Business Review,* Lawrence stated: "The technical aspect of the change is the making of a measurable modification in the physical routines of the job. The social aspect of the change refers to the way those affected by it think it will alter their established relationships in the organization."[40]

According to Lawrence, the variable which determines the result when change is introduced is the social aspect of the change—the changes in the social relationships in the organization. Lawrence further notes that resistance to change can be overcome by getting people involved in the chance to participate in making the change. But he cautions: "Participation, to be of value, must be based on a search for ideas that are seen as truly relevant to the change under consideration. The shallow notion of participation, therefore, still needs to be debunked."[41]

[38]Harold G. Kaufman, "Technical Obsolescence. An Empirical Analysis of Its Causes and How Professionals Cope With It," paper presented at the ASEE Annual Conference, June 19–23, 1978, Vancouver, British Columbia, Canada. August–September 1978, pp. 31–39.
[39]Ibid.
[40]Excerpt from "How to Deal With Resistance to Change," by Paul R. Lawrence (January–February 1969). Reprinted by permission of the *Harvard Business Review.* Copyright © 1969 by the President and Fellows of Harvard College. All rights reserved.
[41]Ibid.

Older people in organizations tend to resist change more **171** than the younger generation. This younger generation—today's 30- to 35-year-olds—were born into a rapidly changing environment; for them change has become a way of life. Today's young engineers were in the universities during the turbulent late sixties and early seventies. Not only did they have the opportunity to observe change, but they participated in bringing about that change. Some insight into the expectations these people have regarding change today is reflected in their attitudes toward their organization's future. Drucker comments: "They expect, indeed they demand, high competence from the boss and genuinely professional management from him. They expect that the organization that employs them actually plans and then carries out its plan."[42]

We cannot help but deal with change, for we cannot of ourselves ever do things quite the same again. Most of us, however, react to a major change with reluctance and find change painful. Why? Don Fabun may have summarized the views of many of us, noting:

"We have," as Kenneth Boulding brilliantly explains in *The Image,* "formed an image of the way things 'should be.'" We have been taught, for instance, that effect follows cause, or that clocks "keep time," or that because people disagree with us they are our enemies. These relationships of things to each other and to us are encoded in our nervous systems and we feel quite strongly that they are not to be bent, spindled or mutilated.

Then we are exposed to some new experience that breaks up the cherished pattern. Our equilibrium is disturbed and it may take considerable psychic energy (and some little pain) to construct a new relationship that will take into account the new evidence.

This may be one reason why the "established order" so vehemently resists change in our times. "Their entire stake of security and status," says Marshall McLunan, "is in a single form of acquired knowledge, so that innovation for them is not novelty, but annihilation."

As Dostoevski said, "Taking a new step, uttering a new word, is what people fear most."

Or Eric Hoffer: "We can never be really prepared for that which is wholly new. We have to adjust ourselves, and every radical adjustment is a crisis in self-esteem. . . . It needs inordinate self confidence to face drastic change without inner trembling."[43]

[42]Peter Drucker, "Report on the Class of '68," *The Wall Street Journal,* February 3, 1978.
[43]Don Fabun, *The Dynamics of Change,* Prentice-Hall, Inc. Englewood Cliffs, N.J., pp. 8–9. Kaiser Aluminum News © 1967.

SUMMARY

An untold number of technological signals are constantly passing through any area of technology. These signals originate in the broad environment in which the organization operates (such as the areas of economic, social, political, technological, and legal matters) within the competitive marketplace, and the organization itself. The anticipation of these signals—and doing something about them—is the key factor in dealing with change.

Innovation, imitation, reaction, and obsolescence are ways of dealing with change. Each of the ways requires some response from the engineering manager ranging from an active to a passive mode. Change is inevitable; this chapter has suggested some ways of dealing with change to make the engineering organization more competitive and supportive of personal and organizational goals.

DISCUSSION QUESTIONS

1   Change should be possible in all aspects of an organization's life. Identify four such aspects that are relevant to an organization.

2   What is the significance of the view that innovation is more often the product of recognition and adaptation than of truly original invention?

3   Can you cite opportunities for change in an organization with which you are familiar?

4   Discuss the differing climates and processes for innovation in a small organization and a large one, respectively.

5   What are some advantages of the project-team or task-force approach to evolving innovation? Are there any drawbacks?

6   For an imitative strategy to work, an organization must recognize and forestall certain pitfalls. Discuss some of these.

7   Reaction is usually passive imitation. But sometimes it is the only mode of change available to an organization. In what kinds of circumstances is the above statement true? What must the organization do to ensure that it achieves an effective reaction?

8   What can be done to prevent obsolescence in an individual?

9   Why are externally generated changes so difficult to anticipate?

10   What is the relationship between organizational structure and technological change?

11   What is there about large technical bureaucracies that may stifle innovation in the organization?

12   How might the project team be used as a means to facilitate change in an organization?

KEY QUESTIONS FOR THE ENGINEERING MANAGER

1 What are the key elements of technological change that have affected this engineering organization in the past 5 years? What insight do these changes give as to what might happen in the next 5 years?

2 What impact has recent technological change had on the structure and management processes of this organization?

3 Is this engineering organization innovative? imitative? reactionary? or obsolete? Why or why not?

4 What have been the most innovative strategies undertaken by this organization in the past 2 years? Have these strategies been successful?

5 Are there any vestiges of bureaucracy in this organization? If so, has this inhibited innovation on the part of the technical people?

6 Has the use of project teams in this organization been tried? Have these teams helped to innovate technology?

7 Has the cultural ambience of this organization been such that the IDEA program (of Texas Instruments) could be applied locally? Why or why not?

8 What imitative strategies has this organization recently pursued? Have these strategies been successful?

9 Who are the most formidable competitors of this organization? What are their strengths and weaknesses?

10 What methods, processes, techniques, etc. have been developed in this organization for maintaining competitive surveillance?

11 Have the necessary interfaces been established with the marketing department to ensure receiving general information on the competitors?

12 Are the engineering managers of this organization approaching obsolescence? What programs have been initiated to reduce obsolescence in engineers and engineering managers in this organization?

13 Am I losing the race between obsolescence and retirement?

BIBLIOGRAPHY

Anshen, Melvin: "The Management of Ideas," *Harvard Business Review,* July-August 1969.

Burck, Charles G.: "How G.M. Turned Itself Around," *Fortune,* Jan. 16, 1978.

Bylinsky, Gene: "The Japanese Spies in Silicon Valley," *Fortune,* Feb. 27, 1978.

Cleland, David I., and William R. King: "Competitive Business Intelligence System," *Business Horizons,* December 1975.

Dalton, Gene W., and Paul H. Thompson: "Accelerating Obsoles-

174 cence of Older Engineers," *Harvard Business Review,* vol. 49, Spring 1971.

Drucker, Peter: "Report on the Class of '68," *The Wall Street Journal,* Feb. 3, 1978.

Fabun, Don: *The Dynamics of Change,* Prentice-Hall, Englewood Cliffs, N.J., 1970.

George, William W.: "Task Teams for Rapid Growth," *Harvard Business Review,* March-April 1977.

Haggerty, Patrick E.: "Long-Term Viability—The Business Problem," *IEEE Student Journal,* September 1968.

———— : "Strategies, Tactics, and Research," *Research Management,* vol. 9, no. 3, 1966.

Hershey, Robert: "Competitive Intelligence for the Smaller Company," *Management Review,* January 1977.

Jasinski, Frank J.: "Adapting Organization to New Technology," *Harvard Business Review,* January-February 1959.

Knight, K. E.: "A Descriptive Model of the Intra-Firm Innovation Process," *Journal of Business,* vol. 40, no. 4, 1967.

Lawrence, Paul R.: "How to Deal with Resistance to Change," *Harvard Business Review on Management,* Harper, New York, 1975.

Levitt, Theodore: "Innovative Imitation," *Harvard Business Review,* September–October 1966.

Loving, Rush, Jr., "W.T. Grant's Last Days—As Seen from Store 1192," *Fortune,* April 1976.

Marquis, D. C.: "The Anatomy of Successful Innovations," *Innovation,* vol. 1, no. 7, November 1969.

Morgan, John S.: *Managing Change,* McGraw-Hill, New York, 1972.

Morton, Jack A.: *Organizing for Innovation,* McGraw-Hill, New York, 1971.

————: "A Systems Approach to the Innovation Process," *Business Horizons,* Summer 1967.

Schoen, Donald R.: "Managing Technological Innovation," *Harvard Business Review,* May-June 1969.

Souder, William E.: *Final Report on: An Exploratory Study of the Coordinating Mechanisms between R&D and Marketing as an Influence on the Innovation Process,* Technology Management Studies Group, Department of Industrial Engineering, University of Pittsburgh, August 1977.

"TI's Magic in Managing Innovation," *Business Week* September **175**
18, 1978.

"Texas Instruments Shows U.S. Business How to Survive in the
1980's, " *Business Week*, September 18, 1978.

White, George R., and Margaret B. W. Graham: "How to Spot a Technological Winner," *Harvard Business Review,* March-April 1978.

PART THREE
MANAGEMENT SCIENCE IN ENGINEERING MANAGEMENT

MANAGEMENT SCIENCE CONCEPTS AND PHILOSOPHY 7

This chapter discusses the basic philosophy of management science and develops a conceptual framework for the use of mathematical models in the decision-making process. Management science is presented as a systematic approach to rational decisions. The stages through which a decision maker goes in making such decisions are explained in the context of engineering management. It is a descriptive chapter to give the reader an appreciation and understanding of what management science is and how it helps the decision maker to sharpen his judgment.

The management science techniques are described but not developed in this chapter. Mathematical models and their solution methodologies are presented in Chaps. 8 and 9.

The similarities and differences between engineering decision making and management decision making are explained. The parallels are drawn between these two types of decisions that the engineering manager makes.

HISTORY OF MANAGEMENT SCIENCE

During the Second World War, the complex decision problems involved in the allocation of resources, scheduling, logistics, and control of the military forces of the United States and allies required an innovative approach to decision making. The problems were unique in their massive size and complexity. Clearly, the time had come to develop rational decision-making methodologies for problems of these enormous proportions. Engineers, scientists, and mathematicians were brought together to form study teams for the development of the concepts, tools, and methods necessary for a scientific approach to decision problems. This was the beginning of a new field

180   based on the mathematical representation and solution of complex decision problems.

About the same time as the mathematical models were being developed for military decisions, another significant development was taking place. The computer was being introduced as a unique tool for the solution of large-scale mathematical problems. The simultaneous developments in both the mathematical formulation and the solution of decision models contributed significantly to the early success and recognition of this new field, called "management science." Among the many synonyms used for this body of knowledge are "operations research," "operational research," "quantitative methods," "systems analysis." The name "management science" will be used in this book to refer to the field of study dealing with the application of scientific methods in a systematic way for the solution of management decision problems.

The earliest examples of the management science applications go as far back as the eighteenth and nineteenth centuries. The analysis of military problems in terms of manpower superiority versus weapons superiority by Lanchester in the 1910–1914 time period was a significant analytical study for major decisions. The statistical antisubmarine warfare study by T. Edison in the mid-1930s was another example of the pre-World War II attempts at a scientific approach to management. The first real, coordinated effort, however, is credited to a team of British engineers and scientists who studied the reasons for the inaccuracy of the British defense against the German submarines in the early 1940s. About the same time, the U.S. Army and Navy started to establish management science groups for the solution of their operational problems. The name "operations research" was popularized by the early emphasis on the use of these newly developed analytical decision tools in military operations.

After World War II, those who had developed management science solely for military purposes started to find new uses for their analytical approach. Several "think tanks," such as RAC (Research Analysis Corporation), RAND Corporation, ONR (Office of Naval Research), etc., were formed. Initially supported by the Army and Navy, they soon began to apply their knowledge to nonmilitary problems. The management science applications became well known in the 1950s, and the universities started offering degree programs in management science in the 1960s. The academic thrust was focused primarily on the techniques and tools of management science, shifting the emphasis from decision orientation to methodology orientation.

This new orientation resulted in many new developments in the state of the art of the solution techniques in several areas, by testing the limitations of these techniques and pushing the boundaries to new limits. The technique orientation of the 1960s, however, became a significant departure from the original motivation for management **181** science. The field came dangerously close to being a tool-generating activity instead of an aid to decision making. An artificial dichotomy was created between theory and applications, and management science ran into identity problems. Maybe this was the natural process that a growing field had to go through. During this period, new solution techniques were developed and existing ones were refined for problems dealing with various types of inventory, scheduling, and queuing systems. Computer programs were improved and the solution efficiency was increased. After about a decade of this growth period, the emphasis again shifted to the problem-solving aspects that had motivated the original contributors to this field.

The new areas analyzed by the management science concepts and methodologies in the 1970s include urban systems, justice systems, health institutions, political districting, and energy planning. The operational problems of the earlier days are now being replaced by much more complicated, multidimensional decision problems in both government and industry.

MATHEMATICAL MODELS FOR DECISION MAKING

A decision can be defined as the removal of uncertainty concerning an alternative course of action. Some decisions are strategic, some operational. Sometimes they deal with alternatives in functional areas; sometimes they have systems effects, impinging on multiple functions of the organization simultaneously.

Decisions can be classified in numerous ways. We will examine here *engineering* and *management* decisions.

ENGINEERING DECISIONS
Mathematical formulation is extensively used for the solution of engineering problems. The engineer develops a mathematical model of the system and uses it to help him make the best decisions concerning the allocation of material resources and the selection of processes. As an example, let us consider the thought process through which the engineer goes in designing a beam under loading $w(x)$, depicted in Fig. 7.1.

First, he determines the conditions affecting the problem. He checks whether the beam is simply supported or has fixity at the ends. He then finds out about the nature of the load $w(x)$. If it is repetitive loading, his approach will be different from the static loading case; he may have to be concerned about vibration problems. If the load is moving, he will again have to use a different analysis, since the critical loading conditions will vary along the length of the beam. After he has oriented himself to the problem by learning as much about it as he can, he formulates a mathematical model. In the model-

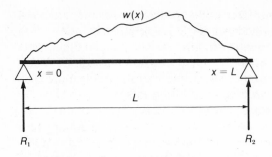

7.1

A simply supported beam under nonuniform loads.

development stage, he goes through a sequence of logically interrelated steps to obtain mathematical expressions for the reactions, shears, and moments on the beam and then combines them. For a simply supported beam under static loading $w(x)$, he develops the following expressions:

$$R_2 = \frac{\int_0^L x w(x)\, dx}{L}$$

$$R_1 = \int_0^L w(x)\, dx - R_2$$

Shear at point l from left:

$$V_l = R_1 - \int_0^l w(x)\, dx$$

Moment at point l from left;

$$M_l = R_1 \cdot l - \int_0^l (l - x) \cdot w(x)\, dx$$

Then, the location l_0 of the maximum moment will be determined as

$$\frac{dM_l}{dl} = 0 \qquad \text{at } l = l_0$$

Once the model is completed, the engineer obtains the numerical values for the variables and parameters used in his model. He needs the length and the functional form and magnitude of the load defined by function $w(x)$. The data he obtains from available sources are substituted into the mathematical expressions, and the model is solved. Using the solution, he can determine the maximum moment and shear and their location on the beam. This information is sufficient to make a decision about the type and amount of material resources that he will allocate to the beam. He chooses the

best size available to support the loads at minimum cost under the **183**
height and width restrictions that should be satisfied. His criterion in
the resource allocation decision is to use the available resources in
the optimal fashion by choosing the proper beam size made of the
proper material. In making that decision he wants to make sure that
the overall cost including the cost of labor is at the minimum level.
He will therefore not only choose the minimum size needed to perform satisfactorily under the given conditions, but also check to see
that the minimum size beam does not require excessive labor cost.

When the resource allocation decision is made, the engineer has solved his problem. He then implements his decision.
Implementation is the transformation of his solution to the final
product. He prepares the engineering drawings, orders the beam,
and sees to it that the beam is constructed according to his specifications.

MANAGEMENT DECISIONS

Just as the engineer uses a rational approach in his technical
decisions by analyzing the mathematical abstractions of the actual
conditions, he could use the same philosophy and methodology in
his engineering management decisions. The basic tools for such
decisions are similar to the tools used in engineering decisions, but
the concepts are broader.

The methodology for rational management decisions has
been developed within the scope of management science. Several
definitions can be given for management science. In simple terms, it
can be viewed as the application of scientific methods in a systematic way to the solution of management decisions. In management
science, the decision systems are represented by rational rules and
interactions even though the actual decisions may or may not be
made under the ideal, rational conditions. The complexities of
human behavior cannot easily be captured by well-defined mathematical relationships. The sudden discontinuities and disruptions in
the external conditions affecting the decision system cannot always
be identified for proper modeling. The conceptual decision elements
and the implicit value structures of the decision makers are difficult
to quantify and incorporate into the models. Despite these difficulties, however, management science offers an alternative to intuitive,
"seat-of-the-pants" decision making. The rational approach to
decisions provides the decision maker with a basis to sharpen his
judgment. The final decisions are, of course, always his prerogative.
Nevertheless, the techniques and methods described herein are
tools to help the manager to be a more rational and, it is hoped, a
better manager.

Management science is not merely a collection of mathe-

184 matical expressions. It is the totality of many philosophies and techniques used in a systematic way to help the decision maker to make decisions. The accuracy of the models used in this approach increases as the complexity of the problem decreases. The operational decisions in areas such as inventory control, production scheduling, etc., can be defined accurately with relative ease compared to the higher-level decisions involved in the overall management of corporations and the allocation of strategic resources under uncertain environmental conditions. The inherent difficulties and shortcomings should be kept in mind in the use of the mathematical models and the interpretation of results obtained from them.

THE MANAGEMENT SCIENCE APPROACH

The scientific approach to problem solving used in management science is characterized by the process shown in Fig. 7.2. The steps shown in the figure are not always followed in an orderly fashion. In many cases, the initial definition of the decision problem is changed several times until the issues are accurately identified. The decision maker, who controls the system by his decisions and directions, may start with the wrong problem definition and seek right answers to that

Management science approach.

7.2

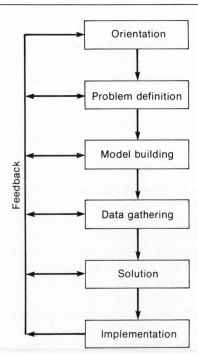

wrong problem. It is not unusual to confuse the symptoms with the **185**
causal factors. As an example, in a materials handling facility, the
delays in the final delivery of materials could easily trigger a study for
the improvement of conveyor belts and result in an increase in their
speed. If the system were closely studied, however, a completely different problem could be discovered. Suppose that the material is arriving at the conveyor belts at a low rate. In that case, an increase in
the belt speed does not improve the system at all. In fact, if the arrival
rate cannot be improved, it would be better to slow down the conveyor belts or to eliminate some of them in order to achieve the same
system performance at a lower cost. Such problem redefinitions are
frequent in complex management decisions. The more the decision
maker gets into the details of his decision problem, the more he discovers the exact cause-effect relationships. As he learns more about
it, he goes back and forth in the steps shown in Fig. 7.2 until he is satisfied with the problem definition and the model developed for its
solution.

In many cases the decision maker and the decision analyst
are not the same person. Analysis is performed by management science specialists working as consultants to provide staff support for
those in control of the system. In such cases the management
science approach becomes a continuous interaction between the
decision maker and the decision analyst.

An engineering manager is both a decision maker and a
decision analyst. By virtue of his position he controls the system by
his decisions both as a technical specialist and as a technical
generalist. Therefore, he is a decision maker. In his technical
decisions he analyzes the problem and evaluates the alternatives
before reaching a conclusion. The quantitative nature of his education and the requirements of his job make him a rational thinker. He is
prepared to justify each of his decisions when challenged. He can do
that by mathematical models supportive of his logic and conclusions.
The basis of engineering decisions is reproducible for someone else
to follow and to evaluate.

When this rational thinker makes a management decision,
he deals with broader issues and a greater degree of uncertainty, but
obviously he cannot be expected to divorce himself from his analytical approach and become a seat-of-the-pants type of decision maker.
His intuition and experience certainly play important roles in his
decisions, but he cannot rely solely upon them. He needs a systematic approach to arrive at an analytical solution to his management
decision problem. Therefore, he is a decision analyst.

The management science approach and methodologies are
powerful tools for the engineering manager to arrive at rational management decisions through an analytical framework. In principle the

186 steps of the management science approach are identical to the steps followed in engineering decisions. There is a one-to-one correspondence between this approach and the thought process through which the hypothetical engineer went when designing the simply supported beam in the earlier example.

Let us assume that an engineering firm has four design tasks to be performed. There are four design engineers to do the work, and the engineering manager is faced with a resource allocation decision. He has to determine which design engineer to assign to each task. Note that the objective is very similar to that in the beam-design example. In both cases, the decision maker wants to allocate the scarce resources available to him in the most effective way possible, and in both cases he can use mathematical-modeling techniques to help him make a rational decision. For the engineering decision in the design example, he needs a knowledge of the methods developed in engineering mechanics and strength of materials. For the management decision in the manpower allocation decision, he needs a knowledge of the methods embodied in management science.

ORIENTATION

Orientation is the first step in the decision process. At this stage, the problem is identified and its characteristics and boundaries are defined in broad terms. If a decision analyst is working for the decision maker, he has to be oriented with the characteristics of the system. For example, the nature of the four design tasks are studied. If any of the tasks should be completed before the others, a sequential decision is needed. If all tasks can be started simultaneously, the decision is simpler, and the decision model is developed accordingly. If some of the tasks require special knowledge and capability, it may be necessary to look for the capability outside the organization, maybe by subcontracting those tasks to another engineering office. This will directly affect the resource allocation decision. Conversely, although four design engineers are available, some of them might have to be eliminated from consideration for some of the tasks because of their specialization in other fields. Another consideration in this orientation stage might be to determine whether or not all four design engineers are indeed available for the new tasks. If they are already assigned to other jobs, it may be difficult, even impossible, to put them on new assignments. The term used in management science for the impossible situation is "infeasible." For example, if all four tasks should start immediately and if one of the four design engineers will not be available for a week, the assignment of that engineer to one of the tasks is infeasible. Information in the orientation stage can be obtained through various contacts with the people who will be affected by the final decision. Interviews are usually very useful. Company records and literature search provide valuable assistance in becoming familiar with the pertinent issues involved in the **187** problem.

PROBLEM DEFINITION

Problem definition follows the orientation stage. In this stage the decision problem is expressed in both qualitative and quantitative terms. The observable results and the underlying causes are identified. Five major characteristics of the system are defined, as follows:

MEASURE OF EFFECTIVENESS —A quantifiable measurement for the comparison of results obtainable under different external conditions and different decisions. Cost, profit, quality, and employee satisfaction are all measures of effectiveness. The overall satisfaction of public priorities can also be defined as a measure of effectiveness. The measure of effectiveness is a numerical value that summarizes the total decision system. Sometimes it is readily identifiable, especially if its units are easy to measure. Cost, profit, and project length are examples of this. In other cases, when quality, satisfaction, and social responsibility issues are involved, the measure of effectiveness is much harder to define because of measurement difficulties. Once the measure of effectiveness is decided upon, it becomes the key criterion for the evaluation of the decision system. It represents the output of the system and reflects the combined effects of all elements involved in that system.

VARIABLE —A characteristic of the system that takes on different values and affects the performance of the system over that range of values. If a variable is controllable by the decision maker, it is called a *decision variable.* The amount of resources allocated to each portion of the system is a decision variable. In the manpower example, the decision variables are the assignment or nonassignment of design engineers to certain tasks. These variables can be defined to take on a value of 1 or 0, representing the respective decision. In general, the variables could have integer or continuous values. If the number of people allocated to a certain project is a decision variable, it will be an integer variable. On the other hand, if the financial resources allocated to different alternatives are defined as the decision variables, there is no need to restrict them to integer values.

The variables could be deterministic or random. *Deterministic variables* are those whose values are not chance-dependent. For example, the variable x_{ij} defining the decision concerning the assignment of engineer i to task j is deterministic. In other types of problems, variables defining demands or customer arrivals depend on a chance-related process. Each comes from a probability distribution and is called a *random variable.*

188   The variables that the decision maker cannot change are *uncontrollable variables.* The demand and customer arrivals are examples of such variables.

PARAMETER —A characteristic of the system that takes on only one value during the foreseeable operation of the system but may change its value when different alternatives are studied. The number of days that engineer i will spend on task j is a parameter; total manpower availability is a parameter; total demand, total capital, and machine hours are all parameters. They have constant values under which the decision system is analyzed. The decision maker could be interested in the performance of the system under different values of the parameters. In such cases, the analysis is repeated with the new values. The typical parameters in probabilistic systems are the means and variances of the probability distributions. For example, the random variable defining the probabilistic demand for a product takes on a different range of values for each set of means and variances of the underlying probability distribution and can be studied separately in each range.

Sometimes the decision system represents a time span over which the parameters change their values. If the demand is expected to vary from one year to another in a decision spanning several years, the different values for the demand parameter are incorporated into the problem formulation at discrete points when they change. Similarly, if the probability distributions or their attributes (such as mean and variance) are expected to change during the time horizon considered in the decision system, those changes are recognized and included in the analysis.

OBJECTIVE —The direction in which the decision maker wants it to move. Quite frequently, multiple objectives have to be satisfied simultaneously, even though some of them may be in conflict. For example, the "design thoroughness" objective competes with the "short completion time" objective, and "commitment to research and development" could be in conflict with the "cost minimization" objective. The nature of the objectives and the tradeoffs among them are specified in quantifiable terms as much as possible in the problem-definition phase. One approach to the solution of multiobjective problems is to obtain subjective measures from the decision maker for the ranking of the various objectives and to combine the weighted objectives in one expression. For example, if cost minimization is considered to be three times as important as the R&D quality, a combined objective can be developed as

$$3O_1 + O_2$$

where O_1 = cost improvement
O_2 = R&D quality

The mathematical representation of the system objective is **189** called the *objective function*. It represents the relationship among the decision variables in terms of their contributions to the overall effectiveness of the decision system. The value of the objective function thus becomes the measure of effectiveness of the total system.

For the assignment of design engineers to the four tasks, one objective could be stated as cost minimization; another objective could be the minimization of total worker-days spent for the four tasks. If the engineers have identical salaries, these two objectives would be the same. If there is a salary differential, cost minimization would not necessarily be the same as the minimization of the total worker-days involved. Still another objective could be the best overall technical performance. It should be noted that for the first two objectives, it is necessary to know how much time each engineer will spend on each task and/or how much it will cost the firm when an engineer is assigned to a specific task. These data are relatively easy to obtain. On the other hand, if the third objective is adopted, then the relative effectiveness of each design engineer in each task will have to be estimated. Such a measure is usually much more difficult to obtain than the directly quantifiable measures.

CONSTRAINTS —The restrictions imposed upon the system by resource availability, the external conditions, the policies of the organization and the technologies involved. For example, the total expenditures for feasibility studies, engineering design, and shop drawings cannot exceed the project budget. This is a constraint as follows:

$$\text{Feasibility cost} + \text{design cost} + \text{drawing cost} \leq \text{budget}$$

Let us suppose that a rule of thumb in such a project is that the man-hours spent for design should be no less than half the time spent for shop drawings. This implies another constraint:

$$\text{Design time} \geq (0.5)(\text{drawing time})$$

Note that these inequalities do not fix the total expenditures or design time in the project. The first constraint is satisfied as long as the expenditures are at or below the upper limit defined by the budget. The second constraint is satisfied as long as the design time is at or above the lower limit. Conceptually, an infinite number of solutions can be found without violating such constraints.

Another type of constraint is the equality constraint. When we define physical laws or strict relationships among the variables, we may have to express them as equations. For example, force is the product of mass and acceleration.

$$F = ma$$

This is an identity that cannot be violated. There is no flexibility in the

190   value of F. It is fixed once m and a are determined. Such an equality constraint puts a greater restriction on the decision problem than the inequality constraints do. If the decisions are restricted by tight constraints, in general, the objective of the system is achieved at a lower level. The release or removal of constraints leads to a gradual improvement in the achievement of system objectives.

In the assignment example, two types of closely related constraints can be developed to define the system:

1   Each engineer has to be assigned to a design task.

2   Each task has to have an engineer assigned to it.

Once the variables, parameters, objectives, and constraints of the decision system have been determined, the problem is formulated and a mathematical model is developed. The correct formulation cannot be overemphasized. If it is not done accurately, the result of the analysis could be the right answer to the wrong problem, or the wrong answer to the right problem. For example, if the objective of the assignment problem is incorrectly defined as maximization of total time spent on the project, the result will be a waste of available resources.

The time and effort necessary for the proper formulation of the problem should never be spared, even if it means the reformulation and the revision of the assumptions several times.

MODEL BUILDING
Model building is the development of a mathematical description of the decision problem to reflect the characteristics of the issues involved. It can be viewed as both science and art. At the model-building stage, relationships among the decision variables are carefully defined for a precise representation of the decision problem. The functional relationships can be linear, nonlinear, deterministic, or probabilistic. The type of model depends on the objectives of the decision problem. Some problems require models for the optimum allocation of resources. Some can be solved by studying the system performance under various policies, asking "what if" questions. Different types of models can be developed for stationary decisions at a given point in time or for multiperiod decisions requiring analysis under dynamic conditions. In some cases, models are used for the control of the organization's ongoing activities; in others, they deal with planning decisions such as plant expansions, product changes, mergers, and acquisitions.

In all cases, a mathematical model is an abstraction of the actual conditions. The level of that abstraction implies certain

tradeoffs. Too much detail could lead to excessively expensive mod-   **191**
els which frequently become ineffective because of the long development time required. Too much abstraction results in ineffective models due to an oversimplification of the issues. The model builder has
to find a happy medium between these two extremes. Another consideration in the model-building stage is the approach to take in the
basic structure of the model. The analyst should decide whether to
use a top-down approach by starting at the top and disaggregating
the system to its components or a bottom-up approach by starting
with the smallest components and aggregating them. Before this
decision is made, it is necessary for the analyst to study the problem
in detail and thoroughly understand its relevant characteristics.
When problems are too complex to be expressed as one large model,
it may be better first to decompose them into submodels and then
aggregate the submodels to obtain an accurate representation of the
total problem.

Extreme care should be exercised in transforming the
problem to a precise mathematical model. The common difficulties in
model building include:

Incorrect functional relationships

Incorrect constraints

Absence of constraints

Absence of variables

Use of incorrect variables

Use of incorrect parameters

In many cases, models can be simplified significantly
without loss of accuracy. In fact, quite frequently minor modifications
could lead to major reductions in computations. Such modifications
could include the following:

Insignificant variables can be eliminated from the model
without an important effect on the final solution.

Some variables can be combined with other similar variables.

Some constraints may dominate the others. The dominated
(redundant) constraints can be removed from the model.

In many cases, integer variables can be treated as continuous variables without a significant inaccuracy.

The nonlinearities in the model may be approximated by
linear relationships with minor inaccuracy but significant
savings in computational efforts.

192   TYPES OF MATHEMATICAL MODELS

There are two major thrusts in mathematical modeling: optimization and simulation. Each represents a different fundamental approach to problem solving and each includes several subcategories, as shown in Figure 7.3.

Optimization is the normative approach to identify the best solution for a given decision problem. The solution typically lies at the minimum or maximum point of the decision function being optimized. If the problem is accurately formulated, the values of the decision variables obtained in the optimum solution result in the highest benefit or the lowest cost as the case may be. After the optimum solution is obtained, the problem is reevaluated for different conditions to obtain a range of solutions. The development of these additional solutions in the neighborhood of the optimum solution is called the "postoptimality" or the "sensitivity" analysis.

Optimization can be used to solve a wide variety of problems, ranging from engineering design to resource allocations in multiplant organizations.

Classical optimization includes closed-form mathematical models solved by differential calculus. This approach is limited to relatively simple problems with a small number of variables.

Mathematical programming is a systematic approach for determining the optimum point in a function under constraints. The function that is optimized (minimized or maximized) is called the "objective function." One special type of mathematical programming problem is the *linear programming* (LP) problem. In an LP formulation both the objective function and the constraints are linear and additive. The solution methods for linear programming models have been well developed. Many computer codes are available for LP

Mathematical models in management science.

7.3

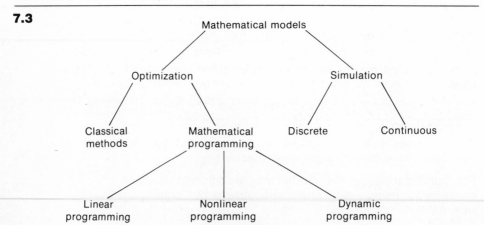

problems. If an optimization problem can be formulated as a linear **193** programming model, its solution is obtained and sensitivity analysis is performed by these codes. If the linearity assumptions cannot be justified, the *nonlinear programming* (NLP) approach is used. NLP is not a unique solution technique similar to LP. Instead, it is the name given to a collection of many techniques developed by several people as the need arose. When decisions are made in multiple time periods or in multiple stages, the policies are studied as a sequential process using *dynamic programming.*

Other types of optimization problems require special solution methodologies, typically developed as extensions of the mathematical programming models described above. For example, if the decision variables take on integer values, as in the case of number of people, number of machines, or yes-no decisions, a technique called "integer programming"[1] is used. When the probabilistic nature of the variables and constraints is to be considered, "stochastic programming"[2] provides the necessary approach for solution. "Goal programming"[3] is another special type of optimization model. It is used for decisions with multiple objectives.

Simulation is the exploratory approach to management decision problems. Components of the decision problem are mathematically defined and related to each other in a series of functional relationships. The result is a mathematical description of the complete decision process. The model is solved repeatedly, using different parameters and different decision variables every time. As those values are changed, a range of solutions is obtained for the problem and the best solution is chosen from that range. This approach is similar in philosophy to the postoptimality analysis, except that it is not restricted to the neighborhood of the optimum point.

The primary difference between optimization and simulation is their starting point. In optimization, we start with a definition of the system objectives and specify the actions that will satisfy those objectives at the optimum level. Once the optimum conditions are established, we further study the vicinity of those points to determine the effect of variations in the system. In simulation, however, we start with the actions and study their effects on the overall system objectives by testing different policies under various external conditions. Simulation is a powerful tool in solving probabilistic problems where the distribution of the final result is more important than a point estimate for that result.

---

[1] R. S. Garfinkel and G. S. Nemhauser, *Integer Programming,* Wiley, New York, 1972; H. M. Salkin, *Integer Programming,* Addison-Wesley, Reading, Mass., 1975.
[2] S. Vajda, *Probabilistic Programming,* Academic, New York, 1972.
[3] James P. Ignizio, *Goal Programming and Extensions,* Lexington Books, Lexington, Mass. 1976; Sang M. Lee, *Goal Programming for Decision Analysis,* Auerbach, Philadelphia, Pa., 1972.

194   *Discrete simulation* deals primarily with operational problems by studying the system at regular intervals and by taking discrete steps in time to observe the changes in a probabilistic sense. Job shop scheduling, queues, and risk analysis are examples of such problems.

Simulation is also applicable to large-scale continuous problems dealing with industry-wide interactions and urban problems. In these problems, the functional relationships are developed to reflect the interactions among decisions and system components and the model is tested under various internal and external conditions to determine the effects of various decisions. This approach is called *continuous simulation*. Econometric modeling,[4] system dynamics studies,[5] models for human behavior,[6] political system,[7] and consumer behavior[8] are a few examples of continuous simulation models.

Certain types of decision problems have become so typical in management science that specific solution methodologies have been developed for them. Examples of such methodologies can be found in project control models (CPM/PERT), queuing models, inventory models, scheduling models, and transportation models. Even though these are specialized model categories, in each case the problem can be solved basically by the optimization approach or the simulation approach, or a combination of the two.

Going back to the assignment example, let us develop a mathematical model for assigning the design engineers to various tasks. Assuming that the objective is to minimize the total worker-days required to perform the necessary job, and defining x_{ij} as the variable which specifies whether or not engineer i is to be assigned to task j, we can formulate the objective function as

$$\min Z = \sum_{J=1}^{4} \sum_{i=1}^{4} c_{ij} x_{ij}$$

where $Z =$ value of objective function
 $c_{ij} =$ time, worker-days, engineer i needs to perform task j
 $x_{ij} =$ decision variable
x_{ij} takes on a value of 1 or 0 depending on whether i is assigned to

[4]G. Fromm and P. Taubman, *Policy Simulations with an Econometric Model*, Brookings Institute, Washington, D.C., 1968; *Wharton Econometric Forecasting Model*, Department of Economics, Wharton School, University of Pennsylvania, Philadelphia, Pa.

[5]J. W. Forrester, *Industrial Dynamics*, MIT Press, Cambridge, Mass., 1961; D. L. Meadows, et al., *The Limits to Growth*, Potomac Associates, Washington, D.C., 1972.

[6]H. Borko, *Computer Applications in the Behavioral Sciences*, Prentice Hall, Englewood Cliffs, N.J., 1962; J. M. Dulton and W. H. Starbuck, *Computer Simulation of Human Behavior*, Wiley, New York, 1971.

[7]I. D. S. Pool, R. Abelson, and S. Popkins, *Candidates, Issues and Strategies*, MIT Press, Cambridge, Mass., 1965.

[8]A. E. Amstutz, *Computer Simulation of Competitive Market Response*, MIT Press, Cambridge, Mass., 1967.

task j or not. In this formulation, the measure of effectiveness of the **195**
system is the total number of worker-days required to complete the
four tasks. The value of the objective function will be the minimum
possible value for that measure.

Assuming that all four design engineers are available to
work on the four tasks, which should all be started immediately, constraints identified in the problem-definition stage can be developed
as follows:

1   Each engineer has to be assigned to a task:

$$\sum_{j=1}^{4} x_{ij} = 1.0 \qquad \text{for } i = 1, \ldots, 4$$

2   Each task has to have an engineer assigned to it:

$$\sum_{i=1}^{4} x_{y} = 1.0 \qquad \text{for } j = 1, \ldots, 4$$

3   The integer constraint, stated as

$$x_{ij} = 0 \quad \text{or} \quad 1 \quad \text{for all } i, j$$

This formulation implicitly assumes another type of constraint:

4   The nonnegativity constraint, i.e.,

$$x_{ij} \geq 0 \qquad \text{for all } i, j$$

Notice that the objective function and the constraints are
both linear and additive functions of the decision variables x_{ij}. The
model presented here is a linear programming model. Furthermore, it
is a special type of LP model where the constraints are formulated in
such a way that the solution will be the assignment of four individuals
to four tasks. Such models are called *assignment models* and their
solution is always an integer solution, i.e., the third constraint is satisfied automatically.

DATA GATHERING

The data are gathered after the model is developed and checked for
internal consistency. Generally there is an abundance of data in organizations. However, the existence of data does not necessarily
mean that the quantitative information is readily available to solve the
model. It is necessary to identify the useful information, to separate
the relevant and irrelevant data, and to concentrate on obtaining the
data required by the model. Depending on the nature of the model,
data can be obtained from the company records, from industry
sources, from competitors, or from government documents. Some
models require data available in quantitative form. Financial models
are an example. Other types of models require measurement and ob-

196   servation over a period of time. For example, when modeling a waiting line, it is necessary to observe the queue characteristics, the arrival patterns, and service time distributions over an extended period. In some cases, data may not be available in quantitative form at all and need to be obtained from experts by using subjective measures. Examples of such data include estimates for job completion times and the probability of success of a proposed project.

In many cases, a complete set of historical data cannot be obtained and statistical curve-fitting techniques are used for estimating the values for missing data points. In other cases, the available data may be in aggregate form and need to be disaggregated; conversely, they may be too disaggregated and need to be aggregated. In summary, the decision process in the majority of decision problems involves a combination of quantitative as well as subjective, intuitive data. The subjective data are quantified with the use of psychometric methods; the quantitative data are obtained from existing records or from observations and measurements.

In the example problem, the data needed for the model are the c_{ij} values, the worker-days required for engineer i to perform task j. Let us assume that after contacting the four design engineers and analyzing their previous performance under similar conditions, the c_{ij} matrix shown in Table 7.1 has been developed. Because of the generic name of the c_{ij} values, this matrix is generally called the "cost matrix."

Using the values given in Table 7.1, the complete model can be developed as follows:

$$\min z = 2x_{11} + 3x_{12} + 10x_{13} + 8x_{14} + 2x_{21} + 7x_{22}$$
$$+ 9x_{23} + 7x_{24} + 2x_{31} + 5x_{32} + 3x_{33} + 4x_{34}$$
$$+ 2x_{41} + 7x_{42} + 4x_{43} + 6x_{44}$$

subject to

Each engineer is assigned:
$$\begin{bmatrix} x_{11} + x_{12} + x_{13} + x_{14} = 1.0 \\ x_{21} + x_{22} + x_{23} + x_{24} = 1.0 \\ x_{31} + x_{32} + x_{33} + x_{34} = 1.0 \\ x_{41} + x_{42} + x_{43} + x_{44} = 1.0 \end{bmatrix}$$

and

Each task has an engineer assigned to it:
$$\begin{bmatrix} x_{11} + x_{21} + x_{31} + x_{41} = 1.0 \\ x_{12} + x_{22} + x_{32} + x_{42} = 1.0 \\ x_{13} + x_{23} + x_{33} + x_{43} = 1.0 \\ x_{14} + x_{24} + x_{34} + x_{44} = 1.0 \end{bmatrix}$$

Since the decision variables x_{ij} will take on integer values 0 or 1 by the nature of the assignment model, it is not necessary to specify the in-

Worker-days required by engineer i to perform task j

7.1

i \ j	Design tasks			
	1	2	3	4
Design engineers				
1	2	3	10	8
2	2	7	9	7
3	2	5	3	4
4	2	7	4	6

teger constraints and the nonnegativity constraints in this problem. Thus the model is completely developed as shown above. Now we can go to the solution step.

SOLUTION
The solution of mathematical models can be obtained in several ways. Graphical solution is the simplest approach for problems with low dimensionality, but it is not a viable method for models with more than two or three variables. Many decision problems require the solution of simultaneous equations. They can be solved manually or by computer, using one of the several reduction techniques available. Calculus is used for the solution of simple optimization models, but special optimization techniques are necessary to solve complex problems. Most optimization methods utilize iterative procedures or systematic search techniques. Computers play an important role in their solution.

Depending upon the nature of the decision model, the solution could be either a specific value or a distribution of values for each variable. The solution could be the description of a process, as in simulation models, or it could represent the best set of decisions for the given conditions, as in optimization models.

The model developed above is an assignment model. It can be solved by the solution technique known as the "simplex method" or by a special procedure taking advantage of the structure of assignment models. The results in both cases are the same, but the special procedure is more efficient for these problems.

The simplex method will be explained in Chap. 9. The special procedure, known as the "assignment algorithm," will be used below to solve the problem.

The principle behind the assignment algorithm is that the minimum solution is not affected by subtracting a constant from any row or column of the cost matrix. The procedure is to subtract a sufficiently large "cost" from the rows and columns so that the optimal solution is determined by inspection.

198   The steps of the assignment algorithm for an $n \times n$ cost matrix are as follows:

1   Subtract the minimum element of each row from each element of that row. At the end of step 1, each row contains at least one element equal to zero.

2   If all columns contain at least one zero element, go to step 3. If any column has all nonzero elements, subtract the minimum element from all elements of that column.

3   Draw the minimum number of straight lines through rows and columns such that all cells with zero costs are covered by these lines. The minimum number of lines needed at this step is equal to the maximum number of assignments that can be made. If that number equals n, go to step 5. If it is less than n, continue with step 4.

4   Select the minimum element that is not covered by the lines. *Subtract it* from all the elements that are *not covered*, and *add it* to the *covered* elements at the intersection of two lines. Go back to step 3.

5   Find the optimal solution by assigning to the cells whose values are zero.

We will apply these five steps to the example problem.

$$
\begin{bmatrix}
2 & 3 & 10 & 8 \\
2 & 7 & 9 & 7 \\
2 & 5 & 3 & 4 \\
2 & 7 & 4 & 6
\end{bmatrix}
\qquad
\begin{bmatrix}
0 & 1 & 8 & 6 \\
0 & 5 & 7 & 5 \\
0 & 3 & 1 & 2 \\
0 & 5 & 2 & 4
\end{bmatrix}
$$

The cost matrix   Step 1

$$
\begin{bmatrix}
0 & 0 & 7 & 4 \\
0 & 4 & 6 & 3 \\
0 & 2 & 0 & 0 \\
0 & 4 & 1 & 2
\end{bmatrix}
\qquad
\begin{bmatrix}
0 & 0 & 7 & 4 \\
0 & 4 & 6 & 3 \\
0 & 2 & 0 & 0 \\
0 & 4 & 1 & 2
\end{bmatrix}
$$

Step 2   Step 3

$$
\begin{bmatrix}
0 & 0 & 7 & 4 \\
0 & 3 & 5 & 2 \\
0 & 2 & 0 & 0 \\
0 & 3 & 0 & 1
\end{bmatrix}
\qquad
\begin{bmatrix}
\text{The solution:} \\
x_{12} = 1 \\
x_{21} = 1 \\
x_{34} = 1 \\
x_{43} = 1 \\
\text{All other } x_{ij} = 0
\end{bmatrix}
$$

Step 4

Step 5

The value of the objective function for this solution is $Z = 13$. **199**
Referring to the model, this means that

Engineer 1 should be assigned to task 2

Engineer 2 should be assigned to task 1

Engineer 3 should be assigned to task 4

Engineer 4 should be assigned to task 3

to complete the four tasks in the minimum total worker-days of 13 days.

Even though the solution has been obtained, it should be pointed out that a single solution is not adequate to understand and control a complex decision system. The sensitivity of the system with respect to changes in parameters should be studied. This analysis is especially important when subjective data are used in the model. It provides the decision maker with a range of values in which the decisions remain the same even if the assured values vary.

Let us assume that the decision maker is not sure about the accuracy of c_{21}; i.e., he feels that engineer 2 has not worked on a design problem similar to task 1 for a long time and it is difficult to estimate how long it will take him to perform that task.

The postoptimality analysis can show us how sensitive the solution is to the value of that parameter. Such an analysis indicates that the assignment decisions will remain unchanged for $0 \leq c_{21} \leq 3$ even though the total worker-days (optimum value of objective function) will vary between 11 and 14 days.

When $c_{21} = 4$, the present solution will still be optimum but the four tasks will take a total of 15 worker days. At this point the engineering manager will also be able to use an alternative allocation pattern as shown in solution 2 for the same 15 worker-day total.

Solution 1:   Solution 2:
Engineer 1 to task 2   Engineer 1 to task 2
Engineer 2 to task 1   Engineer 2 to task 4
Engineer 3 to task 4   Engineer 3 to task 3
Engineer 4 to task 3   Engineer 4 to task 1

Total time = 15 worker-days   Total time = 15 worker-days

If c_{21} is 5 days, the first solution will no longer be optimal. The allocation decisions will have to be made as shown for solution 2 above. The total time will still be 15 worker-days. The reader can verify these results by solving the assignment problem with various values of c_{21}.

In summary, even if the decision maker is not sure how long engineer 2 will take to perform task 1, he knows that the allocation pattern will not be affected by a change in the value of that parameter

200   as long as it is between 0 and 4 days. Unless the manager has reason to believe that it will exceed 4 days, he can allocate the four men to the four tasks as shown in the initial solution.

IMPLEMENTATION

Solving the model is not the end of the management science approach. Unless the results are put to use, the effort spent in developing the model and finding the solution will have been wasted. The most meaningful test of a model lies in its acceptability and implementation. Even highly sophisticated and accurate models would be useless if they were not implemented. Model development by itself is a valuable experience in understanding the system characteristics even without implementation. However, the impact of the model is felt in a system only if it is used in making decisions.

The successful implementation of models is facilitated by bringing the ultimate user into the model-building process at an early stage. If the model is presented to the user as a "black box," there is a high probability that it will not be accepted and its results will not be trusted. However, if the user gets involved in the development of objectives, constraints, assumptions, and the philosophy of the model from the beginning, there will be a considerably higher chance for its acceptance and use in decision making.

As noted earlier, all steps of the management science approach affect each other and are affected by each other. The model builder has to go back and forth among the steps, including the implementation stage, until the final, usable product is developed and put to use.

SUMMARY

Management science is the application of scientific methods in a systematic way for the solution of decision problems. The concepts go back to the eighteenth century, but the first coordinated effort for solving management decision problems by a rational, scientific approach is considered to be the analysis of military decisions during the Second World War.

The management-science approach contains six interrelated steps:

1   Orientation

2   Problem definition

3   Model building

4   Data gathering   **201**

5   Solution

6   Implementation

Orientation consists of the identification of the problem and the description of its characteristics and boundaries in broad terms.

Problem definition involves development and delineation of the system elements, such as:

Measure of effectiveness

Variables

Parameters

Objective(s)

Constraints

Model building is the mathematical description of the decision problem to reflect the economic and technological characteristics.

Data gathering refers to obtaining the numerical values for the elements of the model.

Solution is the process through which the results of the decision system are obtained by analyzing the mathematical model.

Implementation involves putting the results of the decision model into operation.

A recognition of these steps and an acceptance that they make sense can provide the decision maker with a conceptual framework for dealing with the decision aspect of his management responsibilities.

DISCUSSION QUESTIONS

1   Why did the occurrence of the birth of management science and the emergence of the computer as a tool for large-scale problems at about the same time prove beneficial to both?

2   A manufacturing company will develop five products and has five machines to produce them at different costs. The company wants to assign one product to each machine such that the total operating cost will be minimum. If all the alternatives are enumerated, how many alternatives would you have to investigate?

3   Consider the case in which a city district is interested in reducing

202 traffic congestion. It appears to be caused by street parking, at least to the casual observer. Trace the development of a solution beginning with the orientation step.

4 In response to complaints about slow deliveries, a manufacturer installed a sophisticated computerized inventory control and scheduling system. The program functioned marvelously and provided large amounts of data on location of orders and scheduling of shipments. Yet many deliveries were not made. What may have been wrong with the solution? Discuss.

5 Illustrate a situation in which the management science steps are not followed in the orderly fashion of Fig. 7.2. For example, how might the availability of an existing solution procedure affect the model-building step?

6 "Obtaining a solution to a problem is not the end of the management science approach." Comment on this statement.

7 "Ultimately judgment is supreme in the making of management decisions." What is meant by this statement?

8 What are some of the differences between *engineering* and *management* decisions?

9 "An engineering decision is more precise than a management decision." Defend or refute this statement.

10 "In an engineering community, the decision analyst and the decision maker may not be the same person." Explain what is meant by this statement. In your explanation, give some examples.

11 "Defining the problem may be the most important step in the decision process." Defend or refute this statement.

12 "The most meaningful test of a decision model lies in its acceptability and implementation." Explain the rationale behind this statement.

KEY QUESTIONS FOR THE ENGINEERING MANAGER

1 What is my philosophy concerning the "rational manner" by which I make key decisions in this organization?

2 Does my experience in model building in engineering decisions provide any insight into how model building in management decisions can be carried out?

3 Are my engineering management decisions rational in nature?

4 Do I gather all the data necessary to make a rational decision?

5 In analyzing key decisions in this organization, do I think through how those decisions might be best implemented?

6 When a decision analyst presents recommendations to me, am I willing to listen to both the "good" and the "bad" information about that decision?

7   What are my strengths as a decision maker? My weaknesses?   **203**

8   What orientation have I provided for the engineers in my organization on my views on decision techniques?

9   Have I provided adequate information retrieval and processing capability for the use of the engineering community?

10   When key decisions are made in this organization, do I take time to fully discuss the implementation of these decisions with my staff?

11   What key strategic (long-term) and tactical (short-term) decisions have been made in this organization in the past 6 months? Have these decisions been analyzed in a rational manner?

12   What is the most difficult aspect of making decisions in this organization? Can management science help to relieve some of the difficulty?

BIBLIOGRAPHY

Ackoff, R. L., and M. W. Sasieni, *Fundamentals of Operations Research,* Wiley, New York, 1963.

Amstutz, A. E., *Computer Simulation of Competitive Market Response,* MIT Press, Cambridge, Mass., 1967.

Baumol, W. J., *Economic Theory and Operations Analysis,* 3d ed., Prentice-Hall, Englewood Cliffs, N.J., 1972.

Borko, H., *Computer Applications in the Behavioral Sciences,* Prentice Hall, Englewood Cliffs, N.J., 1962.

Churchman, C. W., R. L. Ackoff, and E. L. Arnoff, *Introduction to Operations Research,* Wiley, New York, 1957.

Cohen, Jared L., *Multiobjective Programming and Planning,* Academic, New York, 1978.

Dutton, J. M., and W. H. Starbuck, *Computer Simulation of Human Behavior,* Wiley, New York, 1971.

Eck, R. D., *Operations Research for Business,* Wadsworth, Belmont, Calif., 1976.

Fabrycky, W. J., P. M. Ghare, and P. E. Torgersen, *Industrial Operations Research,* Prentice-Hall, Englewood Cliffs, N.J., 1972.

Forrester, J. W., *Industrial Dynamics,* MIT Press, Cambridge, Mass., 1961.

Fromm, G., and P. Taubman, *Policy Simulations with an Econometric Model,* Brookings Institute, Washington, D.C., 1968.

204 Garfinkel, R. S., and G. L. Nemhauser, *Integer Programming,* Wiley, New York, 1972.

Gaver, D. P., and G. L. Thompson, *Programming and Probability Models in Operations Research,* Brooks/Cole, Belmont, Calif., 1973.

Gupta, S. K., and J. M. Cozzolino, *Fundamentals of Operations Research for Management,* Holden-Day, San Francisco, 1974.

Hillier, Frederick S., and Gerald J. Lieberman, *Introduction to Operations Research,* 2d ed., Holden-Day, San Francisco, 1974.

Huysmans, J. H., *The Implementation of Operations Research,* Wiley, New York, 1976.

Ignizio, James P., *Goal Programming and Extensions,* Lexington Books, Lexington, Mass., 1976.

Lee, Sang M., *Goal Programming for Decision Analysis,* Auerbach, Philadelphia, Pa., 1972.

Meadows, D. L., et. al., *The Limits to Growth,* Potomac Associates, Washington, D.C., 1972.

Miller, D. W., and M. K. Starr, *Executive Decisions and Operations Research,* Prentice-Hall, Englewood Cliffs, N.J., 1969.

Paik, C. M., *Quantitative Models for Managerial Decisions*, McGraw-Hill, New York, 1973.

Phillips, D. T., A. Ravindran, and J. J. Solberg, *Operations Research–Principles and Practice,* Wiley, New York, 1976.

Pool, I. D. S., R. Abelson, and S. Popkins, *Candidates, Issues and Strategies,* MIT Press, Cambridge, Mass., 1965.

Riggs, J. L., and M. S. Inoue, *Introduction to Operations Research and Management Science,* McGraw-Hill, New York, 1975,

Rivett, P., *Principles of Model Building,* Wiley, New York, 1973.

Salkin, H. M., *Integer Programming,* Addison-Wesley, Reading, Mass., 1975.

Shamblin, J. E., and G. T. Stevens, Jr., *Operations Research—A Fundamental Approach,* McGraw-Hill, New York, 1974.

Strum, J. E., *Introduction to Linear Programming,* Holden-Day, San Francisco, 1972.

Taha, Hamdy A., *Operations Research: An Introduction,* Macmillan, New York, 1971.

Thierauf, R. J., and R. C. Klekamp, *Decision Making Through Operations Research,* Wiley, New York, 1975.

Vajda, S., *Probabilistic Programming,* Academic, New York, 1972. **205**

Wagner, Harvey M., *Principles of Operations Research,* 2d ed., Prentice-Hall, Englewood Cliffs, N.J., 1975.

Wharton Econometric Forecasting Model, Department of Economics, Wharton School, University of Pennsylvania, Philadelphia, Pa.

Whitehouse, G. E., and B. L. Wechsler, *Applied Operations Research,* Wiley, New York, 1976.

MATHEMATICAL 8 MODELS

This chapter discusses mathematical models for engineering management decisions. The first part of the chapter is devoted to optimization models, which are presented in the following categories:

1   Classical optimization

2   Linear programming

3   Nonlinear programming

4   Integer programming

5   Dynamic programming

The remainder of the chapter describes simulation methods and briefly mentions special types of mathematical models used for special decision problems.

The objective of this chapter is to introduce the reader to mathematical-modeling concepts. It is intended for the engineering manager or the engineering management student who wants exposure to the available techniques without being concerned with the details of their solution. The techniques presented in this chapter are not a panacea. They cannot solve all decision problems. Some are appropriate for certain situations, some are not. Examples of the problems for which each technique is useful are summarized in Table 8.1.

The typical reader of this book is seen not as a specialist in mathematical programming but rather as the user of management science studies in his role as the manager of an engineering organization. In such a role, he deals with management science at three levels:

1   He recognizes decision situations in which management science can be used.

2   He provides input to management science specialists to solve the problem.

3   He evaluates the results and incorporates them into his decisions.

8.1

TECHNIQUE	TYPICAL PROBLEMS
Classical optimization	Relatively simple engineering design problems
Linear programming	Resource allocations with linear characteristics, CPM networks, product/project mix decisions
Nonlinear programming	Complex design problems and resource allocations with nonlinear characteristics
Integer programming	Assignment of personnel; allocation of equipment to projects
Dynamic programming	Replacement policies for equipment over a period of time
Simulation	Probabilistic decisions such as risky investments, PERT networks, traffic patterns
Special models	Equipment scheduling, transportation of men and equipment between sites, waiting lines, inventories, etc.

This requires that he understand the mathematical models and their results to the extent that he can interpret them and determine their usefulness by extracting the proper information for his decisions. This chapter should provide him with sufficient knowledge and understanding not only of what modeling is but also of how it is used and why it can help to provide the answers that can be used in managing an engineering organization. The linear programming examples presented in the chapter illustrate how the resource allocation problems can be expressed in mathematical terms and how the results are translated into decisions. The typical engineering manager does not solve such problems himself; he manages the decision analysts and engineers who do so. The examples should give the reader the necessary background to ask the right questions and to communicate with the analysts. The simulation example at the end of the chapter is worked out in detail to show how the risks can be incorporated into decisions, especially when it is difficult to develop simple, closed form models.

A detailed treatment of the solution methodology for linear programming is presented in Chap. 9 for readers who are interested in an in-depth study of that technique.

OPTIMIZATION

An engineering manager is a decision maker who has to work with limited resources under numerous constraints. These constraints are imposed upon him by technological requirements, by the capabilities

of his people, by the other departments in the organization, by com-   **209**
petitors, etc. He has to recognize these limiting conditions in making
his decisions to obtain the best possible results. The "best results"
could be maximum benefits, minimum costs, the highest effectiveness of technical personnel, or the most efficient engineering
design.

Engineers are familiar with the optimization approach in the
analysis and design of engineering systems. They allocate material
resources in the best way possible to satisfy acceptable performance
standards for those systems. As managers, they extend the same
thinking process to the allocation of other resources under their control. They decide on the proper mix of products, the proper production levels, the assignment of people to tasks, the selection of projects, and a multitude of similar problems dealing with limited time,
money, workforce, and technology.

Many of these decision problems can be formulated and
solved mathematically. Such formulations are the abstractions of the
real conditions, and typically involve simplifying assumptions. Nevertheless, building mathematical models, solving them, and interpret-
ing their results provide a rational framework that sharpens the engineering manager's judgment and gives him the tools to manage the
system in which he operates.

Some problems are formulated and solved by the classical
optimization methods, using calculus. These are the problems that
can be represented by differentiable functions. More complicated
problems require other, more powerful techniques. One such technique, linear programming, has been successfully applied to a wide
range of problems, such as coal preparation, refinery operations,
scheduling, production planning, workforce allocations, capital budgeting, urban planning, transportation, as well as the operational
problems at the shop level. When problems cannot be represented by
linear mathematical expressions, in many cases they can be solved by
nonlinear techniques involving systematic search methods and differential calculus.

CLASSICAL OPTIMIZATION METHODS

If we can express an optimization problem as a continuous, differentiable function, we can generally solve it by using calculus. Basically,
the first derivate of such a function is zero at a stationary point, and
the sign of the second derivative indicates whether we have a maximum, a minimum, or a point of inflection. This method is easy to use
for simple problems, but it becomes difficult as the problem gets
complicated. For example, if the function has more than one stationary point, each point needs a careful study to determine the optimum value. If the function is not continuous, we have to check its

210   value at each discontinuity. If there are multiple decision variables, the solution is much more involved than the single variable problem. Especially if the function has constraints, the classical optimization methods become exceedingly difficult to use.

Typical problems that can be solved by classical methods include the design of beams for minimum weight, the sizing of containers for minimum cost, and the development of inventory policies for the best combination of storage and ordering costs. It is obvious that the usefulness of these methods is limited to relatively simple problems with a small number of decision variables. Complicated problems require a more sophisticated, systematic approach, such as *mathematical programming.*

MATHEMATICAL PROGRAMMING

This technique provides us with the necessary tools and techniques to determine the optimum feasible point of a constrained function. Mathematical programming takes different forms for different types of problems.

> 1   *Linear programming* is a well-developed, structured method for the solution of problems in which the objective function and the constraints are expressed as linear mathematical statements and the decision variables have continuous values.
>
> 2   *Nonlinear programming* is the general title covering a wide range of methods for solving optimization problems when the objective function and/or the constraints contain nonlinear relationships.
>
> 3   *Integer programming* is the approach used when the decision variables are discrete (1, 2, 3, etc.) or dichotomous (such as 0 or 1).
>
> 4   *Dynamic programming* is the method used to solve problems involving sequential decisions made in multiple stages or in multiple time periods.

By far the most commonly used form of mathematical programming in management decisions is the linear programming (LP). A majority of problems can be formulated and solved by the LP algorithm. Examples of engineering problems include crane configurations for maximum load capacity, design of water-resource systems, control of traffic flow, competitive bidding, manpower assignment, equipment location, resource allocation, and production planning. Even though most of the complex problems usually exhibit nonlinear character-

istics, in many cases they can be linearized and approximated as **211**
linear-programming problems. Because of the importance of linear
programming, it is given special emphasis in this book, while the
other forms of mathematical programming are only briefly described.
The basic concepts underlying the LP models are presented in this
chapter. The following chapter is devoted to a detailed discussion of
the solution methodology.

LINEAR PROGRAMMING

The *decision variables* are the unknowns of the LP problem. Their
values represent the manager's decisions, such as the number of
hours each engineer should spend on each task, the amount of raw
materials to be used in a blending process or the quantity of each
product to be manufactured in a plant. The decision variables are restricted to be nonnegative and continuous, i.e., they are not limited to
be integer values and they cannot be less than zero.

The *cost vector* is a vector of the unit contributions of the
decision variables to the total measure of effectiveness of the system.
If the total cost of the system is the measure of effectiveness, the
hourly salaries of engineering personnel, the unit costs of raw materials or the unit production costs of the final product constitute the
cost vector. On the other hand, if the total profit is the measure of effectiveness, the unit profits, rather than the costs, are used in the cost
vector. Thus the term "cost vector" does not necessarily refer to cost
in the strict sense. The unit contributions, whether they are costs,
profits, opportunity costs, benefits, or other measures of effectiveness, can take on positive, negative, integer, or noninteger values.
Their values are estimated before the model is built.

The *objective function* is the linear combination of the individual contributions of the decision variables to the total measure of
effectiveness of the system. The LP problem is formulated to obtain
either the maximum or the minimum value of the objective function. If
the objective function is to be minimized, it is expressed as follows:

$$\min Z = c_1 x_1 + c_2 x_2 + \cdots + c_n x_n$$

where Z = value of the objective function

c_1 through c_n = costs associated with the n decision variables x_1 through x_n

If the maximum benefit is to be obtained, the c vector represents the
benefits to be received from each unit of the decision variables and
the objective function is expressed as

$$\max Z = c_1 x_1 + c_2 x_2 + \cdots + c_n x_n$$

The *constraints* are the linear expressions that describe the interactions among the decision variables and define the functional relationships of the decision system. Each constraint specifies a resource limitation, a certain technical requirement, or a demand imposed upon the system. If there are m such constraints, the coefficients associated with each of the n decision variables in each of the m constraints constitute an $n \times m$ matrix. It is usually called the A matrix, with each element a_{ij} called the "technological coefficient" or "activity coefficient," representing the effect of the jth decision variable in the ith constraint. The constraints could be expressed as being equal to a constant; less than or equal to a constant; or greater than or equal to a constant. For example, if the first constraint is the budget constraint and if the coefficient a_{ij} is the hourly rate of the jth engineer, the constraint takes the following form:

$$a_{11}x_1 + a_{12}x_2 + a_{13}x_3 + \cdots + a_{1n}x_n \leq B$$

where $x_j =$ number of hours engineer j can be assigned to
the project
$B =$ total budget available

The right-hand side constant of the ith constraint is called a requirement, b_i. All the b_i taken together constitute the requirements vector b.

The general LP problem can now be formulated by using the definitions given above.

Objective function: $\quad$ max (or min) $Z = c_1x_1 + c_2x_2 + \cdots + c_nx_n$

Subject to constraints: $\quad \begin{bmatrix} a_{11}x_1 + a_{12}x_2 + \cdots + a_{1n}x_n \\ a_{21}x_1 + a_{22}x_2 + \cdots + a_{2n}x_n \\ \cdots\cdots\cdots\cdots \\ a_{m1}x_1 + a_{m2}x_2 + = + a_{mn}x_n \end{bmatrix} \begin{matrix} \leq \\ \text{or} \\ = \\ \text{or} \\ \geq \end{matrix} \begin{bmatrix} b_1 \\ b_2 \\ \vdots \\ b_m \end{bmatrix}$

In a more compact form:

$$\text{max or min } Z = \sum_{j=1}^{n} c_j x_j$$

subject to

$$\sum_{j=1}^{n} a_{ij}x_j \begin{bmatrix} \leq \\ = \\ \geq \end{bmatrix} b_i \qquad (i = 1, \cdots, m)$$

The nonnegativity constraints are implicit in LP formulations.

$$x_j \geq 0 \qquad \text{for all}$$

The values of the technological coefficients a_{ij} ($i = 1, \cdots, m$, $j = 1, \cdots, n$), the requirements vector b_i ($i = 1, \cdots, m$), and the cost vector c_j ($j = 1, \cdots, n$) are all constants. The only unknowns are the values of the decision variables x_j ($j = 1, \cdots, n$). The solution procedure assigns values to x_j in such a way that Z reaches the maximum or the minimum possible value under the given conditions.

ASSUMPTIONS IN LP PROBLEMS

There are five major assumptions in the formulation of a decision problem as a linear programming problem:

1   The decision variables are nonnegative (i.e., zero or positive).

2   The objective function is a linear function of the decision variables.

3   The constraints can be expressed as linear equalities or inequalities.

4   The three kinds of parameters used in the formulation have deterministic values. The parameters are:

c_j = unit contributions of decision variables to objective function

b_i = resource levels

a_{ij} = coefficients of decision variables in the constraints

5   The decision variables can take on continuous (noninteger) values.

The name "linear programming" stems from assumptions 2 and 3 above. Assumptions 4 and 5 hold true for the ordinary linear programming problems, but they can be released in extensions of linear programming. Assumption 1 can also be released by modifying the basic LP model to handle nonpositive (zero or negative) or unrestricted (zero, negative, or positive) variables, if such variables exist in the decision system.

As mentioned earlier, linear programming is widely used for solving economic, social, industrial, and military problems.

There are several reasons for such a wide acceptance and use of LP:

1   Linear programming problems can be solved efficiently and systematically with the available solution techniques.

2   Even though the linearity assumptions appear to be too severe, they do not seriously restrict the use of LP because a large number of problems in many different fields can be formulated or approximated as LP problems without a significant departure from reality.

3   Certain types of problems for the transportation and assignment of resources have a special LP formulation that leads to an easy solution and results in integer values for the decision variables.

4   Linear programming lends itself to the analysis of the variability of data. The effect of changes in parameter values can be studied after the optimum solution is obtained.

FORMULATION OF LP MODELS
Linear programming models are formulated in three steps.

1   *Identification of the decision variables,* which are the unknown quantities to be determined, such as the amount of each product to manufacture, the money to allocate to different elements, or the time to spend on various activities.

2   *Development of the objective function* as a linear function of the decision variables.

3   *Mathematical definition of the constraints:*
(*a*)   Description of all system restrictions such as the total time, budget, and workforce availability.
(*b*)   Delineation of the requirements imposed upon the system by the internal and external subsystems; e.g., the minimum total output to be produced, the maximum price that could be charged, etc.
(*c*)   Development of mathematical expressions as equalities and inequalities of linear functions of the decision variables.

These steps will be illustrated in several LP formulation examples.

Example 8.1. You have to decide how to allocate a draftsman, a designer, and an engineer to an engineering project. Their hourly rates are:

Draftsman $8

Designer $12

Engineer $20

Your total budget for the engineering services is $14,400, and the overhead is 60 percent of the salaries. Based on similar jobs in the past, you estimate that the project will need at least 100 hours of analysis by the engineer, 200 hours of design by the designer, and 300 hours of drafting by the draftsman. If the designer does the analysis, he spends twice as much time for it as the engineer. The draftsman cannot be used for analysis, but the design work can be assigned to any of the three men. The engineer and the designer do comparable design work, while the draftsman takes 5 hours to produce what the others could produce in 1 hour. The draftsman is very efficient in drafting but the others can also draft. The designer takes $1\frac{1}{4}$ hours and the engineer 2 hours to do the drafting that the draftsman can finish in 1 hour. Each man is available for up to 220 hours. How many hours should you use each person to minimize the total cost?

Formulation

Decision variables:

x_1 = number of hours the draftsman will design

x_2 = number of hours the draftsman will do drafting

x_3 = number of hours the designer will analyze

x_4 = number of hours the designer will design

x_5 = number of hours the designer will do drafting

x_6 = number of hours the engineer will analyze

x_7 = number of hours the engineer will design

x_8 = number of hours the engineer will do drafting

Objective function: The criterion in this problem is cost minimization. The direct cost can be expressed as

$$8(x_1 + x_2) + 12(x_3 + x_4 + x_5) + 20(x_6 + x_7 + x_8)$$

The objective is to determine $x_1, \ldots, x_8$ such that the direct cost will be at the minimum level.

Therefore, the objective function is

$$\min Z = 8x_1 + 8x_2 + 12x_3 + 12x_4 + 12x_5 + 20x_6 + 20x_7 + 20x_8$$

Constraints:

Six types of constraints can be identified in the problem:

1   Budgetary constraint

2	Analysis constraint
3	Design constraint
4	Drafting constraint
5	Time availability
6	Nonnegativity of the decision variables

Budget: $1.6 [8x_1 + 8x_2 = 12x_3 + 12x_4 + 12x_5 + 20x_6 + 20x_7 + 20x_8] \leq 14,400$

or	$8x_1 + 8x_2 + 12x_3 + 12x_4 + 12x_5 + 20x_6 + 20x_7 + 20x_8 \leq 9,000$
Analysis:	$1/2x_3 \qquad + x_6 \qquad \geq 100$
Design:	$1/5x_1 \qquad + x_4 \qquad + x_7 \qquad \geq 200$
Drafting:	$x_2 \qquad + 1/1.25x_5 \qquad 1/2x_8 \geq 300$
Draftsman's time:	$x_1 + x_2 \qquad \leq 220$
Designer's time:	$x_3 + x_4 + x_5 \qquad \leq 220$
Engineer's time:	$x_6 + x_7 + x_8 \leq 220$
Nonnegativity:	$x_1, \quad x_2, \quad x_3, \quad x_4, \quad x_5, \quad x_6, \quad x_7, \quad x_8 \geq 0$

The last constraint, specifying nonnegative values for the decision variables, merely states that the draftsman, designer, and engineer cannot be assigned to work for less than zero hours. This is an implicit constraint in all LP problems, since the solution is restricted to nonnegative decision variables. Therefore, it is not even necessary to specify this as a separate constraint.

When we solve the problem, we obtain the minimum cost (optimal solution) at

$x_1 = 0$	Design by the draftsman
$x_2 = 220$	Drafting by the draftsman
$x_3 = 0$	Analysis by the designer
$x_4 = 120$	Design by the designer
$x_5 = 100$	Drafting by the designer
$x_6 = 100$	Analysis by the engineer
$x_7 = 80$	Design by the engineer
$x_8 = 0$	Drafting by the engineer

The total cost, including overhead, is $12,800.

The result shows that the draftsman should be assigned to drafting only. This is understandable, since the draftsman's productivity in design work is extremely low. The designer's performance in drafting is marginally below the draftsman and he is only half as efficient as the engineer in analysis. Thus, the designer's time is shared between design work and drafting, but he does not do any analysis. The designer fills in the slack in drafting caused by the fact that 300 hours of the draftsman's time are required, but he is available for only

220 hours. The balance of 80 hours is made up by the designer working for 100 hours at 80 percent efficiency.

The engineer works twice as fast as the designer in doing the analysis, and hence, he does all the analytical work. He also fills in for the slack in design caused by the assignment of the designer to do some of the drafting.

Example 8.2. A chemical plant can use two different processes to manufacture a mixture of two outputs, O_1 and O_2, from two raw materials, R_1 and R_2. The first process uses 7 tons of R_1 and 5 tons of R_2 to produce 2 tons of O_1 and 6 tons of O_2 in one day. The second process uses 5 tons of R_1 and 8 tons of R_2 to produce 5 tons of O_1 and 4 tons of O_2 on a daily basis as shown in Table 8.2.

350 tons of R_1 and 400 tons of R_2 are available and the minimum demands for O_1 and O_2 are 100 tons and 120 tons, respectively.

Even though the same outputs are produced from the same raw materials, the daily net profit is $3000 when process 1 is used and $4000 when process 2 is used.

Assuming that the plant can shift from one process to the other at any point during daily operations, formulate this as an LP problem and determine the best mix of the two processes.

Formulation

Decision variables:

x_1 = number of days to use process 1

x_2 = number of days to use process 2

Objective function:

$$\max Z = 3000x_1 + 4000x_2$$

Constraints:

$$7x_1 + 5x_2 \leq 350$$
$$5x_1 + 8x_2 \leq 400$$

Supply constraints

$$2x_1 + 5x_2 \geq 100$$
$$6x_1 + 4x_2 \geq 120$$

Demand constraints

	INPUT, tons		OUTPUT, tons		**8.2**
	R_1	R_2	O_1	O_2	
1 day of process 1:	7	5	2	6	
1 day of process 2:	5	8	5	4	

218   The optimal solution to this problem is given by

$$x_1 = 25.81 \qquad x_2 = 33.87$$

with

$$\max Z = 212{,}910$$

This solution indicates that we will achieve maximum profit of $212,910 if we use process 1 for 25.81 days and process 2 for 33.87 days.

Let us examine the constraints at the optimum point.

Raw material R_1: $7(25.81) + 5(33.87) = 350$

Raw material R_2: $5(25.81) + 8(33.87) = 400$

Both raw materials are exhausted.

Output O_1: $2(25.81) + 5(33.87) = 324.21$

Output O_2: $6(25.81) + 4(33.87) = 290.34$

Both outputs are produced to satisfy more than the minimum demands.

This result indicates that as long as the market exists for excess production, the company will produce up to the limit of the available materials. However, if the demand is restricted to 100 and 120 units, the corresponding constraints will have to be expressed as equalities rather than inequalities. In that case, the optimal solution becomes

$$x_1 = 9.09 \qquad x_2 = 16.34 \qquad \max Z = \$92{,}730$$

This solution gives

$$R_1 = 145.33$$

$$R_2 = 176.17$$

$$O_1 = 100$$

$$O_2 = 120$$

Notice that in this case the plant will produce its outputs only to satisfy the minimum demand and will not utilize all the raw materials available.

Furthermore, the maximum profit in the earlier version will be obtained in 59.68 days of operation ($25.81 + 33.87$), if the market can absorb the extra production. However, when the demand is limited, the profit of $92,730 will be realized by operating the plant for a total of $9.09 + 16.34 = 25.43$ days.

Example 8.3. A company manufactures its product in two plants, P_1 and P_2, for three markets, M_1, M_2, and M_3. Manufacturing costs and plant capacities are:

	UNIT MANUFACTURING COST	PLANT CAPACITY
P_1	$20	1000 units/month
P_2	$32	2200 units/month

The monthly demands at the three markets are:

M_1	800 units
M_2	1000 units
M_3	1200 units

Unit transportation costs from plants to markets are shown below:

	M_1	M_2	M_3
P_1	$11	$18	$14
P_2	$4	$6	$1

Develop an LP formulation for monthly operations.

Formulation This is a special type of linear programming problem called the "transportation problem." Characteristically, such problems have *m* origins and *n* destinations, with supply constraints at the origins and demand constraints at the destinations. The value of variable x_{ij} represents the amount to be transported from *i* to *j* at a unit cost of c_{ij}.

The problem is formulated as follows:

$$\min Z = \sum_{j=1}^{n} \sum_{i=1}^{m} c_{ij} x_{ij}$$

subject to

$$\sum_{j=1}^{n} x_{ij} \le S_i \qquad i = 1, \cdots, m$$

$$\sum_{i=1}^{m} x_{ij} \ge D_j \qquad j = 1, \cdots, n$$

$$x_{ij} \ge 0 \qquad \text{for all } i, j$$

where S_i = supply at origin *i*
D_j = demand at destination *j*

Decision variables:

x_{ij} = amount of product manufactured at plant *i* and transported to market *j*
$i = 1, 2$
$j = 1, 2, 3$

220   *Objective function:*

Calculate the costs c_{ij} associated with decision variables x_{ij}:

$$c_{11} = 20 + 11 = 31$$
$$c_{12} = 20 + 18 = 38$$
$$c_{13} = 20 + 14 = 34$$
$$c_{21} = 32 + 4 = 36$$
$$c_{22} = 32 + 6 = 38$$
$$c_{23} = 32 + 1 = 33$$

Thus,

$$\min Z = 31x_{11} + 38x_{12} + 34x_{13} + 36x_{21} + 38x_{22} + 33x_{23}$$

Supply constraints:

At P_1: $x_{11} + x_{12} + x_{13} \leq 1000$
At P_2: $x_{21} + x_{22} + x_{23} \leq 2200$

Demand constraints:

At M_1: $x_{11} + x_{21} \geq 800$
At M_2: $x_{12} + x_{22} \geq 1000$
At M_3: $x_{13} + x_{23} \geq 1200$

The LP model developed above can be presented in tabular form as in Table 8.3.

In tabular form, the coefficients of the decision variables are listed in the objective function and the constraints without repeating the variable names.

Solving this problem yields

$$x_{11} = 800$$
$$x_{12} = 200$$
$$x_{13} = 0$$
$$x_{21} = 0$$
$$x_{22} = 800$$
$$x_{23} = 1,200$$
$$\min Z = 102,400$$

The result shows that the constraints are satisfied with x_{11}, x_{12}, x_{22}, and x_{23} having positive values. No products are shipped from plant P_1 to market M_3 or from plant P_2 to market M_1. This result in-

8.3	VARIABLES	x_{11}	x_{12}	x_{13}	x_{21}	x_{22}	x_{23}	
	OF	31	38	34	36	38	33	MINIMIZE
	Constraints:							
	P_1	1	1	1				≤ 1000
	P_2				1	1	1	≤ 2200
	M_1	1			1			≥ 800
	M_2		1			1		≥ 1000
	M_3			1			1	≥ 1200

	Profit Per Day of Operation	
	WITHOUT OVERTIME	WITH OVERTIME
Plant 1	$3000	$2300
Plant 2	$4000	$2900

8.4

dicates that the products with lowest overall cost are to be shipped to the respective markets until the markets are satisfied. If the demand is not satisfied by the lowest-cost product, then the slack is taken up by the more expensive products. This is why x_{11} is preferred over x_{21} for market M_1 and x_{23} is chosen over x_{13} for market M_3.

Example 8.4. Suppose that the two chemical processes described in Example 8.2 are used in two different plants, P_1 and P_2, respectively. The plants are operated 22 days per month without paying overtime. Overtime can be ordered when necessary, but the total working days cannot exceed 30 per month in each plant. The net profits are given in Table 8.4.

The availability of raw materials and demand for outputs are shown in Table 8.5.

The raw materials requirements and output capabilities of the two processes are the same as described for Example 8.2.

Raw materials and outputs can be transported between the two plants as needed. The net transportation costs are given in Table 8.6.

Develop a monthly plan of operations by formulating it as an LP problem.

Formulation:

This problem includes both manufacturing and transportation models. The manufacturing operations in each plant can be modeled independently and then related to each other through the transportation submodel. It should be pointed out that the decision variables in an LP problem need not all have the same units, but their linear combination has to have a consistent measure of effectiveness. For example, the objective function of this problem will be developed in terms of dollars even though some of the decision variables will be expressed in days, some in tons.

	SUPPLY, TONS		DEMAND, TONS	
	R_1	R_2	O_1	O_2
At plant 1	200	220	60	30
At plant 2	150	180	40	90

8.5

8.6

	FROM P_1 TO P_2	FROM P_2 TO P_1
R_1	$5	$6
R_2	$4	$6
O_1	$9	$10
O_2	$12	$13

Decision variables:

x_1 = number of days P_1 to be operated without overtime
x_2 = number of days P_1 to be operated with overtime
x_3 = tons of R_1 to be transported from P_1 to P_2
x_4 = tons of R_1 to be transported from P_2 to P_1
x_5 = tons of R_2 to be transported from P_1 to P_2
x_6 = tons of R_2 to be transported from P_2 to P_1
x_7 = tons of O_1 to be transported from P_1 to P_2
x_8 = tons of O_1 to be transported from P_2 to P_1
x_9 = tons of O_2 to be transported from P_1 to P_2
x_{10} = tons of O_2 to be transported from P_2 to P_1
x_{11} = number of days P_2 to be operated without overtime
x_{12} = number of days P_2 to be operated with overtime

Table 8.7 shows the model developed in tabular form.

8.7

	PLANT 1		TRANSPORTATION				
	x_1	x_2	x_3	x_4	x_5	x_6	x_7
OBJECTIVE FUNCTION	3000	2300	− 5	− 6	− 4	− 6	− 9
Constraints:							
R_1 at P_1	7	7	− 1	+ 1			
R_2 at P_1	5	5			− 1	+ 1	
R_1 at P_2			+ 1	− 1			
R_2 at P_2					+ 1	− 1	
O_1 at P_1	2	2					− 1
O_2 at P_1	6	6					
O_1 at P_2							+ 1
O_2 at P_2							
Total days at P_1	1	1					
P_1 without overtime	1						
Total days at P_2							
P_2 without overtime							

Notice that the nonlinear profit function in this example was linearized by assigning a different decision variable to each linear segment of that function, as shown in Fig. 8.1. Since the objective is to maximize profits, the algorithm loads up x_1 and x_{11} before assigning any values to x_2 and x_{12} in plants P_1 and P_2, respectively. Thus the nonlinearity of the objective function does not cause any difficulty in this LP model.

When this problem is solved, the results are as follows:

$$x_1 = 22$$
$$x_2 = 6.57$$
$$x_3 = 0$$
$$x_4 = 0$$
$$x_5 = 0$$
$$x_6 = 60$$
$$x_7 = 0$$
$$x_8 = 2.86$$
$$x_9 = 0$$
$$x_{10} = 0$$
$$x_{11} = 22$$
$$x_{12} = 8$$
$$Z = \text{maximum profit} = \$191,922$$

TRANSPORTATION			PLANT 2		
x_8	x_9	x_{10}	x_{11}	x_{12}	
−10	−12	−13	4000	2900	MAXIMIZE
					≤ 200
					≤ 120
			5	5	≤ 150
			8	8	≤ 180
+1					≥ 60
	−1	+1			≥ 30
−1			5	5	≥ 40
	+1	−1	4	4	≥ 90
					≤ 30
					≤ 22
			1	1	≤ 30
			1		≤ 22

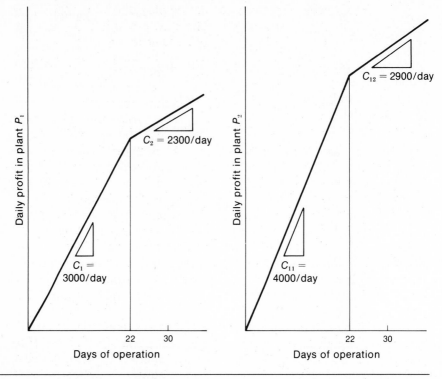

8.1

Nonlinear profit function in the two plants.

One should not permit overtime at plant 1 before plant 2 is fully utilized because the profit with overtime is smaller at the first plant. The results bear this out, with 6.57 days of overtime (x_2) at plant 1 and 8.0 (x_{12}) at plant 2. Also only two of the eight interplant transfers are effected. Since these transfers are not cost-free, it is best to minimize their occurrence as the solution indicates.

NONLINEAR PROGRAMMING

When the assumptions for a linear programming problem are not satisfied, we must look for other ways to solve the optimization problem. If the objective function or the constraint equations are not linear, the problem can be solved by a large number of methods collectively called "nonlinear programming." Any attempt to go through all of them is beyond the scope of this book. We will only give a general description of the various approaches.

If the problem is unconstrained, it can be solved by the gradient methods or the direct-search methods.

Gradient methods can be used when the objective function is differentiable. They move from the current point to another point in the direction determined from the gradient at the current point. The gradient is defined by the partial derivatives of the function.

Direct search methods do not employ derivatives, and they **225** can be applied to objective functions that are not differentiable, provided they are continuous. They are easier to implement and are applicable to a wider variety of objective functions than gradient methods. As an illustration of direct search methods, one particularly effective technique is the Hooke-Jeeves pattern search, which is based on the conjecture that successful adjustments of the decision variables are worth repeating.[1] The method begins cautiously with short excursions from the starting point, but the steps grow with repeated success. Subsequent failure calls for shorter steps, and if a change in direction is required, the technique starts over again with a new pattern. In the neighborhood of an optimal point, the steps become very small to prevent overlooking any promising direction.

When there are equality constraints, it might be possible to use the constraints to eliminate some of the variables from the objective function. This is a good technique, but sometimes it is difficult to identify which variables will enhance the solution by their elimination. One way to solve such problems is to introduce *Lagrange multipliers,* which essentially incorporate the constraints into the objective function.

For inequality constraints, again many methods are available. One effective approach is the use of penalty functions. The basis for this procedure for multivariable nonlinear systems was developed by Fiacco and McCormick and Zangwill.[2] The method augments the objective function by introducing a penalty function for each of the constraints. The augmented objective function is then penalized when one or more constraints are violated. Typically, a penalty function becomes prohibitively large at the constraint barrier, thus keeping the search away from the barrier. The direct methods mentioned under constrained nonlinear programs, earlier in this section, can be used in optimizing the augmented objective function.

The types of problems that can be solved by the nonlinear programming methods include cofferdam design, plant location and resource allocation with nonlinear characteristics.[3]

INTEGER PROGRAMMING

Sometimes one has to make decisions about discrete variables, such as the number of machines to be installed or the number of workers to be employed. The solutions to such problems are integer values for the variables. Or a decision may be dichotomous in the form of "yes"

[1]R. Hooke and T. A. Jeeves, "Direct Search Solution of Numerical and Statistical Problems," *Journal of the Association for Computing Machinery,* vol. 8, April 1961, p. 212.
[2]A.V. Fiacco and G. P. McCormick, "The Sequential Unconstrained Minimization Technique for Nonlinear Programming: A Primal-Dual Method," *Management Science,* vol. 10, 1964; W. I. Zangwill, "Nonlinear Programming via Penalty Functions," *Management Science,* vol. 13, 1970; p. 171.
[3]For other examples, see Robert M. Stark and Robert Nicholls, *Mathematical Models for Design,* McGraw-Hill, New York, 1972.

226   or "no," such as whether or not to undertake a given project. A common practice in such cases is to assign values of 0 and 1 for "no" and "yes," respectively. Problems of this type are referred to as the 0 - 1 problems.

There are several approaches for solving integer programming problems, including Gomory's cutting-plane method,[4] which starts with the optimal LP solution and goes through additional interations until an integer solution is obtained. There is also a method by Land and Doig[5] for mixed-integer programming problems, that is, problems where the integer-value restriction applies only to specified variables. It, too, uses the simplex approach to establish an initial opstriction applies only to specified variables. It, too, uses the simplex algorithm, which is most useful for mixed-integer programming problems. It utilizes implicit enumeration; in other words, it enumerates directly or implicitly all feasible solutions. The branch-and-bound methods are described by Dakin.[6] When 0-1 problems are to be solved, an algorithm developed by Balas can be used. It is sometimes called the additive algorithm.[7]

DYNAMIC PROGRAMMING

Sometimes a problem contains dynamic elements such as time considerations or a sequence of decisions in time. These decisions made at various stages of the evolution of the process are the inputs or decision variables. Together, they constitute a policy. A variety of decision situations fall into this category of multistage problems, including game theory, equipment replacement·policies, economic planning, reliability theory, control theory, chemical reactor design, stochastic processes, and others.

Briefly, the multistage decision processes referred to here are distinguished by the definition of the stages, the nature of decisions at the stages, an objective of the process, and return functions that indicate how the decisions affect the stages. The objective function is usually expressed as the sum of the stage return functions. It can also be expressed in multiplicative form.

If decisions at a later stage do not affect the performance at earlier stages, then the problem can be solved by dynamic programming. We will say here only that it is a method of decomposition incorporating the concept of recursion and Bellman's principle of op-

[4]R. E. Gomory, "Outline of an Algorithm for Integer Solutions to Linear Programs," *Bulletin American Mathematics Society,* vol. 64, 1958, pp. 275–278.
[5]A. H. Land and A. G. Doig, "An Automatic Method of Solving Discrete Programming Problems," *Econometrica,* vol. 28, 1960, pp. 497–521.
[6]R. J. Dakin, "A Tree-Search Algorithm for Mixed-Integer Programming Problems," *Computer Journal,* vol. 8, 1965, p. 250.
[7]E. Balas, "An Additive Algorithm for Solving Linear Programs with Zero-One Variables," *Operations Research,* vol. 13, 1965, pp. 517–546.

timality.[8] The return functions mentioned above are recursive rela-   **227**
tionships. The method essentially provides a framework for formulating an optimization problem, and any appropriate optimization technique can then be utilized to obtain a solution.

SIMULATION

A common characteristic of the optimization models is their prescriptive nature. We start with an objective, express it in terms of specific
actions, and determine which of those actions we should take in
order to achieve the objective at the best possible level. The output of
those models is a prescription of strategies.

Simulation, on the other hand, is a descriptive approach.
There are many situations that cannot be expressed in the form of a
prescriptive mathematical model because of the probabilistic nature
of the problem, the complexity of the issues involved in it, or the nature of the interactions among the decision elements. In such situations, we can identify the system components, develop a model
describing their internal relationships, and test that model for the effects of various actions that can be taken. The output of this type of
model is a simulation of the system under different external conditions and different decisions. Phillips, Ravindran, and Solberg
describe simulation as the process of creating the essence of reality
without ever actually attaining that reality itself.[9]

One of the best definitions is given by Naylor[10] as follows:
"Simulation is a numerical technique for conducting experiments on
a digital computer, which involves certain types of mathematical and
logical models that describe the behavior of business or economic
systems (or some component thereof) over extended period of time."
Simulation is generally considered as the approach to be used when
everything else fails. The availability of simulation languages and
powerful computers have made it a widely accepted method in management science. It provides us with insights into complex problems
that cannot be obtained by any other methodology. The two major
categories of simulation are discrete and continuous simulation.

DISCRETE SIMULATION

Discrete simulation, sometimes also called the "Monte Carlo" technique, is used primarily for studying operational decisions, such as
inventory problems, waiting lines, financial risk analysis, project
control networks, etc. Probability distributions of the events are developed by using random numbers in these models. The rela-

[8]R. Bellman, *Dynamic Programming*, Princeton, N.J., 1957.
[9]D. T. Phillips, A. Ravindran and J. J. Solberg, *Operations Research*, Wiley, New York, 1976, p. 359.
[10]T. H. Naylor, J. L. Balintfy, D. S. Burdick, and K. Chu, *Computer Simulation Techniques*, Wiley, New York, 1966, p. 3.

228   tionships among the events are simulated with their respective probabilities in order to obtain the distribution of the final output under those combined effects. The basic steps of discrete simulation models are the following:

1   Develop system interactions among the controllable and uncontrollable variables

2   Identify the probability distributions for uncontrollable variables

3   Use random numbers to simulate the probabilistic events

4   Simulate the total system by combining the probabilistic events

5   Develop probabilistic statements for the system behavior after repeating steps 1 to 4 a sufficient number of times

These steps are illustrated in the following example.

Suppose that a company has developed a new technology and is planning to market it. Because of the absence of any market data, subjective probabilities have been obtained on the annual demand. The company is concerned about competitors' actions and it will adjust its selling price according to the competitors' pricing strategy. To complicate the matter, the unit cost will depend on whether or not the labor union will get higher wages next year. The current tax rate is 50 percent, but legislation is pending to provide tax breaks for new technologies. For simplicity, assume that the losses incurred in a year cannot be carried to the following year for tax reduction.

We will simulate the company's operations to determine its IAT (income after tax) over the next 3 years.

SYSTEM INTERACTIONS   The model can be developed as follows for year i:

$$IAT = (\text{demand})_i \times (\text{unit price}_i - \text{unit cost}_i) \times \alpha_i$$

where $\alpha = 1 - \text{tax rate}$ if unit price $>$ unit cost
$\alpha = 1$ if unit price $\leq$ unit cost

PROBABILITY DISTRIBUTIONS

Demand   Assume that the company has obtained the best, worst, and most likely estimates for the demand and decided to use a triangular density for probabilities as shown in Fig. 8.2.

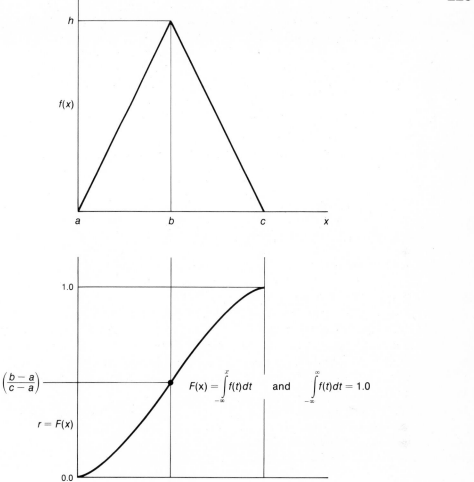

8.2

Triangular density function.

For simplicity, the demand will be considered to be a continuous variable rather than a quantity in discrete units.

Unit Price The subjective probabilities are:

$$p(\$10) = 0.50$$
$$p(\$15) = 0.30$$
$$p(\$20) = 0.20$$

230 *Unit Cost*

p(Labor pressure resulting in a cost of $12) = 0.40

p(No labor dispute, thus $8 cost)   = 0.60

Tax Rate on Profit

p(Tax = 25 percent) = 0.20

p(Tax = 40 percent) = 0.30

p(Tax = 50 percent) = 0.50

SIMULATION OF PROBABILITY DISTRIBUTIONS   Random numbers are numbers generated via a random process in such a way that each digit has the same probability as any other digit. There are numerous methods for generating them. One source is the random number tables, as shown in Table 8.8.

Random Numbers

8.8	02946	96520	81881	56247	17623	47441	27821	91845
	85697	62000	87957	07258	45054	58410	92081	97624
	26734	68426	52067	23123	73700	53730	06111	64486
	47820	32353	95941	72169	58374	03905	06805	95353
	76603	99330	40571	41186	04981	17531	97372	39558
	47526	26522	11045	83565	45639	02485	43905	01823
	70100	85732	19741	92951	98832	38188	24080	24519
	86819	50200	50889	06493	06638	03619	90906	95370
	41614	30074	23403	03656	77580	87772	86877	57085
	17930	26194	53836	53692	67125	98175	00912	11246
	24649	31845	25736	75231	83808	98997	71829	99430
	79899	34061	54308	59358	56462	58166	97302	86828
	76801	49594	81002	30397	52728	15101	72070	33706
	02567	08480	61873	63162	44873	35302	04511	38088
	40723	15275	09399	11211	67352	41526	23497	75440
	42658	70183	89417	57676	35370	14915	16569	54945
	65080	35569	79392	14937	06081	74957	87787	68849
	02906	38119	72407	71427	58478	99297	43519	62410
	75153	86376	63852	60557	21211	77299	74967	99038
	14192	49525	78844	13664	98964	64425	33536	15079
	32059	11548	86264	74406	81496	23996	56872	71401
	81716	80301	96704	57214	71361	41989	92589	69788
	43315	50483	02950	09611	36341	20326	37480	34626
	27510	10769	09921	46721	34183	22856	18724	60422
	81782	04760	36716	82519	98272	13969	12429	03093

We will use this table in our example, but we could also generate random numbers by writing simple computer programs or using the software packages available for this purpose.

The use of random numbers to simulate probabilistic events involves a mapping of the random number on the probability distribution of the event under consideration. The basic principle behind this mapping will be illustrated by the inverse transformation method.

The inverse transformation method is based on the observation that the area under the probability density curve can be obtained by integrating the density function, and it is related to the random number scale as follows:

The area under the probability density curve is given by

$$F(x) = \int_{-\infty}^{x} f(t) \, dt$$

such that $F(x^*)$ is the cumulative probability for the random variable x to take on a value up to x^*, i.e.,

$$F(x^*) = p(x \le x^*)$$

On the other hand, since all random numbers have the same probability of being selected, if we pick a random number R ($0.0 \le R < 100$), we can make a statement that the probability of picking a number up to R is in fact equal to $r = R/100$, that is,

$$p(\text{Random number} = R) = R/100 = r$$

In other words, if the random number is 0.86, we can say that the probability of picking any number between 0 and 0.86 in a span of 0 to 99 is equal to 86 percent. We then find the point on the cumulative distribution where $F(x^*) = 0.86$ and transform it to x^*. This transformation is made by the inverse-transformation calculations. Notice that the probability that the random variable x will take on a value up to x^* is also 0.86. Thus we can use the value x^* corresponding to R on the $F(x)$ scale as the value of the random variable. Formally stated: If we set $r = F(x^*)$, we can say

$$p(\text{Random number} \le R) = p(x \le x^*)$$

This completes our mapping.

We will now explain the transformation of $F(x^*)$ to x^* first using a triangular density function and then discrete probabilities. Consider the function shown in Fig. 8.3.

We first determine the ratio $(b - a)/(c - a)$ and then select a two-digit random number R between 0 and 99 and calculate $r = R/100$. If $0 \le r \le [(b - a)/(c - a)]$, we calculate x as

$$x = a + \sqrt{r(c - a)(b - a)}$$

If $[(b - a)/(c - a)] \le r < 1.0$, we calculate

$$x = c - \sqrt{(1.0 - r)(c - a)(c - b)}$$

232

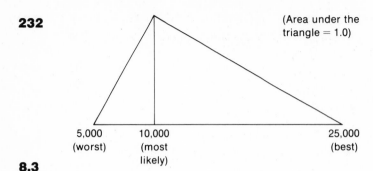

8.3

Demand estimates.

x is the value of the random variable from the given triangular density function. (Derivation of the expressions for *x* are left to the reader.)

Discrete probabilities: When we are dealing with discrete probabilities such as the distributions given for the unit cost, unit price, and tax rate in our example, we first develop the cumulative distribution and then determine the ranges in that distribution and define the random numbers which correspond to each range. The value of the random variable is determined according to the range to which the random number belongs. This is illustrated in Tables 8.9–8.11.

If *R* is between 0 and 49, we use $10 for unit price. If it is between 50 and 79, we use $15; and for *R* greater than 79, we use $20.

STIMULATION OF THE SYSTEM We will use the first two digits of the random numbers in Table 8.8, starting at the left uppermost corner and going downward. One cycle of the simulation is shown in Table 8.12.

Explanation For the triangular demand distribution with

$$a = 5000, \ b = 10,000, \text{ and } c = 25,000,$$

$$\frac{b-a}{c-a} = \frac{5,000}{20,000} = 0.25$$

YEAR	RN 1	DEMAND	RN 2	PRICE	RN 3	COST
1	.02	6,414	.85	$20	.26	$12
2	.76	16,515	.47	$10	.70	$ 8
3	.41	11,696	.17	$10	.24	$12

In year 1, $R_1 = 2$, $r = 0.02 < 0.25$:

$$x = \text{demand} = 5000 + \sqrt{0.02\,(20{,}000)\,(5000)} = 6414 \text{ units}$$

Then

$$R_2 = 85 \rightarrow \text{Price} = \$20$$
$$R_3 = 26 \rightarrow \text{Cost} = \$12$$
$$R_4 = 47 \rightarrow \text{Tax} = 40\%$$

In the second year, $R_1 = 76$, $r = 0.76 > 0.25$:

$$\therefore \text{Demand} = 25{,}000 - \sqrt{(1 - 0.76)\,(20{,}000)\,(15{,}000)} = 16{,}515$$

$$R_2 = 47 \rightarrow \text{Price} = \$10$$

$$R_3 = 70 \rightarrow \text{Cost} = \$8$$

$$R_4 = 79 \rightarrow \text{Tax} = 50\%$$

8.9

UNIT PRICE	PROBABILITY	CUMULATIVE PROBABILITY	RANGE FOR RANDOM NUMBER
$10	0.50	0.50	0–49
$15	0.30	0.80	50–79
$20	0.20	1.00	80–99

Random Numbers for Unit Cost

8.10

UNIT COST	PROBABILITY	CUMULATIVE PROBABILITY	RANGE FOR RANDOM NUMBER
$12	0.40	0.40	0–39
$ 8	0.60	1.00	40–99

Random Numbers for Tax Rate

8.11

TAX RATE	PROBABILITY	CUMULATIVE PROBABILITY	RANGE FOR RANDOM NUMBER
25%	0.20	0.20	0–19
40%	0.30	0.50	20–49
50%	0.50	1.00	50–99

Risk Analysis by Monte Carlo Simulation

8.12

RN 4	TAX RATE	PRICE − COST	IBT	TAX	IAT	CUM IAT
.47	0.4	8	51312	20525	30787	30787
.86	0.5	2	33030	16515	16515	47302
.79	0.5	− 2	− 23392	———	(23392)	23910

234

The third year is simulated the same way. Since the cost will be higher than the price, there will be a loss in year 3. The cumulative IAT for 3 years is $23,910. This is the end of the first cycle of simulation.

RESULTS The simulation process illustrated above is repeated for many cycles.[11] After a sufficient number of cycles (e.g., 1000), the results are summarized. We can obtain the probability distribution for the 3-year cumulative IAT; the probability that the company will end up with a loss in 3 years, or the probability that the total net profit in 3 years will be more than $50,000, etc. These probabilistic statements can be derived directly from the frequency distributions of the appropriate values.

CONTINUOUS SIMULATION
In contrast to the primary emphasis of discrete simulation on operational decisions, continuous simulation models focus on the total systems at the aggregate level. Social, urban, political, and energy systems can be studied by this approach. In the context of engineering management decisions, the effects of technological innovation, productivity, and technical obsolescence on long-term policies can be evaluated. System components are identified and interrelated by logical functional relationships that describe their impact on each other. As a change is introduced into one or more of the components, that change is followed through the total system and results in a directional shift in the system's overall behavior. These shifts are observed over an extended period of time to determine the long-range implications of major changes that can be introduced by the external conditions or the internal policies.

SPECIAL TYPES OF MATHEMATICAL MODELS
Some decision problems are encountered so frequently that special models have been developed to solve them. One of these is the queueing model. The arrival of customers is studied as the Poisson process and decisions are made about the number of service stations and their spacing. This approach lends itself readily to the planning of turnpike facilities and the scheduling of production lines, etc.

Another group of special models is used for inventory problems. The tradeoffs among inventory costs, shortage costs, and ordering costs are analyzed and balanced to obtain the strategies leading to the minimum overall costs. The output of inventory models includes the economic order quantity (EOQ), the frequency of orders,

---

[11] A discussion of the stopping rules is beyond the scope of this book. They are discussed in any good book on simulation—for example, Naylor, et al. (Footnote 10).

and the point at which the order should be placed. The stochastic na- **235**
ture of demands is incorporated into these models.

Other special models include networks, scheduling, and transportation models.

SUMMARY

Mathematical models are used for the analysis of management decisions at various levels. Some decision problems can be expressed as optimization models and solved for a specification of the action that leads to the optimum results. The most widely used optimization model is the linear programming model. It is fully developed and formalized to the extent that computer codes are now available to solve a broad range of problems fitting into the linear programming category. When linearity assumptions are not satisfied, it may be necessary to use other optimization techniques. Such techniques include classical optimization, nonlinear programming, integer programming, and dynamic programming.

When decision problems cannot be formulated for prescriptive solutions offered by the normative approach of optimization methods, we can use simulation in an exploratory sense. It is a useful tool for analyzing complex relationships among probabilistic events and is used both for the analysis of discrete changes in decision problems and for the evaluation of continuous impacts of policies and external conditions on total systems.

In addition to these two major approaches, a number of special models can also be developed for frequently encountered decision problems. Even though the overall objective of these special models may again be to optimize or to simulate the system, we recognize them as separate models because of their common characteristics and frequent use. Best-known in this special group are the queueing, inventory, network, scheduling, and transportation models.

DISCUSSION QUESTIONS

1 In the introduction to this chapter, the authors identify certain mathematical models and suggest typical problems for which these techniques are useful. Identify some additional problems where mathematical models can be used. Be prepared to defend your choices.

2 An engineering manager is a user of management science. As a user he must understand its advantages and disadvantages in his

managerial role. Identify and discuss your choice of advantages and disadvantages.

3   Mathematical programming takes different forms for different problems. What are these forms? In what circumstances can each be used?

4   What are the major assumptions to be considered in the formulation of a decision situation in the context of a linear programming problem?

5   "Luck plays a role in the outcome of an engineering management decision." Is this statement irrelevant in a discussion of the role of mathematical models in the decision process?

6   Simulation has been described as creating a form of reality without ever attaining that reality. What is meant by this statement?

7   Mathematical modeling emerged about 30 years ago. Engineering managers dealt with complex decision problems prior to that time. How did these engineers cope with such complex problems? Discuss.

8   A consulting firm has entered competitive bidding for the design and construction of a large facility. The job will be awarded to the lowest bidder. Discuss the firm's objective(s), constraints, assumptions, and information requirements involved in arriving at a bid price.

9   A major city is considering building a high-speed mass transportation system between its suburbs and the downtown area. Because of budget restrictions, the system cannot be extended to all the suburbs. It will serve the metropolitan area population to varying degrees, depending on which suburbs are included in the network. Discuss how mathematical modeling can help to decide where to build the system.

10   The characteristics of steel are determined by its mineral content. Consequently, the upper and lower limits of minerals such as carbon, chrome, manganese, and silicon are specified as the percentage of the final product in accordance with the user's specifications. The materials available for producing steel include pig iron, scrap metal, various alloys, ferrosilicon, carbide, etc. They typically have varying unit costs and varying mineral contents per pound. Some materials may be in short supply at any given time. Discuss how steel production decisions can be formulated as an LP model.

11   A company is considering eight projects to choose from. Each project may be selected at a fractional amount from zero to one. If project j is chosen at the 100 percent level, its net return will be rj. Assume that the costs and returns are proportional for each project; that is, if it is decided to invest 30 percent of the total cost of a project, the expected return of that project will be 30 percent of its maximum return. Let lj be the investment required for project j chosen at the 100 percent level and assume that the company has a total budget of B. Formulate a linear programming model to examine the total return.

12 Suppose the project selection example of the above question **237** has been changed so that:

Projects should either be chosen at 100 percent level or be rejected

Projects 1, 2, and 3 are mutually exclusive; i.e., if one is chosen, the others cannot be included in the final package of projects

Projects 4 and 5 are dependent, i.e., if one is chosen, the others should also be chosen

Project 6 cannot be chosen unless projects 7 and 8 have both been chosen.

Discuss how these modifications can be incorporated into the model.

KEY QUESTIONS FOR THE ENGINEERING MANAGER

1   What are the five or six key decisions in my organization where a management science approach could be used?

2   Do I understand enough about management science to be able to communicate with the specialists in that field concerning management problems and opportunities in this organization?

3   A "management model" is a representation of the real world which can facilitate the decision process. What management models might be useful for me to use in this organization?

4   Am I sufficiently qualified to evaluate the recommendation of management scientists who assist me in making management decisions?

5   Ultimately all decisions rest on an integration of judgment, experience, and intuition. Is this statement compatible with the use to which I might put management scientists in this organization?

6   What is the nature of the input which I should be prepared to provide to management scientists in this organization?

7   How might optimization be used to assist me in making management decisions involving the allocation of resources?

8   What application might nonlinear programming have in management decisions in this organization?

9   How might simulation help me in making strategic (long-term) decisions in this organization?

10   How should I evaluate the results of the use of management science in this organization?

11   What is the most critical management opportunity facing this

238 organization. Can (should) management science be used in this decision?

REFERENCES

Balas, E., "An Additive Algorithm for Solving Linear Programs with Zero-One Variables," *Operations Research,* vol. 13, 1965, pp. 517–546.

Bellman, R., *Dynamic Programming,* Princeton, Princeton, N.J., 1957.

Dakin, R. J., "A Tree-Search Algorithm for Mixed-Integer Programming Problems," *Computer Journal,* vol. 8, 1965, p. 250.

Fiacco, A. V., and G. P. McCormick, "The Sequential Unconstrained Minimization Technique for Nonlinear Programming: A Primal-Dual Method," *Management Science,* vol. 10, 1964.

Gomory, R. E., "Outline of an Algorithm for Integer Solutions to Linear Programs," *Bulletin American Mathematics Society,* vol. 64, 1958, pp. 275–278.

Hooke, R., and T. A. Jeeves, "Direct Search Solution of Numerical and Statistical Problems," *Journal of the Association for Computing Machinery,* vol. 8, April 1961, p. 212.

Land, A. H., and A. G. Doig, "An Automatic Method of Solving Discrete Programming Problems," *Econometrica,* vol. 28, 1960, pp. 497–521.

Naylor, T. H., J. L. Balintfy, D. S. Burdick, and K. Chu, *Computer Simulation Techniques,* Wiley, New York, 1966.

Phillips, D. T., A. Ravindran, and J. J. Solberg, *Operations Research,* Wiley, New York, 1976.

Stark, Robert M., and Robert L. Nicholls, *Mathematical Foundations for Design,* McGraw-Hill, New York, 1972.

Zangwill, W. I., "Nonlinear Programming via Penalty Functions," *Management Science,* vol. 13, 1970, p. 171.

LINEAR PROGRAMMING 9

The preceding chapter presented mathematical models at a conceptual level and described the kinds of problems that are handled by different techniques. In addition, the problem formulation and representative results for the linear programming approach were provided, but the details of the solution were not discussed.

This chapter discusses the simplex algorithm of linear programming. The material is provided for the engineer and analyst who plan to use linear programming to solve decision problems. An engineering manager who has the responsibility for supervising engineers needs to understand the methodology and the types of problems to which the technique can be applied. His level of understanding should be such that he can ask the right kinds of questions and know if he is getting the right answers. If he is to accept the decisions facilitated by linear programming, then he must have a conviction that this is the way to do things. By scanning this chapter, the reader should be able to determine if an in-depth reading is required.

The mathematical concepts are explained when first presented in this chapter. An understanding of vector algebra is assumed. The intent is to teach linear programming to engineering management students at graduate and advanced undergraduate levels. The reader should first understand the framework of linear programming in Chap. 8.

Fundamental concepts and definitions are introduced at the beginning of the chapter. A graphical treatment of problems with two variables is next offered to illustrate the important features of the solution procedure. The presence of *alternative* solutions and *unbounded* condition are investigated. *Redundant* constraints and *infeasible* problems are illustrated.

The chapter then delves into the general solution procedure of the simplex algorithm. The details of the algorithm are presented using the production planning example of the previous chapter.

In formulating a problem, particular values are assigned to the appropriate parameters. These values may be subject to variations. What then happens to the optimal solution? This necessitates the analysis of the problem and the optimal solution to see if and how

240   it changes as the parameter values vary. This analysis, called the "sensitivity," or "postoptimality" analysis, is just as important as obtaining the optimal solution in the first place. The chapter ends with a discussion of sensitivity analysis.

LP PROBLEMS

Recall that a linear programming model is formulated as follows:

$$\text{max or min } Z = \sum_{j=1}^{n} c_j x_j \tag{9.1}$$

subject to

$$\sum_{j=1}^{n} a_{ij} x_j \begin{bmatrix} \leq \\ = \\ \geq \end{bmatrix} b_i \qquad i = 1, \ldots, m \tag{9.2}$$

and

$$x_j \geq 0 \qquad j = 1, \ldots, n \tag{9.3}$$

After the problem is formulated, it can be solved systematically by following a sequence of steps until the result is obtained. The desired result of the LP problem is the *optimum feasible solution,* which is a set of x_j resulting in the optimum value for the objective function Z given in expression (9.1), while satisfying all constraints of expression (9.2) and the nonnegativity conditions of expression (9.3). It can be shown that if feasible solutions exist for a linear programming problem, they form a convex set.[1] It is the *feasible region* for the problem. Each extreme point of that set represents a *feasible solution* with no more than m variables taking positive values and the rest remaining at zero level. Such solutions are called *basic feasible solutions.* One of the basic feasible solutions is the optimum feasible solution.[2]

The convex set described above is defined by the intersections of constraints expressed as equalities. The constraints are not restricted to be equalities in the original formulation of the problem, but they can all be converted to equalities by adding new variables to the system.

Thus, the optimum feasible solution of a linear program-

---

[1]"Convex set" describes a region such that a straight line joining any two points in that region also lies in it; i.e., for x_1 and x_2 in the convex region, $x^* = \lambda x_1 + (1 - \lambda)x_2$ is also in that region for $0 \leq \lambda \leq 1$ and $x_1 \neq x_2$. For example:

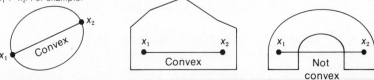

[2]For a discussion of these concepts and the proofs of basic LP theorems, see the books listed at the end of this chapter.

ming problem is at an extreme point of a convex set of solutions, **241**
defined by the constraint equations. This observation leads to an interesting conclusion. If we study all basic feasible solutions, we
could select the optimum among them. This could be done by setting $n - m$ variables equal to zero and solving the m constraint
equations for the remaining m variables, for all possible combinations of these m variables. At the outset, this sounds like a workable
procedure, but it should be kept in mind that for n variables (including the variables added to convert inequality constraints to equalities) and m constraints ($n > m$), the upper bound for the number of
possible solutions is

$$\binom{n}{m} = \frac{n-}{[m - (n - m) -]}$$

Although this is a finite number, the complete enumeration involves
an extremely high number of solutions before the optimum is determined. For example, even for a small problem with $n = 10$, $m = 4$,
there could be as many as

$$\frac{10\ !}{[4\ !\ (6\ !)]} = 210 \text{ solutions}$$

Instead of enumerating all these solutions, a systematic
approach called the "simplex method" is used to solve LP problems.
It is an algorithm developed in the late 1940s almost simultaneously
by George B. Dantzig[3] and by A. Charnes and W. W. Cooper,[4] working
independently.

The simplex algorithm can be used for solving any size and
type of LP problem. Obviously, for procedures coded for computer
use, the problem size must not exceed the computer limits.

In addition to the general solution methodology of the
simplex algorithm, there are other methods that can be used for solving simple LP problems or special types of problems. The simplest
solution is by the graphical method. If the problem has only two
decision variables, the constraints and the objective function can be
plotted and the solution is obtained by visual inspection or by simple
calculations. For problems with more than two variables, this method
becomes extremely difficult to use. Even for a three-dimensional
problem, the graphical method is very cumbersome.[5]

In some cases, the characteristics of the problem structure
can be used for special solution methods. A class of LP problems
called the "transportation problem" is an example. The optimum
resource allocation pattern for transporting goods or materials from

[3]*Linear Programming and Extensions*, Princeton, N.J., 1963
[4]*Management Models and Industrial Application of Linear Programming*, vols. 1 and 2, Wiley, New York, 1960.
[5]The dimension of an LP problem is determined by the number of decision variables.

m origins to n destinations can be determined by a convenient technique called the "transportation algorithm." A more specialized version of the transportation problem is the "assignment problem," which involves the distribution of exactly n resources to exactly n positions, such as the allocation of 4 engineers to 4 tasks, where each task has to have an engineer and each engineer has to be assigned to a task. The solution of such a problem was illustrated in Chap. 7. It should be pointed out that all these special problems can be solved by the generalized solution methodology, the simplex algorithm.

Before simplex is introduced, the graphical solution approach will be presented to clarify some of the salient characteristics of the LP problems.

THE GRAPHICAL SOLUTION

Although linear programming problems with only two variables are very rare in actual situations, the discussion in this section will be limited to two-dimensional problems in order to illustrate the principles of the graphical method.

Example 9.1 Let us consider the problem given in Example 8.2. (The cost coefficients are in thousands for convenience.)

$$\max Z = 3x_1 + 4x_2$$

subject to

$$\left. \begin{array}{l} 7x_1 + 5x_2 \leq 350 \\ 5x_1 + 8x_2 \leq 400 \end{array} \right\} \text{Supply constraints}$$

$$\left. \begin{array}{l} 2x_1 + 5x_2 \geq 100 \\ 6x_1 + 4x_2 \geq 120 \end{array} \right\} \text{Demand constraints}$$

$$\left. \begin{array}{l} x_1 \geq 0 \\ x_2 \geq 0 \end{array} \right\} \text{Nonnegativity constraints}$$

The solution of this problem should specify the values of x_1 and x_2 which will give the objective function Z its maximum value without violating any of the constraints.

The procedure to obtain such a solution by the graphical method follows the steps below:

1   Formulate the linear programming problem.

2   Represent the feasible region in a graph by plotting all constraints in the first quadrant.

3   Plot the objective function for an arbitrary value of Z.

4   Move the objective function parallel to the arbitrary line plotted in step 3 until it passes through a point in the feasible region that represents its optimal value.

Note that in step 4, every line drawn parallel to the original, arbitrary line of step 3 has a different value for the objective function. In a maximization problem where both coefficients are positive (i.e., both variables make a positive contribution to the objective function), the optimal value is obtained when the line reaches a feasible point farthest from the origin. Conversely, in a minimization problem (with c_1 and c_2 both positive), the optimal solution is found when the objective function passes through the point closest to the origin. It is not necessary to extend the graph beyond the first quadrant because of the nonnegativity constraint. The values $x_1 \geq 0$ and $x_2 \geq 0$ can be found only in the first quadrant.

Figure 9.1 shows the graphical representation of the example problem. The constraints are plotted at their limiting values, by expressing them as equalities instead of inequalities. The sense of each constraint is indicated by arrows pointing in the appropriate

Graphical solution for Example 1.

9.1

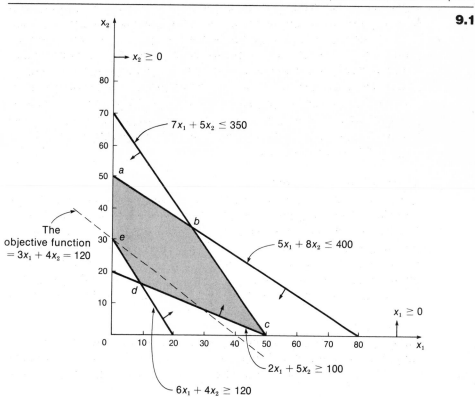

244   direction. The feasible region is the shaded area defined as *abcde* in the figure. Any combination of x_1 and x_2 in that region is an acceptable (feasible) solution, although not necessarily the optimum solution.

Setting $Z = 120$ arbitrarily, the objective function is drawn as shown by the dotted line. When that line moves away from the origin, the last point in the feasible region through which it passes is *b*. Therefore, the optimal solution is at point *b*, the intersection of the first two constraints. The values of x_1 and x_2 can either be read directly from the graph or be determined by solving the two constraint equations simultaneously.

$$7x_1 + 5x_2 = 350$$
$$5x_1 + 8x_2 = 400$$

This gives the solution as

$$x_1 = 25.81$$
$$x_2 = 33.87$$

The solution above results in the optimal value of

$$25.81\,(3) + 33.87\,(4) = 212.91$$

Optimality can be verified by evaluating the objective function at each extreme point of the feasible region as shown in Table 9.1.

The optimum solution for a given set of constraints depends on the ratio of the coefficients in the objective function. For

$$Z = c_1 x_1 + c_2 x_2$$

the ratio $r = c_1/c_2$ is the negative of the slope of the objective function, as shown in Fig. 9.2.

The ratios of the coefficient of x_1 to the coefficient of x_2 in the example are given in Table 9.2.

The effect of a change in the ratio r on the optimum solution will be as follows:

$$0 \le r < \frac{5}{8} \qquad \text{the solution is at point } a$$

Solutions of Example 9.1

9.1	POINT	x_1	x_2	Z
	a	0	50	200.00
	b	25.81	33.87	212.91 ← max
	c	50	0	150.00
	d	9.09	16.36	92.71
	e	0	30	120.00

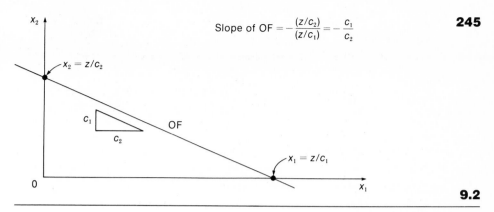

$$\text{Slope of OF} = -\frac{(z/c_2)}{(z/c_1)} = -\frac{c_1}{c_2}$$

9.2

Slope of objective function.

$r = \dfrac{5}{8}$ the solution is between points *a* and *b*

$\dfrac{5}{8} < r < \dfrac{7}{5}$ the solution is at point *b*

$r = \dfrac{7}{5}$ the solution is between points *b* and *c*

$\dfrac{7}{5} < r \leq \infty$ the solution is at point *c*

When the optimum solution is at an extreme point (corner) of the feasible region, it is called the *unique solution.* When it falls on a line segment, it indicates that there are *alternative solutions.*

Assuming that the objective function is modified as

$$\max Z = 5x_1 + 8x_2$$

Slope of Objective Function and Constraints

9.2

	COEFFICIENT OF		RATIO OF COEFFICIENTS
	x_1	x_2	
Objective function	3	4	3/4 ← *r*
Constraint 1	7	5	7/5
Constraint 2	5	8	5/8
Constraint 3	2	5	2/5
Constraint 4	6	4	6/4
Constraint 5	1	0	∞
Constraint 6	0	1	0

246 its value will be

$$5(25.81) + 8(33.87) = 400 \qquad \text{at point } b \text{ and}$$

$$5(0) + 8(50) = 400 \qquad \text{at point } a$$

The same value will also be obtained at any point on the straight line connecting those two points, even though the values of x_1 and x_2 will vary. These will be the alternative optima. Note that any point on a straight-line segment can be represented as a linear combination of the two end points of that line.

Since the optimal value remains unchanged on the line segment ab, we can say that any point p between a and b results in the optimal solution

$$p = \lambda a + (1 - \lambda)b \qquad \text{for } 0 \le \lambda \le 1$$

Point p can be expressed as

$$(x_1^{(p)}, x_2^{(p)}) = \lambda(x_1^{(a)}, x_2^{(a)}) + (1 - \lambda)(x_1^{(b)}, x_2^{(b)})$$

Thus

$$x_1^{(p)} = \lambda(0) + (1 - \lambda)(25.81) = 25.81 - 25.81\lambda$$

$$x_2^{(p)} = \lambda(50) + (1 - \lambda)(33.87) = 33.87 + 16.13\lambda$$

The value of the objective function at point p is

$$z_p = 5x_1^{(p)} + 8x_2^{(p)} = 5(25.81 - 25.81\lambda) + 8(33.87 + 16.13\lambda)$$
$$= 400 - 0\lambda = 400 \qquad \text{regardless of the value of } \lambda$$

MINIMIZATION PROBLEM

If this were a minimization problem, i.e., if the objective function were

$$\min Z = 3x_1 + 4x_2$$

the optimum solution would be found at the corner of the feasible region closest to the origin. By sliding the dotted line in Fig. 9.1 toward the origin, we can obtain the minimum solution at point d, where $x_1 = 9.09$, $x_2 = 16.36$, and $Z = 92.71$.

The optimum solution will again be different as the ratio of the coefficients of the decision variables in the objective function changes.

The effect of r, the ratio of c_1/c_2, on the minimum solution is as follows:

$$0 \le r < \frac{2}{5} \qquad \text{the solution is to point } c$$

$$r = \frac{2}{5} \qquad \text{the solution is between points } c \text{ and } d$$

$$\frac{2}{5} < r < \frac{6}{4} \qquad \text{the solution is at point } d$$

$$r = \frac{6}{4} \qquad \text{the solution is between points } d \text{ and } e$$

$$\frac{6}{4} < r < \infty \qquad \text{the solution is at point } e$$

SIGN OF c_j

So far, the behavior of the LP problem has been discussed only with nonnegative cost coefficients c_j in the objective function. The non-negativity of c_j is not a requirement. The decision variables x_j are limited to nonnegative values; their unit contributions to the objective function could be zero, negative, or positive. With both c_j non-negative, we showed that the maximum solution is obtained by sliding the objective function as far away from the origin as possible. Since both variables have positive contributions, the maximum combination is obtained when they both reach their highest possible values without violating the constraints. Conversely, the objective function attains its minimum value when both variables are at their minimum points, thus at a point as close to the origin as possible.

When c_j takes on negative values, the above discussion needs modification. The location of the optimum point under various combinations of unit costs is shown in Table 9.3. It should be noted that in Table 9.3, the optimum solution of the maximization problem has the same location as the optimum solution of the negative of the minimization problem. Formally stated:

$$Z_{\max} = \max \sum_{j=1}^{n} c_j x_j = -Z_{\min} = -\min \sum_{j=1}^{n} -c_j x_j$$

In the example, if the objective function is changed to

$$\min Z = -3x_1 - 4x_2$$

the optimum solution will still be at

$$x_1 = 25.81, \ x_2 = 33.87$$

with the value of the objective function equal to $Z_{\min} = -212.91$.

ADDITIONAL CONSTRAINTS

The addition of more constraints to a problem generally has a detrimental effect on the value of the objective function. Thus, the current solution would be an upper bound if new constraints were to be added to the example problem, i.e.,

$$Z_{\max}^{\text{more constraints}} \leq Z_{\max}^{\text{fewer constraints}}$$

and

$$Z_{\min}^{\text{more constraints}} \geq Z_{\min}^{\text{fewer constraints}}$$

Location of the Optimum Point

9.3

MAXIMIZATION PROBLEM		OPTIMUM POINT	MINIMIZATION PROBLEM	
$Z = \max c_1x_1 + c_2x_2$			$Z = \min c_1x_1 + c_2x_r$	
SIGN OF c_1	SIGN OF c_2		SIGN OF c_1	SIGN OF c_2
+	+	max x_1 and x_2 Farthest away from the origin	−	−
+	−	max x_1 and min x_2	−	+
−	+	min x_1 and max x_2	+	−
−	−	min x_1 and x_2 Closest to the origin	+	+

Suppose the following constraint is added:

$$x_1 + x_2 \leq 30$$

The graphical representation will be as shown in Fig. 9.4. The new feasible region is much smaller than before, as shown by the shaded area. The optimum solution is now at point ϵ and the maximum value of Z is 120.0, which is less than the maximum value of the original problem.

Suppose that the problem is not to be restricted by $x_1 + x_2 \leq 30$, but by two less stringent constraints:

$$x_1 \leq 30$$
$$x_2 \leq 30$$

The graphical representation will be as shown in Fig. 9.3. The new feasible region in this case is shown as *eihgd*. The solutions at points *h* and *i* are given in Table 9.4.

The original optimal point *b* is no longer a feasible point. The maximum value is obtained at point *i*, which results in a smaller Z_{max} than before.

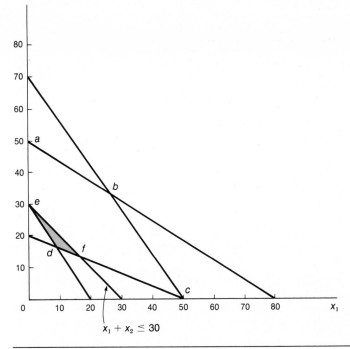

9.3

Example 1 with additional constraint.

Example 1 with two additional constraints.

9.4

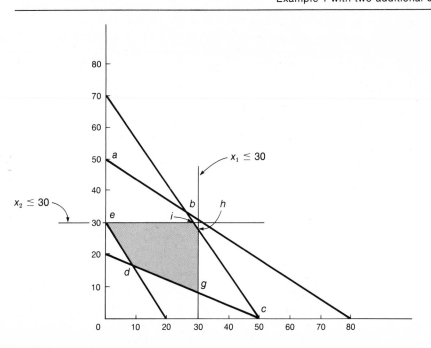

Solutions at New Extreme Points h and i

9.4	POINT	x_1	x_2	Z
	h	30	28	202.00
	i	28.57	30	205.71

In both of these cases, the additional constraints were "active," i.e., the optimal solution was located on the new constraints. This was the reason for the change in the optimal value of the objective function.

Sometimes a new constraint can be added without affecting the solution of the problem. For example, if the inequality in expression (9.4) is reversed, the constraint $x_1 + x_2 \geq 30$ will not alter the optimum solution of the original problem because it will not be active, i.e., the optimum solution will still be at point b and will still result in $Z_{max} = 212.91$.

Some constraints remain outside the feasible region. They do not contain any usable information about the problem because the solution is already restricted to the region defined by the other constraints. They are called "redundant" constraints and can be

Inactive constraint.

9.5

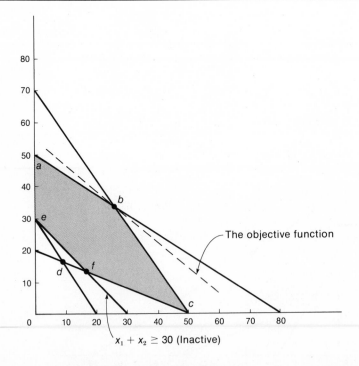

The objective function

$x_1 + x_2 \geq 30$ (Inactive)

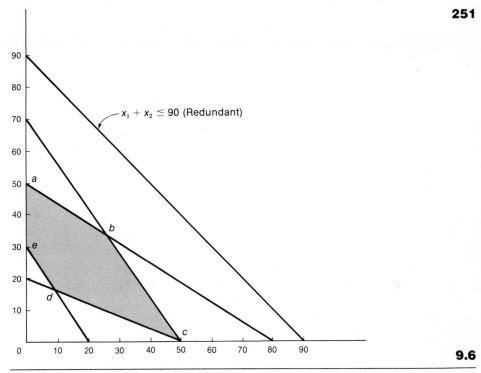

90

80

$x_1 + x_2 \leq 90$ (Redundant)

70

60

50 *a*

40

b

30 *e*

20

d

10

c

0   10   20   30   40   50   60   70   80   90

9.6

Redundant constraint.

dropped from the problem formulation. For exampᵢᵤ, consider the constraint

$$x_1 + x_2 \leq 90$$

in the example problem. The new constraint is outside the feasible region *abcde*, as shown in Fig. 9.6. Its elimination has no effect on the problem at all.

INFEASIBLE PROBLEM
If constraints are inconsistent in a problem, no feasible region may exist for a solution. These problems are called "infeasible." In such cases it is necessary to check the problem formulation and to remove the inconsistencies.

Suppose in the example problem the inequalities are reversed in all four constraints:

$$7x_1 + 5x_2 \geq 350$$
$$5x_1 + 8x_2 \geq 400$$
$$2x_1 + 5x_2 \leq 100$$
$$6x_1 + 4x_2 \leq 120$$

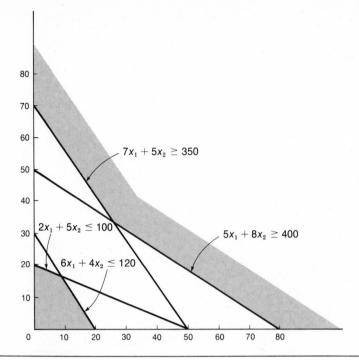

9.7

Infeasible problem.

The graphical representation is shown in Fig. 9.7. Notice that no feasible region exists and no solution can be found for the problem because there is no possible way to satisfy all the constraints.

UNBOUNDED PROBLEM
Another error in the formulation of the problem could lead to the "unbounded" case. The objective function can reach $+\infty$ in maximization and $-\infty$ in minimization problems if no bounds are defined for the resources. This case is illustrated in Fig. 9.8, which uses only the first two constraints of the infeasible case. Since both x_1 and x_2 can increase indefinitely, the optimum solution would be obtained at

$$x_1 = \infty$$

$$x_2 = \infty$$

$$Z_{max} = \infty$$

Obviously, there is no unbounded problem in reality. Limits are always imposed on the resources. An unbounded case is the indication of an error that needs to be corrected by a careful examination of the model.

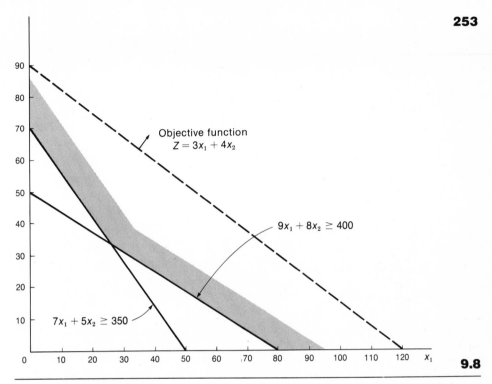

9.8

Unbounded case.

GENERAL SOLUTION PROCEDURE

The graphical solution presented in the previous section is a good learning tool because of its pictorial display of the behavior of LP problems. Its use is limited, however, because of the low dimensionality of problems to which it can be applied. In reality, problems with only two decision variables do not even require the LP technique, since many of them can be solved by observation or simple calculations.

The general solution procedure for large LP problems is called the "simplex method." It can solve any size problem with a systematic search, moving from one extreme point of the convex set of solutions to another point, always improving the solution, thus eliminating the need to study all possible solutions.

Some operations on the LP problem may be necessary in order to put it into the simplex format. The most important operations are explained below.

CHANGING THE SENSE OF THE OPTIMIZATION
A maximization problem can be converted to a minimization problem by multiplying the objective function by -1. The optimum solution

254   vector (values of x_j) will not change with this operation, but the value of the objective function will reverse its sign, i.e.,

$$\max \sum_{j=1}^{n} c_j x_j = -\min \sum_{j=1}^{n} (-c_j) x_j$$

CHANGING THE SENSE OF A CONSTRAINT

The sense of inequality is reversed if a constraint is multiplied by -1, i.e.,

$$\sum_{j=1}^{n} a_{ij} x_j \leq b_i \doteq \sum_{j=1}^{n} (-a_{ij}) x_{ij} \geq -b_i$$

For example, the inequalities

$$2x_1 + 3x_2 - x_3 \geq -4 \qquad \text{and} \qquad -2x_1 - 3x_2 + x_3 \leq 4$$

are equivalent.

CONVERTING INEQUALITIES TO EQUALITIES

The basis for the simplex algorithm is the solution of a set of simultaneous linear equations. Since many of the constraints are in the form of inequalities, they need to be converted to equalities before the solution process can start.

An inequality can be transformed into an equality by adding or subtracting a nonnegative quantity from it as follows:

$$\sum_{j=1}^{n} a_{ij} x_j \leq b_i \doteq \sum_{j=1}^{n} a_{ij} x_j + s_i = b_i \qquad (9.5)$$

and

$$\sum_{j=1}^{n} a_{ij} x_j \geq b_j \doteq \sum_{j=1}^{n} a_{ij} x_j - s_i = b_i \qquad (9.6)$$

Notice that the inequality in expression (9.5) indicates that the system has a resource at a certain level b_i but it may or may not be completely used up. In other words, the optimum solution may require a smaller amount of that resource than the total available. Thus a *slack* could exist in the system. s_i represents that slack, and it is called the "slack variable." If the resource is completely used up, s_i takes on a zero value; if there is an unused amount of b_i, then the slack variable has a positive value. On the other hand, if the optimum solution satisfies a constraint at more than the minimum level, the system has a *surplus* in it. In that case, s_i in expression (9.6) is called the "surplus variable," and gets a positive value. If expression (9.6) is satisfied at the minimum acceptable level, there will be no surplus and s_i will be equal to zero.

Slack and surplus variables are added to the LP model as regular variables, since they satisfy the nonnegativity condition stipulated for the decision variables. If they appear in the final solution, they indicate unused capacities and extra outputs.

Going back to Example 9.1, the constraints can all be converted to equalities by adding a new variable to each one as follows:

$$7x_1 + 5x_2 + x_3 = 350$$
$$5x_1 + 8x_2 + x_4 = 400$$
$$2x_1 + 5x_2 - x_5 = 100$$
$$6x_1 + 4x_2 - x_6 = 120$$

In this formulation x_3 and x_4 are slack variables and x_5 and x_6 are surplus variables.

CONVERTING NEGATIVE VARIABLES TO NONNEGATIVE VARIABLES

Linear programming formulations are restricted to nonnegative variables. However, in some cases a variable may have to be defined as a negative quantity. When this situation occurs, that variable is replaced by a new, nonnegative variable. This is accomplished by defining

$$x_j' = -x_j \qquad \text{for negative } x_j$$

where $x_j \leq 0$

$$x_j' \geq 0$$

Suppose x_1 in the previous example is a negative variable. The model is modified by defining $x_1' = -x_1$ and changing the original formulation as follows:

$$\max Z = -3x_1' + 4x_2$$

subject to

$$-7x_1' + 5x_2 \leq 350$$
$$-5x_1' + 8x_2 \leq 400$$
$$-2x_1' + 5x_2 \geq 100$$
$$-6x_1' + 4x_2 \geq 120$$
$$x_1', x_2 \geq 0$$

After the optimal solution is obtained, the value of x_1 is determined as

$$x_1 = -x_1'$$

Thus, if x_1' is in the final solution at a positive level, it indicates the existence of x_1 at a negative level in the system.

256 CONVERTING NONRESTRICTED VARIABLES
TO NONNEGATIVE VARIABLES

In some cases it may be desirable to define a variable without restricting it to nonnegative or nonpositive values. This occurs when a variable is used to indicate both profit (at positive level) and loss (at negative level) or both incoming amounts (positive) and outgoing amounts (negative).

There are several methods to convert such a variable to a nonnegative variable. The most common method is to define it as the difference between two nonnegative variables.

If x_j is nonrestricted, it is replaced by $x_j' - x_j''$ everywhere it appears in the model. The problem is solved with both of the new variables being nonnegative. For example, suppose x_2 is a nonrestricted variable in Example 9.1. The revised model will be as follows:

$$\max Z = 3x_1 + 4x_2' - 4x_2''$$

subject to

$$7x_1 + 5x_2' - 5x_2'' \leq 350$$
$$5x_1 + 8x_2' - 8x_2'' \leq 400$$
$$2x_1 + 5x_2' - 5x_2'' \geq 100$$
$$6x_1 + 4x_2' - 4x_2'' \geq 120$$
$$x_1, x_2', x_2'' \geq 0$$

It should be pointed out that the two variables x_2' and x_2'' in the model above are not linearly independent. The vectors

$$\begin{bmatrix} 4 \\ 5 \\ 8 \\ 5 \\ 4 \end{bmatrix} \quad \text{and} \quad \begin{bmatrix} -4 \\ -5 \\ -8 \\ -5 \\ -4 \end{bmatrix}$$

consisting of the coefficients of x_2' and x_2'' in the objective function, and the constraints are parallel. However, according to a basic theorem of linear programming, the solutions of the problem contain only linearly independent vectors.[6] It follows, then, that at most one of those variables could appear in the solution at a positive level, with the other being equal to zero. After the solution is obtained, the value of the original variable x_2 is determined as follows: Since $x_2 = x_2' - x_2''$:

If $x_2' = 0$	and	$x_2'' = 0$	then $x_2 = 0$
If $x_2' > 0$	and	$x_2'' = 0$	then $x_2 > 0$
If $x_2' = 0$	and	$x_2'' > 0$	then $x_2 < 0$

---

[6]For proof of this theorem, see Hadley or Gass, both listed in the Bibliography at the end of this chapter.

This is a convenient way of converting a nonrestricted variable to a **257** nonnegative variable, but it should be pointed out that *one* additional variable is introduced into the model for *each* nonrestricted variable.

STANDARD FORM OF LP

Using the operations discussed above, one can convert all LP formulations to the *standard form.* It is a representation of the LP model in such a way that all constraints are expressed as equations, all variables are nonnegative, and all right-hand-side constraints are also nonnegative. The objective function could be the maximization or the minimization type. Standard form is a convenient representation of the model for the simplex procedure.

THE SIMPLEX METHOD

After the LP problem is formulated and modified as necessary, it is solved by an iterative procedure that finds successive, improved feasible solutions and arrives at the optimum feasible solution in a finite number of steps if such a solution exists. This procedure, based on the solution of simultaneous linear equations, is called the "simplex method."

Since the number of variables n is not equal to the number of constraints m, the solution at each step sets $n - m$ variables equal to zero and solves the m equations for the remaining m variables. These are called the *basic variables;* and the solution vector consisting of the basic variables is called the *basis.* Each basis represents an extreme point of the feasible region in the multidimensional space. The algorithm moves from one basis to a neighboring basis if such a move improves the solution without violating any of the constraints. The algorithm stops when it reaches a solution that cannot be improved upon. It is the optimal solution. As long as the constraints have not been violated, it is the *optimal feasible solution.*

INITIAL SOLUTION

The algorithm needs an initial solution to start the procedure. Such a solution can sometimes be obtained by observation, but it is very inefficient to rely on observations. Instead, the origin ($x_j = 0$; all x_j) is selected as the initial solution.

The origin does not necessarily represent a feasible solution for the problem, especially if some of the constraints are equalities and the "greater than" ($\geq$) type. Furthermore, since all the decision variables are equal to zero at the origin, the initial solution does not contain any decision variables at all. Under these circumstances, the initial solution is meaningless for the original formulation. It becomes meaningful only if we modify the problem, solve the modified version first, and then go back to the original problem. This modification is

258   done by augmenting the original problem with an identity matrix[7] consisting of *slack* and *artificial* variables. Slack variables were explained earlier. They have a physical meaning in the problem, and they constitute a part of the identity matrix with their $+1$ coefficients. The surplus variables are similar, but their coefficients are -1, instead of $+1$; thus they do not belong to the identity matrix. In order to complete the identity matrix, an *artificial variable* is added to each constraint which does not contain a surplus variable.

Unlike the slack and surplus variables, the artificial variables have no physical interpretation. They are used merely as a convenience to initiate the solution process. They appear in all constraints expressed as equalities or in the ($\geq$) form in the original problem.

As an example, consider the following constraints:

$$x_1 + x_2 - x_3 \leq 5 \tag{9.7}$$

$$2x_1 - 5x_2 + 3x_3 \geq 10 \tag{9.8}$$

$$3x_1 + 2x_2 + 5x_3 = 8 \tag{9.9}$$

First, they are converted to equalities by adding a slack variable x_4 to expression (9.7) and subtracting a surplus variable x_5 from expression (9.8).

$$x_1 + x_2 - x_3 + x_4 \qquad = 5 \tag{9.10}$$

$$2x_1 - 5x_2 + 3x_3 \qquad - x_5 = 10 \tag{9.11}$$

$$3x_1 + 2x_2 + 5x_3 \qquad = 8 \tag{9.12}$$

Notice that the augmented part consists of x_4 and x_5 but it is not enough for an identity matrix yet.

Next, artificial variables x_6 and x_7 are added to expressions (9.11) and (9.12), respectively:

$$x_1 + x_2 - x_3 + x_4 \qquad\qquad = 5$$

$$2x_1 - 5x_2 + 3x_3 \qquad - x_5 + x_6 \qquad = 10$$

$$3x_1 + 2x_2 + 5x_3 \qquad\qquad + x_7 = 8$$

In this formulation, x_4, x_6, and x_7 form an identity matrix as shown in Table 9.5.

The augmented problem has seven unknowns and three equations. By letting $x_j = 0$ for $j = 1, 2, 3$, and 5, we can obtain the initial solution as

$$x_4 = 5$$

$$x_6 = 10$$

$$x_7 = 8$$

Since these variables have nonnegative values, we now have

---

[7]An identity matrix is a square matrix whose diagonal elements are $+1$; all other elements are zero.

Constraint Coefficients in the Augmented
Problem

9.5

x_1	x_2	x_3	x_5	x_4	x_6	x_7
1	1	−1	0	1	0	0
2	−5	3	−1	0	1	0
3	2	5	0	0	0	1

a feasible solution for the augmented problem but not necessarily for the original problem.

When new variables are added to the constraints, they also have to be added to the objective function. The costs (relative contributions to the objective function) associated with the slack and surplus variables are zero. The costs of the decision variables typically include all unit costs (or profits) involved in the system, including the cost of idle resources. Consequently, the cost of an unused resource or an excess capacity does not have to be considered again when the slack-surplus variables are used.

The artificial variables also do not have any cost associated with them. They are included in the augmented model only as a mathematical convenience and they are not intended to remain in the solution. It is desirable to get rid of them as soon as possible. There are two approaches to remove them from the solution.

1   The big *M* method, which assigns a penalty function to the artificial variables and forces them out of the solution in the early iterations.

2   The two-phase simplex method, which first minimizes the sum of artificial variables, then solves the original problem after all artificial variables are reduced to zero.

We will discuss the big *M* method in this book. In this method, the penalty associated with each artificial variable is a large cost in the opposite sense to the objective of the problem. For example, if the objective is to maximize, the cost of each artificial variable is a very large negative value ($-M$); if we are minimizing, the cost is a very large positive value ($+M$). This mechanism assures the removal of those variables from the solution as soon as possible.

In manual calculations, the cost M can be used without a numerical value, recognizing its high value throughout the solution process. In computer codes, M is generally given a value as high as 1000 times the maximum cost of the decision variables.

Going back to Example 9.1, the augmented problem will be:

$$\max Z = 3x_1 + 4x_2 + 0x_3 + 0x_4 + 0x_5 - Mx_6 + 0x_7 - Mx_8$$

260 Subject to

$$7x_1 + 5x_2 + x_3 = 350$$
$$5x_1 + 8x_2 + x_4 = 400$$
$$2x_1 + 5x_2 - x_5 + x_6 = 100$$
$$6x_1 + 4x_2 - x_7 + x_8 = 120$$
$$x_1, \ldots x_8 \geq 0$$

THE SIMPLEX TABLEAU

The linear programming problem is represented in tabular form at each iteration of the simplex algorithm. The elements of the simplex tableau are shown in Table 9.6 and defined below.

The Elements of the Simplex Tableau

9.6

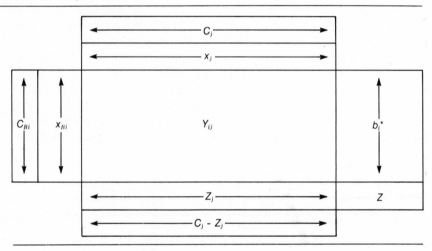

x_j = variables

C_j = relative contribution of X_j to the objective function (the unit "cost" of variable X_j)

x_{Bi} = ith basic variable (variable in the solution in this iteration)

C_{Bi} = unit "cost" of the ith basic variable

b_i^* = quantity of the ith basic variable in the solution in this iteration

Y_{ij} = rate of substitution between X_j and X_{Bi}; i.e., the amount of X_{Bi} to be removed from the solution when one unit of X_j is introduced without violating the feasibility conditions

Z = value of the objective function at this iteration, calculated as

$$Z = \sum_{i=1}^{m} C_{Bi} \, b_i^*$$

$Z_j =$ "loss" in the value of the objective function caused by the removal of X_{Bi} from the solution due to the introduction of one unit of X_j, calculated as

$$Z_j = \sum_{i=1}^{m} C_{Bi} \, Y_{ij} \qquad \text{for each } j$$

$C_j - Z_j =$ difference between the unit contribution of X_j to the objective function and the loss caused by the introduction of one unit of X_j into the solution; i.e., the *net contribution* of one unit of X_j to the objective function

INITIAL TABLEAU In the initial tableau, X_{Bi} are the slack and artificial variables, Y_{ij} are the same as the a_{ij} coefficients of the constraints, the b_i^* are the right-hand-side constants, b_i, as shown below.

The Initial Tableau

C_{Bi}	Basis	C_j	3	4	0	0	0	$-M$	0	$-M$	
			x_1	x_2	x_3	x_4	x_5	x_6	x_7	x_8	b_i^*
0	x_3		7	5	1	0	0	0	0	0	350
0	x_4		5	8	0	1	0	0	0	0	400
$-M$	x_6		2	5	0	0	-1	1	0	0	100
$-M$	x_8		6	4	0	0	0	0	-1	1	120
	Z_j		$-8M$	$-9M$	0	0	M	$-M$	M	$-M$	$-220M$
	$C_j - Z_j$		3 $+8M$	4 $+9M$	0	0	$-M$	0	$-M$	0	

9.7

The solution vector in the initial tableau consists of x_3, x_4, x_6, and x_8, the variables in the identity matrix embedded in the augmented problem. Since all decision variables are at zero level, the solution is at the origin.

The $C_j - Z_j$ row indicates the net contribution of variable x_j to the objective function when it is introduced into the solution. Since this is a maximization problem, the variable x_j for which $C_j - Z_j > 0$ is a candidate for the improvement of the solution. It shows that the inclusion of one unit of that variable in the next basis will result in a net increase in the value of the objective function.

OPTIMAL TABLEAU The $C_j - Z_j$ values are studied at the end of each iteration.

If all $C_j - Z_j \leq 0$, the maximum has been reached.

If some $C_j - Z_j = 0$ for x_j not in the optimal basis, there are alternative optimum solutions.

If one or more $C_j - Z_j > 0$, and at least one column j^* for

262   which $C_j{}^* - Z_j{}^* > 0$, has all $Y_{ij}{}^* \leq 0$, the problem is un-bounded.

NONOPTIMAL TABLEAU   If, at any iteration, one or more $C_j - Z_j > 0$ and each column with this condition has $Y_{ij} > 0$ for at least one row i, the solution can be improved. In that case, the variable x_k to enter the *basis* is selected as $C_k - Z_k = \max(C_j - Z_j)$. If k is not unique, an arbitrary selection is made.

Since $C_j - Z_j > 0$ indicates unit contribution of x_j to the objective function, we want to include as many units of x_j as possible in the solution without violating the constraints. Notice that we have $b_i{}^*$ amount of each basic variable x_{Bi} in the solution, and we have to remove Y_{ij} amount from it when we introduce one unit of x_j. Therefore, for each variable x_j, we can calculate the maximum permissible amount without causing any of the current basic variables to be negative, as follows: $b_i{}^*/Y_{ij}$ is the maximum amount of x_j that can be substituted for x_{Bi}. Additional units of x_j in the solution require more resources than available, thus forcing the value of some of the variables to be negative.

This observation leads to the rule for the selection of the basic variable that will be removed from the solution. Variable r is selected as

$$\frac{b_r}{Y_{rk}} = \min \frac{b_i{}^*}{Y_{ik}} \qquad \text{for } Y_{ik} > 0$$

If r is not unique, the problem is *degenerate*. It indicates that some of the variables in the next solution will be at zero level. Degeneracy could potentially lead to continuous cycling without ever achieving the optimum solution. Although this could theoretically lead to a major problem, it has never occurred in actual situations. When the minimum ratio $b_i{}^*/Y_{ik}$ is the same for several basic variables, the first one can be selected to leave the basis.[8]
(Note: The discussion above of the optimal and nonoptimal tableaus applies to maximization problems, but the same rules apply also to minimization problems by reversing the inequality in the $C_j - Z_j$ values. In other words, the rules given above are converted to $C_j - Z_j < 0$ or ≤ 0 when $>$ or $\geq$ is indicated.)

IMPROVEMENT OF THE SOLUTION
The procedure for the selection of variables entering and leaving the basis in each iteration can be summarized as two rules.

1   The nonbasic variable that shows the maximum contribution to the objective function is chosen to enter the basis as follows:

---

[8] For a discussion of degeneracy and the mathematically sound tie breaker, see Hadley, cited in the end-of-chapter Bibliography.

For x_k to enter:

$$C_k - Z_k = \max \ (C_j - Z_j) \qquad C_j - Z_j > 0 \text{ in a maximization problem}$$

or

$$C_k - Z_k = \min \ (C_j - Z_j) \qquad C_j - Z_j < 0 \text{ in a minimization problem}$$

2   The basic variable that will be depleted before any other basic variable leaves the basis is chosen as follows:

For x_{Br} to leave:

$$\frac{b_i^*}{Y_{rk}} = \min\frac{b_i^*}{Y_{ik}}; \qquad Y_{ik} > 0 \text{ in both maximization and minimization problems}$$

Using rule 1 in the example, x_2 is chosen as the new basic variable. Note that both x_1 and x_2 will have net positive contributions to the objective function. One unit of x_1 will increase Z by $3 + 8M$; one unit of x_2 will increase Z by $4 + 9M$. x_2 is selected because its unit contribution is higher.

Using rule 2, the maximum amount x_2 in the next iteration is calculated as follows:

Change of Basis

9.8

C_{Bi}	Basis	x_1	x_2	x_3	x_4	x_5	x_6	x_7	x_8	b_i^*	b_i^*/Y_{ik}
			4								
	x_3		5							350	$350/5 = 70$
	x_4		8							400	$400/8 = 50$
$-M$	x_6		5							100	$100/5 = 20 \longleftarrow r$
	x_8		4							120	$120/4 = 30$
	Z_j		$-9M$							$-220M$	
	$C_j - Z_j$		$4+9M$								

k

The minimum b_i^*/Y_{ik} is 20, corresponding to the basic variable x_6. If an attempt is made to put more than 20 units of x_2 into the next solution, it will be necessary to remove more than 100 units of x_6 from the basis. Thus the value of x_6 will be negative. Since this will violate the feasibility condition, x_2 in the subsequent tableau will have to be limited to 20. This change will reduce x_6 to 0, change the amount of x_3, x_4, and x_8 in the solution, and increase the value of the objective function by an amount equal to $20 (4 + 9M) = 80 + 180M$.

The identity matrix will now consist of x_3, x_4, x_2, and x_8, and

264   the third constraint will play the key role in the transformation for the first iteration.

We can obtain the tableau at iteration 1 by solving the four equations for x_3, x_4, x_2, and x_8. This is done by:

1   Solving the third constraint for x_2

2   Substituting the expression that is obtained for x_2 in step 1 into constraint equations 1, 2, and 4.

The results are summarized as follows for the elements of the new tableau:
In column k:

$$Y_{rk}^{new} = 1$$

$$Y_{ik}^{new} = 0 \qquad \text{for all } i \neq r$$

In row r:

$$Y_{rj}^{new} = \frac{Y_{rj}^{old}}{Y_{rk}^{old}}$$

The other rows

$$Y_{ij}^{new} = Y_{ij}^{old} - \frac{Y_{rj}^{old}}{Y_{rk}^{old}} Y_{ik}^{old}$$

$$(C_j - Z_j)^{new} = (C_j - Z_j)^{old} - \frac{(C_k - Z_k)^{old}}{Y_{rk}^{old}} Y_{rj}^{old}$$

$$Z^{new} = Z^{old} + \frac{(C_k - Z_k)^{old}}{Y_{rk}^{old}} b_r^{*old}$$

These transformations can be performed by completing a loop around each element of the tableau, as shown below:

Pivot Operations for the Change of Basis.

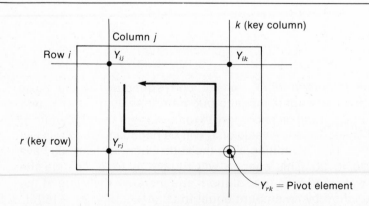

Tableau at Iteration 1

9.9

C_{Bi}	Basis	C_j	3	4	0	0	0	$-M$	0	$-M$		
			x_1	x_2	x_3	x_4	x_5	x_6	x_7	x_8	b_i^*	b_i^*/Y_{ik}
0	x_3		5	0	1	0	1	-1	0	0	250	50
0	x_4		9/5	0	0	1	8/5	$-8/5$	0	0	240	400/3
4	x_2		2/5	1	0	0	$-1/5$	1/5	0	0	20	50
$-M$	x_8		22/5	0	0	0	4/5	$-4/5$	-1	1	40	100/11
	Z_j		1.6	4	0	0	-0.8	0.8	M	$-M$	80	
			$-4.4M$				$-0.8M$	$+0.8M$			$-40M$	
	$C_j - Z_j$		1.4	0	0	0	0.8	-0.8	$-M$	0		
			$+4.4M$				$+0.8M$	$-1.8M$				

$r \longrightarrow$ points to x_8 row. k points to x_1 column.

The new Z_j are calculated after the $C_j - Z_j$ values are determined.

$$Z_j^{new} = C_j - (C_j - Z_j)^{new}$$

After these transformations are completed the tableau shown in Table 9.9 is obtained.

This is not the optimal tableau yet, as indicated by the positive values in the $C_j - Z_j$ row. x_1 is chosen to enter the basis, replacing x_8 in the next tableau.

The tableau in iteration 2 is obtained as shown in Table 9.10.

Tableau at Iteration 2

9.10

C_{Bi}	Basis	C_j	3	4	0	0	0	$-M$	0	$-M$		
			x_1	x_2	x_3	x_4	x_5	x_6	x_7	x_8	b_i^*	b_i^*/Y_{ik}
0	x_3		0	0	1	0	1/11	$-1/11$	25/22	$-25/22$	2250/11	2250
0	x_4		0	0	0	1	14/11	$-14/11$	9/22	$-9/22$	2460/11	1230/7
4	x_2		0	1	0	0	$-3/11$	3/11	10/11	$-10/11$	180/11	—
3	x_1		1	0	0	0	2/11	$-2/11$	$-5/22$	5/22	100/11	50
	Z_j		3	4	0	0	$-6/11$	6/11	$-65/22$	65/22	1020/11	
	$C_j - Z_j$		0	0	0	0	6/11	$-M$	65/22	$-M$		
								6/11		$-65/22$		

$r \longrightarrow$ points to x_1 row. k points to x_5 column.

Tableau at Iteration 3

9.11

C_{Bi}	Basis	C_j	3	4	0	0	0	$-M$	0	$-M$		
			x_1	x_2	x_3	x_4	x_5	x_6	x_7	x_8	b_i^*	b_i^*/y_{rk}
0	x_3		$-1/2$	0	1	0	0	0	$5/4$	$-5/4$	200	250
0	x_4		-7	0	0	1	0	0	2	-2	160	80
4	x_2		$3/2$	1	0	0	0	0	$-1/4$	$1/4$	30	—
0	x_5		$11/2$	0	0	0	1	-1	$-5/4$	$5/4$	50	—
	Z_j		6	4	0	0	0	0	-1	1	120	
	$C_j - Z_j$		-3	0	0	0	0	$-M$	1	$-M$ -1		

$r \longrightarrow$ (points to x_4 row)

k (points to x_7 column)

The artificial variables have been removed from the basis in the tableau in Table 9.10. Because of their high penalty, $-M$, they will not reenter any more. x_5 will be in the next basis and x_1 will leave. Note that the b_j^*/Y_{ik} value is negative. The tableau at iteration 3 is shown in Table 9.11.

This is still not the optimal tableau. Variable x_7 will replace x_4 in the next iteration, Table 9.12.

We have not reached the optimal solution yet, since the introduction of one unit of x_1 improves the objective function by 0.5 units.

The next basis consists of x_1, x_7, x_2, and x_5, as shown in Table 9.13.

Tableau at Iteration 4

9.12

C_{Bi}	Basis	C_j	3	4	0	0	0	$-M$	0	$-M$		
			x_1	x_2	x_3	x_4	x_5	x_6	x_7	x_8	b_i^*	b_i^*/Y_{ik}
0	x_3		$31/8$	0	1	$-5/8$	0	0	0	0	100	800/31
0	x_7		$-7/2$	0	0	$1/2$	0	0	1	-1	80	—
4	x_2		$5/8$	1	0	$1/8$	0	0	0	0	50	80
0	x_5		$9/8$	0	0	$5/8$	1	-1	0	0	150	400/3
	Z_j		$5/2$	4	0	$1/2$	0	0	0	0	200	
	$C_j - Z_j$		$1/2$	0	0	$-1/2$	0	$-M$	0	$-M$		

$r \longrightarrow$ (points to x_3 row)

k (points to x_1 column)

Optimal Tableau

9.13

		C_j	3	4	0	0	0	$-M$	0	$-M$	
C_{Bi}	Basis		x_1	x_2	x_3	x_4	x_5	x_6	x_7	x_8	b_i^*
3	x_1		1	0	8/31	$-5/31$	0	0	0	0	800/31
0	x_7		0	0	28/31	$-2/31$	0	0	1	-1	5280/31
4	x_2		0	1	$-5/31$	7/31	0	0	0	0	1050/31
0	x_5		0	0	$-9/31$	25/31	1	-1	0	0	3750/31
Z_j			3	4	4/31	13/31	0	0	0	0	6600/31
$C_j - Z_j$			0	0	$-4/31$	$-13/31$	0	$-M$	0	$-M$	

This is the optimal tableau because all $C_j - Z_j < 0$ for the nonbasic x_j, indicating that the inclusion of the nonbasic variables into the solution will result in a net decrease in the value of the objective function.

INTERPRETATION OF THE OPTIMAL TABLEAU
The final result is obtained in the optimal tableau as follows:

$$x_1 = \frac{800}{31} = 25.81$$

$$x_2 = \frac{1050}{31} = 33.87$$

$$x_5 = \frac{3750}{31} = 120.97$$

$$x_7 = \frac{5280}{31} = 170.32$$

$$x_3, x_4, x_6, x_8 = 0$$

$$Z_{optimal} = \frac{6600}{31} = 212.91$$

This is the same result as we obtained in the graphical solution previously. It indicates that the maximum profit of $212,910 can be realized if process 1 and process 2 are used 25.81 and 33.87 days, respectively. This is the best solution, even though it results in more production than the minimum needed. The positive values of x_5 and x_7 indicate that the demands have been oversatisfied. The result is realistic as long as the market can support this overproduction. The demand constraints expressed as inequalities ($\geq$) rather than equalities describe a market where the extra products can be sold at the same unit profit. Therefore, as long as the supplies last, the best decision is, indeed, to produce as much as possible.

It should be pointed out that the artificial variables never entered the basis once they left. They are equal to zero in the optimal

268   solution. This is what we expect. As explained earlier, the artificial variables have no physical meaning, and their presence in the final solution cannot be interpreted in terms of system characteristics. They remain in the final solution only if the problem is infeasible.

SHADOW PRICES

The Z_j associated with the original basic variables have a special economic meaning in the optimal tableau. By definition, Z_j represents the amount to be removed from the objective function in order to accommodate one unit of variable x_j in the solution.[9] Remember that the initial solution consisted of x_3, x_4, x_6, and x_8. The Z_j values of these variables in the optimal tableau are 4/31, 13/31, 0, and 0, respectively. Using the definition of Z_j, we conclude that if one unit of x_3 were introduced into the solution, it would reduce Z, the optimum value of the objective function, by 4/31 units. Keep in mind that the introduction of x_3 into the solution is the same as leaving some of the first raw material R_1 unused. Currently, x_3 equals zero in the optimal solution; therefore all of the available supply of R_1 is being utilized. This is indicated by the fact that x_3, the slack variable associated with the first constraint, is not in the basis. If one unit of unused resource R_1 reduces the objective function by 4/31 units, we could conclude that the opposite would also be true if an additional unit of R_1 were made available. This statement can be made because of the linearity assumption. Therefore, Z_3 represents the unit contribution of resource R_1 to the objective function. This is an important observation. It provides the decision maker with a major piece of information about the profitability of each resource. For example, if resource R_1 can be obtained at a premium price over and above the price paid for the original amount, the decision maker could still consider obtaining it as long as the additional payment per unit is less than 4/31.

A similar discussion can also be presented for x_4. In that case, x_4 is to be viewed in conjunction with the second raw material R_2, associated with the second constraint.

The Z_j associated with the other two initial basic variables, x_6 and x_8, are equal to zero. The reason for this is that the demands expressed in constraints 3 and 4 are already more than satisfied and the surplus variables related to those constraints are in the optimal solution. Thus a change in the demands will not affect the optimal value of the objective function as long as there is a surplus in the system. These values,

$$Z_3 = \frac{4}{31}$$

---

[9] $Z_j = \sum_{i=1}^{m} C_{Bi} Y_{ij}$ for all j, as defined earlier.

$$Z_4 = \frac{13}{31}$$

$$Z_6 = 0$$

$$Z_8 = 0$$

are the *shadow prices* of variables x_3, x_4, x_6, and x_8, respectively. They are a measure of the relative contribution of each resource (in this case, the supply and demand conditions) to the objective function.

This statement can be verified by checking the total contribution of the resources, as follows:

(1) RESOURCE i	(2) QUANTITY b_i	(3) Z_j ASSOCIATED WITH i	(2) × (3) CONTRIBUTION
R_1 (supply in constraint 1)	350	4/31	1400/31
R_2 (supply in constraint 2)	400	13/31	5200/31
O_1 (demand in constraint 3)	100	0	0
O_2 (demand in constraint 4)	120	0	0
Total			6600/31

Notice that the total contributions of the resources equal the optimal value of the objective function. Thus

$$Z^o = \sum_{i=1}^{m} C_{Bi}^o b_i^{*o} = \sum_{i=1}^{m} Z_{i,\,\text{init}}^o b_{i,\,\text{init}}$$

where $0 = $ values in optimal tableau
$B = $ basis
i, init $= $ variable i in initial basis
Stated differently:

$$\Delta Z^o = Z_{i,\,\text{init}}^o \Delta b_{i,\,\text{init}}$$

or

$$Z_{i,\,\text{init}}^o = \frac{\Delta Z^o}{\Delta b_{i,\,\text{init}}} \qquad (9.13)$$

Expression (9.13) shows that the shadow price Z_i^o of a resource i represents the rate of change in the objective function attributed to that resource.

The analysis of the optimal results gives significant insight into the system. The solution describes the best decisions under the assumed values of the parameters, but those values are only estimates in most cases. The actual costs c_j, resource levels b_i, and technological coefficients a_{ij} could be different from the values used

270   in the LP formulation. It is conceivable that an additional decision variable (for example, a new product) not included in the original problem may have to be considered, or more constraints may be discovered after the optimal solution is obtained. The decision maker wants to determine how sensitive the solution is in view of such changes—for example, how much a cost c_j can vary without affecting the optimal solution. One way to answer such questions is, of course, to solve the problem again, under the new conditions. This can always be done, but it requires a major effort to reformulate and solve the problem before a conclusion is reached. This is not the best approach. Alternatively, bounds can be placed on parameters to define a range for the solution without changing optimality. These bounds give the decision maker the flexibility he needs in his decisions. For example, he may conclude that, even though the initial parameters may not be accurate, the optimal decision will remain the same within foreseeable variations in those values. Or, he might see that the decision system is more sensitive to changes in some elements than in others. Such an interpretation is at least as important as the optimal solution itself in making decisions about the system.

We can conclude that the LP solution specifies the best decisions under the given conditions, but it is by no means a complete specification of "what can be done" with the system under study. A more complete understanding becomes available only after the sensitivity of the decision system is studied.

SENSITIVITY (POSTOPTIMALITY) ANALYSIS

Sensitivity analysis is the study of postoptimality conditions to determine the effects of changes in the values of the parameters. More specifically, it is a study to determine the amount by which each parameter can change without altering optimality.

The parameters of the LP problem are the c_j's, the b_i's, and the a_{ij}'s. This section will explain how a change in these values for basic and nonbasic variables could affect the optimal solution. The discussion will be presented for a maximization problem.

THE COSTS c_j

If the cost of a variable changes, it could change the optimal solution by replacing a nonbasic variable with a basic variable. In other words, the $C_j - Z_j$ values may no longer be all less than or equal to zero. The reason for this can be seen in the definition of $C_j - Z_j$:

$$C_j - Z_j = C_j - \sum_{i=1}^{m} C_{Bi} \, Y_{ij}$$

Thus,

$$C_j - Z_j = C_j - (C_{B1}, C_{B2}, \ldots C_{Bm}) \begin{bmatrix} Y_{1j} \\ Y_{2j} \\ \cdot \\ \cdot \\ \cdot \\ Y_{mj} \end{bmatrix}$$

CHANGE IN THE COST OF A NONBASIC VARIABLE   From the definition of $C_j - Z_j$ above, we can see that a change in the cost of a nonbasic variable will affect the $C_j - Z_j$ of that variable only. If the change is large enough to make $C_j - Z_j > 0$, that variable will be a candidate for a new basis, otherwise the optimal solution will not be changed.

Suppose the cost of x_j is changed from C_j to $C_j + P_j$, where P_j is any value (negative or positive). The optimal solution remains unaltered as long as

$$(C_j - Z_j)^{new} = (C_j + P_j) - Z_j \leq 0$$

Thus,

$$P_j \leq Z_j - C_j$$
$$P_j \leq - (C_j - Z_j)$$

Therefore, the optimal solution is not affected as long as the change in C_j is less than the negative of $C_j - Z_j$ for the nonbasic variable under consideration. Referring to the optimal tableau of the example problem, x_3 will not be a candidate to improve the optimal solutions if $P_3 \leq -(C_3 - Z_3)$; that is $P_3 \leq 4/31$. Since $C_3^{new} = C_3 + P_3$, and $C_3 = 0$, the optimal solution will not be affected if C_3^{new} remains in the range

$$-\infty \leq C_3^{new} \leq \frac{4}{31}$$

Similarly, the optimality will not be affected if C_4^{new} remains in the range

$$-\infty \leq C_4^{new} \leq \frac{13}{31}$$

C_6 and C_8 need not be studied, since they are associated with the artificial variables.

CHANGE IN THE COST OF A BASIC VARIABLE   From the definition of $C_j - Z_j$, it can also be seen that a change in the cost of a basic variable C_{Bi} will affect all $C_j - Z_j$ values.

In the example problem, suppose C_1 is changed by an amount ΔC_1. The following changes will occur in the optimal tableau (refer to Table 9.13).

$$C_1^{new} = 3 + \Delta C_1$$

$$(C_3 - Z_3)^{new} = \frac{-4}{31} - \left(\frac{8}{31}\right)(\Delta C_1)$$

$$(C_4 - Z_4)^{new} = \frac{-13}{31} + \left(\frac{5}{31}\right)(\Delta C_1)$$

$$Z^{new} = \frac{[6600 + 800\,(\Delta C_1)]}{31}$$

All other elements will remain the same. As a result of these changes, if any $(C_j - Z_j)$ becomes positive, x_1 will be removed from the solution and a new variable will enter. In order not to change the optimal solution mix, the following conditions have to be satisfied:

$$(C_3 - Z_3)^{new} = \frac{-4}{31} - \frac{8}{31}\,(\Delta C_1) \leq 0 \qquad (9.14)$$

$$(C_4 - Z_4)^{new} = \frac{-13}{31} + \frac{5}{31}\,(\Delta C_1) \leq 0 \qquad (9.15)$$

From expression (9.14):

$$-0.5 \leq \Delta C_1 \leq \infty \qquad \text{for } x_3 \text{ not to enter}$$

From expression (9.20):

$$-\infty \leq \Delta C_1 \leq 2.6 \qquad \text{for } x_4 \text{ not to enter}$$

Thus, as long as ΔC_1 is between -0.5 and 2.6, (or C_1 between 2.5 and 5.6) the optimal basis will remain as it is. If C_1 is less than 2.5, it will leave the basis and x_3 will enter. If C_1 is greater than 5.6, x_4 will replace it in the solution. These are summarized in Table 9.14.

Sensitivity of the Cost Vector

9.14	CHANGE IN C_1	NEW VALUE OF C_1	OPTIMAL BASIS	OPTIMAL Z
	$-0.5 < \Delta C_1 < 2.6$	$2.5 < C_1 < 5.6$	$x_1,\ x_2,\ x_5,\ x_7$	$\dfrac{6600 + 800\,(\Delta C_1)}{31}$
	$\Delta C_1 = -0.5$	$C_1 = 2.5$	$x_1,\ x_2,\ x_5,\ x_7$ or $x_3,\ x_2,\ x_5,\ x_7$ (alternative optima)	$\dfrac{6600 + 800\,(-0.5)}{31}$ $= 200$
	$\Delta C_1 = 2.6$	$C_1 = 5.6$	$x_1,\ x_2,\ x_5,\ x_7$ or $x_4,\ x_2,\ x_5,\ x_7$ (alternative optima)	$\dfrac{6600 + 800\,(2.6)}{31}$ $= 280$
	$\Delta C_1 < -0.5$	$C_1 < 2.5$	$x_3,\ x_2,\ x_5,\ x_7$	New value to be calculated
	$\Delta C_1 > 2.6$	$C_1 > 5.6$	$x_4,\ x_2,\ x_5,\ x_7$	New value to be calculated

If the optimal mix changes (as in $C_1 < 2.5$ or $C_1 > 5.6$), the **273** new optimal solution can be obtained by developing one or more postoptimal tableaus.

When changes in C_2, C_5, and C_7 are studied one at a time in a similar way, the following ranges are defined for the optimal basis to remain optimal.

$$2.5 \leq C_1 \leq 5.6$$

$$2.14 \leq C_2 \leq 4.8$$

$$-0.52 \leq C_5 \leq 0.44$$

$$-0.14 \leq C_7 \leq 6.5$$

Notice that these are the ranges for one cost parameter at a time. When multiple changes are considered, they should be studied simultaneously.

For example, if C_1 and C_2 are both changed by amounts ΔC_1 and ΔC_2, respectively, the new $C_j - Z_j$ values for x_3 and x_4 will depend on those changes as follows:

$$(C_3 - Z_3)^{\text{new}} = \frac{-4}{31} - \frac{8(\Delta C_1)}{31} + \frac{5(\Delta C_2)}{31} \tag{9.16}$$

$$(C_4 - Z_4)^{\text{new}} = \frac{-13}{31} + \frac{5(\Delta C_1)}{31} - \frac{7(\Delta C_2)}{31} \tag{9.17}$$

Expressions (9.16) and (9.17) are plotted in Fig. 9.10. The optimal basis remains optimal as long as the point $(\Delta C_1, \Delta C_2)$ falls into the shaded area.

For any pair of ΔC_1 and ΔC_2 in the shaded area, the new Z is calculated as follows:

$$Z^{\text{new}} = Z^0 + \sum_{i=1}^{m} C_{Bi}^0 \cdot b_i^{*o}$$

The optimal values indicated by the superscript (o) can be obtained from the optimal tableau shown on Table 9.13.

$$Z^{\text{new}} = \frac{6600}{31} + \left(\frac{800}{31}\right) \Delta C_1 + \left(\frac{5230}{31}\right) \Delta C_7 + \left(\frac{1050}{31}\right) \Delta C_2 + \left(\frac{3750}{31}\right) \Delta C_5$$

Since we are considering changes only in C_1 and C_2,

$$Z^{\text{new}} = \frac{6600}{31} + \left(\frac{800}{31}\right) \Delta C_1 + \left(\frac{1050}{31}\right) \Delta C_2$$

This analysis can be extended to any number of changes, but it is cumbersome to consider multiple changes simultaneously. It may be easier to solve the problem with new estimates if multiple changes are anticipated.

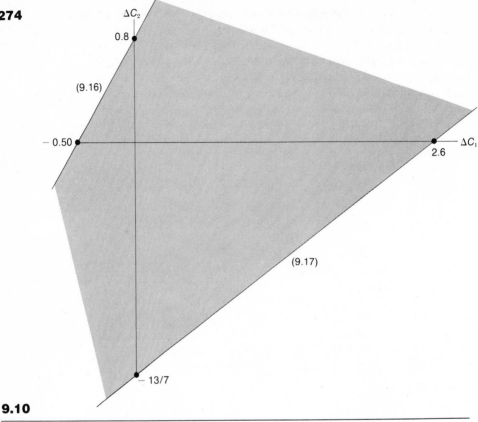

9.10

Effect of simultaneous changes in two cost coefficients.

CHANGES IN THE RIGHT-HAND-SIDE VALUES (THE REQUIRE-MENTS VECTOR **b**)

If a requirement level b_i is changed by an amount Δb_i, it may change
the feasibility of the problem. If the modification in b_i causes one or
more of the basic variables take on negative values in the optimum
tableau, a new optimum solution is needed. However, as long as the
values of the variables in the optimal solution remain nonnegative,
the optimal basis is unaltered even though their values and the optimum value of the objective function may be different.

CHANGE IN A REQUIREMENT b_i WHOSE SLACK OR SURPLUS VARI-ABLE IS IN THE OPTIMAL SOLUTION

If the slack (or surplus) vari-
able associated with a constraint is in the optimal basis, there is an
idle capacity of the resource requirement expressed in that constraint
(supply or demand in the example problem). If a slack variable is in
the solution, the excess amount can be removed from the system
without making the problem infeasible; in case of a surplus variable,

the respective constraint can be made more stringent by the amount equal to the optimal value of the surplus.

In our example, the two surplus variables in the solution are x_5 and x_7. They are associated with the demand constraints. Referring to Table 9.13, the optimal solution oversatisfies the demand by producing 3750/31 units more of O_1 than the minimum requirement of 100 units; similarly, 5280/31 units of O_2 are produced above the minimum required amount of 120 units. Thus the optimal solution will remain optimal even if the values of b_3 and b_4 are increased by these amounts.

$$-\infty \leq b_3^{new} \leq 100 + \frac{3750}{31} = \frac{6850}{31}$$

$$-\infty \leq b_4^{new} \leq 120 + \frac{5280}{31} = \frac{9000}{31}$$

Mathematically, this result can be obtained by using the relationship between the initial tableau and the other tableaus of the simplex algorithm.

Any column in a specific simplex tableau can be obtained by premultiplying the corresponding column in the initial tableau by the "inverse of the basis matrix," in that specific tableau.[10]

Therefore, if the inverse of the basis matrix, denoted by $\mathbf{B}^{-1}$ is available for a given tableau, all elements of that tableau, including the right-hand-side values b_i^*, can be determined directly from the original problem formulation.

Matrix inversion will not be discussed in this book. The interested reader is referred to books on linear algebra. It is interesting to note that detailed calculations are not necessary to obtain $\mathbf{B}^{-1}$, because it can be obtained directly from the simplex tableau itself. It is found under the initial basic variables in each tableau.

Since the initial basis consisted of x_3, x_4, x_6, and x_8 in the example, $\mathbf{B}^{-1}$ is the square (4×4) matrix under those variables in the optimal tableau.

$$\mathbf{B}^{-1} = \begin{bmatrix} \dfrac{8}{31} & \dfrac{-5}{31} & 0 & 0 \\[2mm] \dfrac{28}{31} & \dfrac{-2}{31} & 0 & -1 \\[2mm] \dfrac{-5}{31} & \dfrac{7}{31} & 0 & 0 \\[2mm] \dfrac{-9}{31} & \dfrac{25}{31} & -1 & 0 \end{bmatrix}$$

[10]For proof, see D. T. Phillips, A. Ravindran, and J. J. Solberg, *Operations Research* Wiley, New York, 1976.

Using $\mathbf{B}^{-1}$ and the original right-hand-side values b_i of 350, 400, 100, and 120, the b_i^* in the optimal tableau can be calculated as follows:

$$\left(\frac{8}{31}\right)(350) + \left(\frac{-5}{31}\right)(400) + 0 + 0 = \frac{800}{31}$$

$$\left(\frac{28}{31}\right)(350) + \left(\frac{-2}{31}\right)(400) + 0 + (-1)(120) = \frac{5280}{31}$$

$$\left(\frac{-5}{31}\right)(350) + \left(\frac{7}{31}\right)(400) + 0 + 0 = \frac{1050}{31}$$

$$\left(\frac{-9}{31}\right)(350) + \left(\frac{25}{31}\right)(400) + (-1)(100) + 0 = \frac{3750}{31}$$

The b_i^* are the same as those in the optimal tableau. If a change in the original b_i values is considered, the new $\mathbf{b}^*$ vector can be calculated by substituting the modified b_i into the calculations above.

For example, if b_3^{new} and b_4^{new} are used instead of 100 and 120, respectively, the new $\mathbf{b}^*$ vector becomes:

$$\mathbf{b}^{*\text{new}} = \mathbf{B}^{-1}\mathbf{b}^{\text{new}} = \begin{bmatrix} \frac{8}{31} & \frac{-5}{31} & 0 & 0 \\ \frac{28}{31} & \frac{-2}{31} & 0 & -1 \\ \frac{-5}{31} & \frac{7}{31} & 0 & 0 \\ \frac{-9}{31} & \frac{25}{31} & -1 & 0 \end{bmatrix} \begin{bmatrix} 350 \\ 400 \\ b_3^{\text{new}} \\ b_4^{\text{new}} \end{bmatrix} = \begin{bmatrix} \frac{800}{31} \\ \frac{9000}{31} - b_4^{\text{new}} \\ \frac{1050}{31} \\ \frac{6850}{31} - b_3^{\text{new}} \end{bmatrix}$$

Note that b_1^* and b_3^* are the same as before, but the b_2^* is changed to $9000/31 - b_4^{\text{new}}$ and b_4^* to $6850/31 - b_3^{\text{new}}$. In order to maintain feasibility, we must have

$$\frac{6850}{31} - b_3^{\text{new}} \geq 0 \quad \text{and} \quad \frac{9000}{31} - b_4^{\text{new}} \geq 0$$

That is,

$$-\infty \leq b_3^{\text{new}} \leq \frac{6850}{31} \quad \text{and} \quad -\infty \leq b_4^{\text{new}} \leq \frac{9000}{31}$$

These, of course, are the same ranges as calculated before.

CHANGE IN THE REQUIREMENT b_i WHOSE SLACK OR SURPLUS VARIABLE IS NOT IN THE OPTIMAL SOLUTION If the slack or surplus variable associated with a constraint is not in the optimal solution, there is no idleness in the system regarding that constraint.

In other words, the resource represented by that constraint is fully utilized. A change in such a resource could affect all decision variables and also the optimal solution.

Suppose b_1 is changed from 350 to $350 + \Delta b_1$ in the example. Using the inverse of the basis matrix, we can calculate the new $\mathbf{b}^*$ as follows:

$$\mathbf{b}^{*\text{new}} = \mathbf{B}^{-1}\mathbf{b}^{\text{new}} = \begin{bmatrix} \dfrac{8}{31} & \dfrac{-5}{31} & 0 & 0 \\[2mm] \dfrac{28}{31} & \dfrac{-2}{31} & 0 & -1 \\[2mm] \dfrac{-5}{31} & \dfrac{7}{31} & 0 & 0 \\[2mm] \dfrac{-9}{31} & \dfrac{25}{31} & -1 & 0 \end{bmatrix} \begin{bmatrix} 350 + \Delta b_1 \\[2mm] 400 \\[2mm] 100 \\[2mm] 120 \end{bmatrix}$$

$$b_1^{*\text{new}} = x_1^{\text{new}} = \frac{800}{31} + \frac{8}{31}\Delta b_1$$

$$b_2^{*\text{new}} = x_7^{\text{new}} = \frac{5280}{31} + \frac{28}{31}\Delta b_1$$

$$b_3^{*\text{new}} = x_2^{\text{new}} = \frac{1050}{31} - \frac{5}{31}\Delta b_1$$

$$b_4^{*\text{new}} = x_5^{\text{new}} = \frac{3750}{31} - \frac{9}{31}\Delta b_1$$

Thus, the net changes in the decision variables are

$$\begin{aligned}
\Delta x_1 &= \left[\frac{8}{31}\Delta b_1 \right. \\[2mm]
\Delta x_7 &= \quad \frac{28}{31}\Delta b_1 \\[2mm]
\Delta x_2 &= \quad \frac{-5}{31}\Delta b_1 \\[2mm]
\Delta x_5 &= \left. \quad \frac{-9}{31}\Delta b_1 \right]
\end{aligned}$$

The coefficients of Δb_1 are the same as the Y_{ij} values of the column under x_3 in the optimal tableau. Recall that x_3 is the surplus variable associated with the first constraint whose resource b_1 is changed by an amount equal to Δb_1. The new optimal value of Z is calculated as

$$Z^{\text{new}} = \sum_{i=1}^{m} C_{Bi}\, b_i^{*\text{new}}$$

$$= (3, 0, 4, 0) \begin{bmatrix} \dfrac{800}{31} + \dfrac{8}{31} \Delta b_1 \\[2ex] \dfrac{5280}{31} + \dfrac{28}{31} \Delta b_1 \\[2ex] \dfrac{1050}{31} - \dfrac{5}{31} \Delta b_1 \\[2ex] \dfrac{3750}{31} - \dfrac{9}{31} \Delta b_1 \end{bmatrix}$$

$$Z^{new} = \frac{6600}{31} + \frac{4}{31} \Delta b_1$$

Again, the net change in Z is the product of Δb_1 and the Z_j value under x_3 in the optimal tableau.

Interpretation Keep in mind that Y_{ij} is the rate of substitution between the nonbasic variable x_j and the basic variable x_{Bi}. It represents the amount by which x_{Bi} should be reduced if one unit of x_j is introduced into the solution.

In this example x_3 is the slack variable associated with the first constraint. Since it is not in the solution, the first constraint is fully satisfied. In other words, the available supply of raw material R_1 is all used up in the optimal solution. If x_3 were in the solution at a positive level, that much of R_1 would have been unused. Conversely, if we were to obtain additional amounts of that material, we could use it to produce more outputs. Therefore Y_{ij} can also be interpreted to represent the amount by which each x_{Bi} can be *increased* if the constraint, whose slack or artificial variable is x_j, is relaxed by one unit.

Using this economic interpretation, we observe that a unit increase in raw material R_1 relaxes the first constraint and affects the basic variables by amounts shown under x_3, which is the slack variable associated with that constraint. Hence, we can determine the new values of the basic variables directly from the optimal tableau without going through the matrix multiplication above.

THE RANGE OF b If b_1 is changed from 350 to $350 + \Delta b_1$, the optimal solution remains feasible as long as the new values of the basic variables are still nonnegative.

$$x_1^{new} = \frac{800}{31} + \frac{8}{31} \Delta b_1 \geq 0 \rightarrow \Delta b_1 \geq -100$$

$$x_7^{new} = \frac{5280}{31} + \frac{28}{31} \Delta b_1 \geq 0 \rightarrow \Delta b_1 \geq \frac{-1320}{7}$$

$$x_2^{new} = \frac{1050}{31} - \frac{5}{31} \Delta b_1 \geq 0 \rightarrow \Delta b_1 \leq 210$$

$$x_5^{\text{new}} = \frac{3750}{31} - \frac{9}{31} \Delta b_1 \geq 0 \rightarrow \Delta b_1 \leq \frac{1250}{3}$$

In the range $-100 \leq \Delta b_1 \leq 210$, the optimal solution is unaltered. This range corresponds to

$$250 \leq b_1 \leq 560$$

For this range of b_1, the solution is still made up of x_1, x_2, x_5, and x_7 but the optimal value of Z changes:

$$Z^{\text{new}} = Z + Z_3\Delta b$$

$$= \frac{6600}{31} + \frac{4}{31} \Delta b$$

Thus using -100 and 210 for the lower and upper bound of Δb,

$$\frac{6200}{31} \leq Z^{\text{new}} \leq \frac{7440}{31}$$

or

$$200 \leq Z^{\text{new}} \leq 240$$

Similarly, if b_2 is changed from 400 to $400 + \Delta b_2$, the new values of the basic variables become:

$$x_1 = \frac{800}{31} - \frac{5}{31} \Delta b_2$$

$$x_7 = \frac{5280}{31} - \frac{2}{31} \Delta b_2$$

$$x_2 = \frac{1050}{31} + \frac{7}{31} \Delta b_2$$

$$x_5 = \frac{3750}{31} + \frac{25}{31} \Delta b_2$$

resulting in a new value of Z equal to

$$\frac{6,600}{31} + \frac{13}{31} \Delta b_2$$

The limiting values of Δb_2 can be calculated.

$$\frac{800}{31} - \frac{5}{31} \Delta b_2 \geq 0 \rightarrow \Delta b_2 \leq 160$$

$$\frac{5280}{31} - \frac{2}{31} \Delta b_2 \geq 0 \rightarrow \Delta b_2 \leq 2640$$

$$\frac{1050}{31} + \frac{7}{31} \Delta b_2 \geq 0 \rightarrow \Delta b_2 \geq -150$$

$$\frac{3750}{31} + \frac{25}{31} \Delta b_2 \geq 0 \rightarrow \Delta b_2 \geq -150$$

280   Therefore,

$$-150 \leq \Delta b_2 \leq 160$$

or

$$250 \leq b_2 \leq 560$$

In this range of b_2, the optimal solution will not be affected but the value of Z will again vary because

$$Z^{\text{new}} = Z + Z_4 \Delta b_2$$

$$= \frac{6600}{31} + \frac{13}{31} \Delta b_2$$

or

$$150 \leq Z^{\text{new}} \leq 280$$

If both b_1 and b_2 are changed simultaneously, the combined effect can be studied by considering the Y_{ij} values for x_3 and x_4 together:

$$\left.\begin{array}{l} b_1^{\text{new}} = 350 + \Delta b_1 \\ b_2^{\text{new}} = 400 + \Delta b_2 \end{array}\right\} \quad \text{simultaneously}$$

The new values of the basic variables will be:

$$x_1^{\text{new}} = \frac{800}{31} + \frac{8}{31} \Delta b_1 - \frac{5}{31} \Delta b_2 \geq 0 \tag{9.18}$$

$$x_7^{\text{new}} = \frac{5280}{31} + \frac{28}{31} \Delta b_1 - \frac{2}{31} \Delta b_2 \geq 0 \tag{9.19}$$

$$x_2^{\text{new}} = \frac{1050}{31} - \frac{5}{31} \Delta b_1 + \frac{7}{31} \Delta b_2 \geq 0 \tag{9.20}$$

$$x_5^{\text{new}} = \frac{3750}{31} - \frac{9}{31} \Delta b_1 + \frac{25}{31} \Delta b_2 \geq 0 \tag{9.21}$$

with the corresponding value of $Z_{\text{opt}}^{\text{new}}$ being

$$Z_{\text{opt}}^{\text{new}} = \frac{6600}{31} + \frac{4}{31} \Delta b_1 + \frac{13}{31} \Delta b_2$$

Expressions (9.18) through (9.21) are summarized in graphical form in Fig. 9.11. The optimal basis remains unaltered as long as the point $(\Delta b_1, \Delta b_2)$ falls into the shaded area.

CHANGES IN THE A MATRIX

ADDING A NEW VARIABLE x_p If a new variable x_p (such as a new product, a new project, etc.) improves the optimal value of the objective function when introduced into the solution, the optimal solution will be changed. However, as long as its unit contribution to Z does

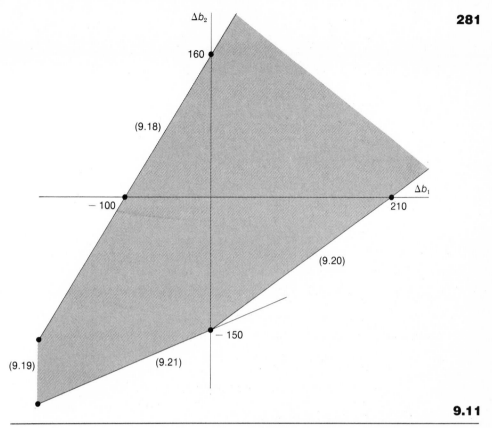

9.11

Effect of simultaneous changes in two resource levels.

not exceed its unit cost, the new variable will not enter the solution, and the optimality will not be altered.

Suppose a third process is under consideration in the example. The daily profit is estimated to be $2000 when that process is used. It uses 6 tons of raw material R_1 and 10 tons of R_2 to develop 1 ton of output O_1 and 3 tons of output O_2 per day. Should we use the new process?

Let us represent the daily use of the new process by x_9 and add the following vector to the LP model:

$$c_9 = 2$$

$$a_{19} = 6$$

$$a_{29} = 10$$

$$a_{39} = 1$$

$$a_{49} = 3$$

282   Since this is a new column in the initial tableau, we can calculate the corresponding values in the optimal tableau as follows:

$$\mathbf{Y}_9 = \mathbf{B}^{-1}\,\mathbf{a}_9$$

$$= \begin{bmatrix} \dfrac{8}{31} & \dfrac{-5}{31} & 0 & 0 \\[2mm] \dfrac{28}{31} & \dfrac{-2}{31} & 0 & -1 \\[2mm] \dfrac{-5}{31} & \dfrac{7}{31} & 0 & 0 \\[2mm] \dfrac{-9}{31} & \dfrac{25}{31} & -1 & 0 \end{bmatrix} \begin{bmatrix} 6 \\[2mm] 10 \\[2mm] 1 \\[2mm] 3 \end{bmatrix} = \begin{bmatrix} \dfrac{-2}{31} \\[2mm] \dfrac{55}{31} \\[2mm] \dfrac{40}{31} \\[2mm] \dfrac{165}{31} \end{bmatrix}$$

Therefore, in the optimal tableau, the column under x_9 appears as follows:

C_{Bi}	X_{Bi}	C_j	$\ldots\ldots x_9$
3	x_1		$-2/31$
0	x_7		$55/31$
4	x_2		$40/31$
0	x_5		$165/31$

We can now calculate Z_9 and $C_9 - Z_9$:

$$Z_9 = (3,\ 0,\ 4,\ 0) \begin{bmatrix} \dfrac{-2}{31} \\[2mm] \dfrac{55}{31} \\[2mm] \dfrac{40}{31} \\[2mm] \dfrac{165}{31} \end{bmatrix} = \frac{154}{31}$$

$$C_9 - Z_9 = 2 - \frac{154}{31} = \frac{-92}{31}$$

Since $C_9 - Z_9$ is negative, the new process will not increase the value of the objective function; thus it does not need to be considered at all.

If C_9 were greater than Z_9, i.e., $C_9 > 154/31$, it would mean that the new process would be profitable. We would then consider x_9 as a candidate for the basis and perform one postoptimal iteration to determine the new solution.

Economic Interpretation Remember that the shadow prices were **283** explained as the unit contributions of the available resources (or other requirements represented by the constraints). If we use the new process, we will need 6 units of supply 1 and 10 units of supply 2 to produce 1 unit for the first demand and 3 units for the second demand. Notice that these are the amounts to be *used up* by the new process in each constraint. In other words, one day of the new process will make each constraint tighter by these amounts if we decide to use it. Thus, those quantities of the requirements vector will be removed from the system in exchange for the net profit of one day's use of this process.

Using the shadow prices and the a_9 vector, we can calculate the loss in the value of the objective function induced by one day's use of the new process.

$$\text{Loss in } Z = \sum_{i=1}^{4} (a_{i9}) \text{ (shadow price)}_i$$

$$= 6 (Z_3) + 10 (Z_4) + 1 (Z_6) + 3 (Z_8)$$

$$= 6 \left(\frac{4}{31}\right) + 10 \left(\frac{13}{31}\right) + 1 (0) + 3 (0)$$

$$= \frac{154}{31}$$

which is the same value as calculated before.

Therefore, as long as C_9, the gain per unit of x_9, is not greater than \$154/31, the new process will not be utilized and the optimal solution will not be affected.

CHANGE IN THE TECHNOLOGICAL COEFFICIENTS a_{ij} FOR A NON-BASIC VARIABLE x_j If the technological coefficients of a nonbasic variable are changed, the effect will be similar to the introduction of a new variable into the system.

For example, suppose we found out that the input requirements for the new process in the example are not accurate. We want to determine when this process could be a valid alternative to the other two that we are considering.

Let us say that the vector for the new variable is as follows:

$$c_9 = 2$$
$$a_{19} = ?$$
$$a_{29} = ?$$
$$a_{39} = 1$$
$$a_{49} = 3$$

284   This process would be profitable if $C_9 - Z_9 \geq 0$:

$$C_9 - Z_9 = 2 - \left[a_{19} \frac{4}{31} + a_{29} \frac{13}{31} + 1(0) + 3(0) \right]$$

$$= 2 - \frac{4}{31} a_{19} - \frac{13}{31} a_{29} \geq 0 \tag{9.22}$$

$$= 62 - 4a_{19} - 13a_{29} \geq 0$$

For any combination of a_{19} and a_{29} below the boundary shown in Fig. 9.12, the new process will be profitable.

In the example we had $a_{19} = 6$, $a_{29} = 10$. This point is outside the shaded area and it makes the new process unprofitable. However, if we consider $a_{29} = 2$ instead of 10, we obtain

$$C_9 - Z_9 = 2 - \left[6 \frac{4}{31} + 2 \frac{13}{31} + 1 \ (0) + 3 \ (0) \right]$$

$$= 2 - \frac{24}{31} - \frac{26}{31} - 0 - 0$$

$$= \frac{22}{31} > 0$$

This indicates that the new process will be profitable under these conditions and that x_9 will be a candidate to enter the solution.

CHANGE IN THE TECHNOLOGICAL COEFFICIENTS a_{ij} FOR A BASIC VARIABLE x_j This case requires an involved procedure. It will not be

Sensitivity of technological coefficients, a_{ij}.

9.12

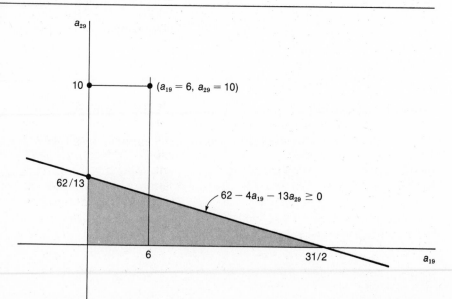

discussed here. A treatment of this case can be found in books such as Simmonard, Gass, etc.[11]

A NEW CONSTRAINT

After the optimum solution is obtained, we might realize that an additional constraint has to be considered in the problem. If this constraint has been violated, the solution is infeasible; we need a new solution. If it is not violated, then we can ignore it.

When a constraint is added, we check the solution by substituting the optimal values of the decision variables into that constraint. If the constraint is satisfied, we consider it redundant and use the optimal solution as it is.

For example, suppose we found out that the two processes considered in the problem cannot be used for more than a total of 60 days. Let us check this new constraint.

$$x_1 + x_2 \leq 60$$

$$x_1^0 = \frac{800}{31}$$

$$x_2^0 = \frac{1050}{31}$$

$$x_1^0 + x_2^0 = \frac{1850}{31} < 60 \qquad \text{OK}$$

If, however, we were limited to a total of 30 days, the solution would be infeasible. In such cases, a reverse procedure called the "dual simplex method" is used. In dual simplex algorithm, the optimality is maintained in each iteration but feasibility is not. The objective of the method is to reduce the infeasibility at each step until reaching the feasible solution while maintaining optimality conditions at all times.

DUAL SIMPLEX ALGORITHM

The infeasible optimality is called the "dual feasible" condition.

When one or more of the b_i^* are negative and all $C_j - Z_j$ are nonpositive in a maximization problem (or all $C_j - Z_j$ are nonnegative in a minimization problem), we have dual feasibility. In such cases, we proceed to make the right-hand-side elements (the b_i^* values) nonnegative in subsequent iterations without altering the sign of the $C_j - Z_j$ values. "Primal feasibility" is reached when all b_i^* become nonnegative. Since the optimality condition is also satisfied at that point the optimal solution is thus reached.

[11]M. Simmonard, *Linear Programming*, W. Jewell (trans.), Prentice-Hall, Englewood Cliffs, N.J., 1966; S. I. Gass, *Linear Programming*, 4th ed., McGraw-Hill, New York, 1975.

The selection of the pivot row and pivot column is done in reverse order of the regular simplex method. In dual simplex algorithm, first the basic variable is chosen to leave the basis, then the nonbasic variable is selected to enter the next basis, as follows:

1   Selection of the variable x_r to leave the basis: Choose a basic variable that is making the present solution infeasible, i.e., a basic variable that is at a negative level. If several basic variables have negative values b_i^*, select x_r, the variable to leave the basis, as

$$b_r^* = \min_i (b_i^*) \quad \text{for } b_i^* < 0$$

If the most negative b_i^* is not unique, select any one of them and identify row r as the pivot row.

2   Selection of the nonbasic variable x_k to enter the basis: There are two conditions to be satisfied:
   (a)   Infeasibility should be reduced. In other words, we should at least have the negative b_i^* take on a positive value in the next iteration.
   (b)   The next tableau should remain dual feasible, i.e., the $C_j - Z_j$ values should not change sign.

The first condition is satisfied by considering only those nonbasic variables x_j with negative coefficients in row r as candidates to enter the basis, i.e., consider x_j for basis only if $Y_{rj} < 0$.

The second condition requires that the nonbasic variable x_k to enter the basis is selected as follows (without considering the artificial variables):

$$\left| \frac{C_k - Z_k}{Y_{rk}} \right| = \min \left| \frac{C_j - Z_j}{Y_{rj}} \right| \quad \text{for } Y_{rj} < 0$$

Let us show that this rule guarantees the dual feasibility of the next tableau. Note that

$$(C_j - Z_j)^{\text{new}} = \underbrace{(C_j - Z_j)^{\text{old}}}_{1} - \underbrace{Y_{rj}}_{2} \underbrace{\frac{C_k - Z_k}{Y_{rk}}}_{3} \tag{9.23}$$

In a maximization problem

$$\frac{C_k - Z_k}{Y_{rk}} \geq 0$$

because $C_k - Z_k \leq 0$ and $Y_{rk} < 0$.

For $Y_{rj} \geq 0$: $(C_j - Z_j)^{\text{new}}$ shown in expression (9.23) will always be less than or equal to zero because quantity 1 is nonpositive and quantities 2 and 3 are nonnegative. Thus, the product of quantity 2 and quantity 3 is greater than or equal to zero; and the new value of $(C_j - Z_j)$ is less than or equal to the old value, which was nonpositive to start with.

For those $Y_{rj} < 0$: Rewrite $(C_j - Z_j)^{\text{new}}$ as follows:

$$(C_j - Z_j)^{\text{new}} = \underbrace{Y_{rj}}_{1} \left[\underbrace{\frac{(C_j - Z_j)^{\text{old}}}{Y_{rj}}}_{2} - \underbrace{\frac{C_k - Z_k}{Y_{rk}}}_{3} \right] \tag{9.24}$$

Note that the difference between quantities 2 and 3 in Eq. (9.24) is always nonnegative, because

$$\frac{(C_j - Z_j)^{\text{old}}}{Y_{rj}} \geq 0$$

$$\frac{C_k - Z_k}{Y_{rk}} \geq 0$$

and

$$\frac{C_k - Z_k}{Y_{rk}} = \min \frac{C_j - Z_j}{Y_{rj}}$$

Therefore $(C_j - Z_j)^{\text{new}}$ is the product of a negative (Y_{rj}) and a nonnegative quantity.

So, $(C_j - Z_j)^{\text{new}}$ is always less than or equal to zero.
In a minimization problem:

$$\frac{C_k - Z_k}{Y_{rk}} \leq 0$$

because $C_k - Z_k \geq 0$ and $Y_{rk} < 0$.
Thus, for $Y_{rj} \geq 0$, we have

$$(C_j - Z_j)^{\text{new}} = \underbrace{(C_j - Z_j)^{\text{old}}}_{\geq 0} - \underbrace{Y_{rj}}_{\geq 0} \underbrace{\frac{C_k - Z_k}{Y_{rk}}}_{\leq 0}$$

$$(C_j - Z_j^{\text{new}} \geq (C_j - Z_j)^{\text{old}}$$

Since $(C_j - Z_j)^{\text{old}} \geq 0$, it follows that $(C_j - Z_j)^{\text{new}}$ is always nonnegative.

For $Y_{rj} < 0$:

$$(C_j - Z_j)^{\text{new}} = \underbrace{Y_{rj}}_{< 0} \left[\underbrace{\frac{(C_j - Z_j)^{\text{old}}}{Y_{rj}}}_{\leq 0} - \underbrace{\frac{C_k - Z_k}{Y_{rk}}}_{\leq 0} \right] \tag{9.25}$$

$$\frac{(C_j - Z_j)^{\text{old}}}{Y_{rj}} \leq 0$$

and

$$\frac{C_k - Z_k}{Y_{rk}} \leq 0$$

but

$$\left| \frac{C_k - Z_k}{Y_{rk}} \right| = \min \left| \frac{C_j - Z_j}{Y_{rj}} \right|$$

288   Therefore, the quantity in brackets in expression (9.25) is less than or equal to zero.

Since $(C_j - Z_j)^{\text{new}}$ is the product of a negative quantity and a nonpositive quantity, it is always greater than or equal to zero.

EXAMPLE Suppose, in our problem, that the new constraint states that the two processes cannot add up to more than 30 days:

$$x_1 + x_2 \leq 30$$

Since $x_1^o + x_2^o = 1850/31 > 30$, we know that the solution will no longer be feasible. We can convert the new constraint to equality by adding a slack variable x_9, and incorporate it into the final tableau as shown in Table 9.15.

Adding a New Constraint to Optimal Tableau

9.15

		C_j	3	4	0	0	0	−M	0	−M	0	
C_{Bi}	Basis		x_1	x_2	x_3	x_4	x_5	x_6	x_7	x_8	x_9	b_i^*
3	x_1		1	0	8/31	−5/31	0	0	0	0	0	800/31
0	x_7		0	0	28/31	−2/31	0	0	1	−1	0	5280/31
4	x_2		0	1	−5/31	7/31	0	0	0	0	0	1050/31
0	x_5		0	0	−9/31	25/31	1	−1	0	0	0	3750/31
0	x_9		1	1	0	0	0	0	0	0	1	30
Z_j			3	4	4/31	13/31	0	0	0	0	0	6600/31
$C_j - Z_j$			0	0	−4/31	−13/31	0	−M	0	−M	0	

The new tableau does not contain an identity matrix, but it can be modified easily, using row operations. Subtract row 1 and row 3 from row 5 to obtain the tableau shown as Table 9.16.

Dual Feasibility

9.16

		C_j	3	4	0	0	0	−M	0	−M	0	
C_{Bi}	Basis		x_1	x_2	x_3	x_4	x_5	x_6	x_7	x_8	x_9	b_i^*
3	x_1		1	0	8/31	−5/31	0	0	0	0	0	800/31
0	x_7		0	0	28/31	−2/31	0		1	−1	0	5280/31
4	x_2		0	1	−5/31	7/31	0	0	0	0	0	1050/31
0	x_5		0	0	−9/31	25/31	1	−1	0	0	0	3750/31
0	x_9		0	0	−3/31	−2/31	0	0	0	0	1	−920/31	← r
Z_j			3	4	4/31	13/31	0	0	0	0	0	6600/31
$C_j - Z_j$			0	0	−4/31	−13/31	0	−M	0	−M	0	

k

As expected, this tableau is dual feasible with optimality condition satisfied but primal feasibility violated because x_9 is equal to $-920/31$. Since x_9 is the only nonpositive basic variable, it is the only candidate for removal from the basis. Thus x_9 is x_r.

x_3 and x_4 are candidates to enter the basis because Y_{53} and Y_{54} are both negative.

$$\frac{C_3 - Z_3}{Y_{53}} = \frac{(-4/31)}{(-3/31)} = \frac{4}{3}$$

$$\frac{C_4 - Z_4}{Y_{54}} = \frac{(-13/31)}{(-2/31)} = 13/2$$

Since $\quad \dfrac{C_3 - Z_3}{Y_{53}} < \dfrac{C_4 - Z_4}{Y_{54}}$

x_3 is the entering variable.

The next tableau is shown in Table 9.17. This tableau is also dual feasible. x_7 will leave the basis and x_4 will enter. x_7 will be removed because it has the most negative value. However, notice that the ratio b_i^*/Y_{ik} is equal to 160 for both $i = 1$ and $i = 2$ (i.e., for basic variables x_1 and x_7). This indicates degeneracy.

All basic variables are nonnegative and all $C_j - Z_j$ are less than or equal to zero in the tableau shown in Table 9.18. Therefore, the optimal solution has been reached. Basic variable x_1 is at zero level. This is because of the degeneracy indicated in the previous tableau.

The solution with the new constraint $(x_1 + x_2 \leq 30)$ is $x_1 = 0$, $x_2 = 30$, $Z = 120$. Thus, if the total use of the two processes is limited to 30 days, we should utilize the second process to the full extent because of its higher profitability, and not use process 1 at all.

Postoptimal Tableau 1 in Dual Simplex Algorithm

9.17

C_{Bi}	Basis	C_j	3	4	0	0	0	$-M$	0	$-M$	0	
			x_1	x_2	x_3	x_4	x_5	x_6	x_7	x_8	x_9	b_i^*
3	x_1		1	0	0	$-1/3$	0	0	0	0	8/3	$-160/3$
0	x_7		0	0	0	$(-2/3)$	0	0	1	-1	28/3	$-320/3$ ← r
4	x_2		0	1	0	1/3	0	0	0	0	$-5/3$	250/3
0	x_5		0	0	0	1	1	-1	0	0	-3	210
0	x_3		0	0	1	2/3	0	0	0	0	$-31/3$	920/3
Z_j			3	4	0	1/3	0	0	0	0	4/3	520/3
$C_j - Z_j$			0	0	0	$-1/3$	0	$-M$	0	$-M$	$-4/3$	

$\uparrow$
k

Postoptimal Tableau 2 in Dual Simplex
Algorithm

9.18

C_{Bi}	Basis	C_j	x_1	x_2	x_3	x_4	x_5	x_6	x_7	x_8	x_9	b_i^*
			3	4	0	0	0	$-M$	0	$-M$	0	
3	x_1		1	0	0	0	0	0	$-1/2$	$1/2$	$22/3$	0
0	x_4		0	0	0	1	0	0	$-3/2$	$3/2$	$-14/3$	160
4	x_2		0	1	0	0	0	0	$1/2$	$-1/2$	$-19/3$	30
0	x_5		0	0	0	0	1	-1	$3/2$	$-3/2$	-17	50
0	x_3		0	0	1	0	0	0	1	-1	$-59/3$	200
Z_j			3	4	0	0	0	0	5	-5	$-10/3$	120
$C_j - Z_j$			0	0	0	0	0	$-M$	-5	$-M-5$	$10/3$	

Notice that the 30-day limit has resulted in the underutilization of raw materials. This is indicated by $x_3 = 200$ and $x_4 = 160$. Since x_3 and x_4 are the slack variables associated with the supply constraints, the new optimum solution shows that there will be 200 units of the first raw material and 160 units of the second raw material left over at the end of 30 days.

It is interesting to note that $x_7 = 0$ in this solution, indicating that there will be no extra units of output 2, which the previous solution had produced. Since x_5 and x_7 are both zero now, both outputs O_1 and O_2 will be produced exactly as demanded; neither will be overproduced.

Let us look at the graphical solution of the problem for the point represented in this tableau. The new feasible region, efd, is shaded in Fig. 9.13, making two of the constraints redundant. The solution obtained in postoptimal tableau 2 is at point e. Notice that three constraints:

$$x_1 \geq 0$$
$$x_1 + x_2 \leq 30$$
$$6x_1 + 4x_2 \geq 120$$

intersect at point e. When an extreme point of the convex region where the solution lies is represented by more than two constraints, a degenerate solution is indicated. When degeneracy occurs, one or more of the variables appear in the basis at zero level. If a basic variable has a zero value, it can be replaced by a nonbasic variable without changing the solution. Conceptually, the solution shifts from one description of a point to another description. In the example problem, the basic solution is at point e, represented as the intersection of constraints $x_1 \geq 0$ and $x_1 + x_2 \leq 30$. If x_7 replaces x_1, the values of the variables and the value of Z remain the same but the basic solution contains X_7 instead of x_1 at zero level. In that case, a

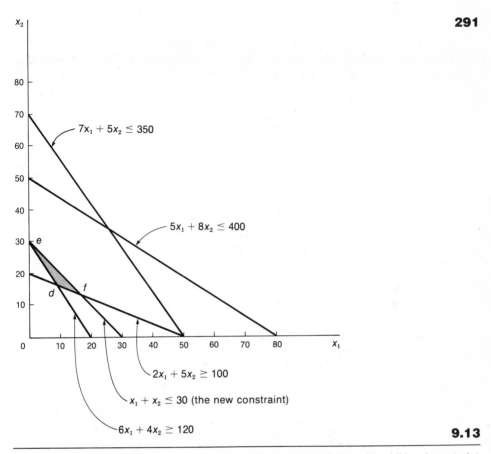

9.13

Postoptimal solution with additional constraint.

representation of the same point e is obtained as the intersection of constraints $x_1 \geq 0$ and $6x_1 + 4x_2 \geq 120$, instead of $x_1 + x_2 \leq 30$.

For our example, however, the optimum has already been reached, and degeneracy is academic at this point.

SUMMARY

Linear programming is one of the most widely used methods for the allocation of scarce resources. It is a well developed technique employing a systematic method to find the optimum point in problems expressed as linear expressions.

Linear programming formulations consist of an objective function and several constraints. The constraints can be in the form of equalities and inequalities depending on the system characteristics.

A linear programming problem can be solved graphically if only two variables are involved. However, such an approach cannot be used for problems with higher dimensions. The solution technique for the general LP problem is called the "simplex" method, which studies the extreme points of the surface defined by the constraints. The optimum solution is located at one of the extreme points, and it can be found efficiently by the simplex method.

The problem may become infeasible when constraints are inconsistent. In such cases, no solution can be found. Sometimes the problem has no bound and one can go to infinity in searching for the optimum. This is a case with improper formulation. This condition can be identified in the simplex method.

After the solution is obtained, the neighboring region of the optimum point is studied for determining the effect of changes in assumed parameter values on the solution of the problem. This postoptimal study is called the "sensitivity analysis."

DISCUSSION QUESTIONS

1 Write this problem in standard form:

$$\max Z = 5x_1 + 12x_2$$

Subject to

$$2x_1 + 7x_2 \geq 14$$
$$3x_1 + 5x_2 \geq 15$$
$$x_1, x_2 \geq 0$$

2 Write this problem in standard form:

$$\max Z = 6x_1 + 10x_2$$

Subject to

$$1.5x_1 - x_2 \geq 1.5$$
$$4x_1 - 4x_2 \leq 16$$
$$3x_1 + 5x_2 \geq -15$$
$$x_1 \text{ is unrestricted}$$
$$x_2 \leq 0$$

3 Consider the following constraints

$$2x_1 + x_2 \leq 40$$
$$-2x_1 + x_2 \leq 20$$
$$-x_1 + x_2 \geq -10$$

(a) What is the upper bound for the number of possible solutions?
(b) Identify the m basic and $(n - m)$ nonbasic variables at each possible solution point (feasible or infeasible) by complete enumeration.

4   Answer the question above after changing the third constraint to

$$-x_1 + x_2 \geq 10$$

5   Use the constraints in question 3 with objective function

$$\max Z = x_1 - 2x_2$$

Find the optimum solution for the following cases:

(a) $x_1, x_2 \geq 0$
(b) $x_1, x_2 \leq 0$
(c) $x_1 \geq 0, x_2 \leq 0$
(d) $x_1 \leq 0, x_2 \geq 0$
(e) x_1, x_2 unrestricted

6   Answer question 5 with the objective function

$$\min Z = x_1 - 2x_2$$

7   Answer question 5 with the following additional constraints:

$$-5 \leq x_1 \leq 5$$
$$-5 \leq x_2 \leq 5$$

8   Solve the following problems and comment on the solution in each case.

(a) max $Z = -3x_1 + 2x_2$
Subject to

$$x_1 \leq 3$$
$$x_1 - x_2 \leq 0$$
$$x_1, x_2 \geq 0$$

(b) max $Z = 3x_1 + 2x_2$
Subject to

$$2.5x_1 + 2x_2 \leq 5$$
$$x_1 + x_2 \leq 2$$
$$7x_1 + 3x_2 \leq 21$$
$$x_1, \quad x_2 \geq 0$$

(c) max $Z = 3x_1 - 2x_2$
Subject to

$$x_1 + x_2 \leq 1$$
$$2x_1 + 2x_2 \geq 4$$
$$x_1, \quad x_2 \geq 0$$

9 Solve the following LP problem

$$\max Z = x_1 + x_2 + 3x_3 + 2x_4$$

Subject to

$$2x_1 + 2x_2 + 4x_3 - 3x_4 \leq 16$$
$$x_1 + 4x_2 + x_3 + x_4 \geq 16$$
$$-4x_1 + 3x_2 + x_3 + 5x_4 = 16$$
$$x_j \geq 0 \qquad j = 1, \ldots, 4$$

10 For the linear programming problem

$$\max Z = 2x_2 - 5x_3$$

Subject to

$$x_1 + x_3 \geq 2$$
$$2x_1 + x_2 + 6x_3 \leq 6$$
$$x_1 - x_2 + 3x_3 = 0$$
$$x_1, x_2, x_3 \geq 0$$

the final tableau giving the optimal solution is as follows:

C_{Bi}	Basis	C_j	0	2	-5	0	0	-M	-M	
			x_1	x_2	x_3	x_4	x_5	x_6	x_7	b_i^*
2	x_2		0	1	0	0	1/3	0	-2/3	2
0	x_4		0	0	2	1	1/3	-1	1/3	0
0	x_1		1	0	3	0	1/3	0	1/3	2
Z_j			0	2	0	0	2/3	0	-4/3	$Z = 4$
$C_j - Z_j$			0	0	-5	0	-2/3	-M	-M + 4/3	

Note:

x_4 = surplus variable in constraint 1

x_{5x} = artificial variable in constraint 1

x_6 = slack variable in constraint 2

x_7 = artificial variable in constraint 3

(a) What are the shadow prices?
(b) Give an interval for each of the objective function coefficient (c_1, c_2 and c_3) such that the optimal solution remains optimal.
(c) Give an interval for each right-hand-side constant (b_1, b_2, and b_3) such that the optimal solution remains optimal.

(d) Assume that the costs c_1 and c_2 are Δc_1 and $2 + \Delta c_2$. Derive a set of inequalities for the pair $(\Delta c_1, \Delta c_2)$ such that the optimal solution will remain optimal. Draw a graph of the region implied by this set of inequalities.

(e) Repeat part c for the right-hand-side constants of the first and second constraints. Use $2 + \Delta b_1$ for the first constraint, $6 + \Delta b_2$ for the second constraint.

11   A new variable x_8 is added to the model given in question 10. It has no contribution to the objective function $(c_8 = 0)$ and its constraints coefficients are 3, 2, and 5. What happens to the optimal solution?

12   The coefficient of x_1 in the second constraint of the model in question 10 is changed to $(2 + a)$. What are the limits of a for the optimal solution to remain optimal.

13   The following constraint is added to the LP model given in question 10. $x_1 + x_2 + x_3 \leq 3$

What happens to the optimal solution?

KEY QUESTIONS FOR THE ENGINEERING MANAGER

1   Am I comfortable with my level of understanding of linear programming as it might be applied in this organization?

2   Can I ask the right questions about the solution of linear programming problems?

3   How might sensitivity analysis be applied to some of the current decisions in my department?

4   How can linear programming be used to facilitate strategic planning process in this organization?

5   What kinds of linear programming problems exist in this organization where only two principal variables exist? Do I have adequate understanding to apply the graphical solution to these problems?

6   What sorts of large linear programming problems exist in this organization which requires a computer code based on the general solution procedure?

7   Is my understanding of the simplex adequate to supervise its use by a technical staff?

8   Can I communicate with my general manager on the results of linear programming models applied to our tactical problems?

9   What scarce resource problems in this organization currently limit the progress toward accomplishing our goals?

10   Can linear programming provide insight into how these problems might be alleviated?

11   What are the five or six most serious problems/opportunities fac-

296 ing this organization? Can I identify the objectives and constraints for them?

12   Have I identified the managers to be from among the professionals in this organization? What self-development strategies have I encouraged for them in the area of mathematical skills for decision making?

BIBLIOGRAPHY

Charnes, A., and W. W. Cooper:.*Management Models and Industrial Application of Linear Programming,* vols. 1 and 2, Wiley, New York, 1960.

Dantzig, G.B.: *Linear Programming and Extensions,* Princeton, Princeton, N.J., 1963.

Gass, S. I.: *An Illustrated Guide to Linear Programming,* McGraw-Hill, New York, 1976.

————: *Linear Programming,* 4th ed., McGraw-Hill, New York, 1975.

Gupta, S. K., and J. M. Cozzolino: *Fundamentals of Operations Research for Management,* Holden-Day, San Francisco, 1974.

Hadley, G.: *Linear Programming,* Addison-Wesley, Reading, Mass., 1962.

Hillier, F. S., and G. J. Lieberman: *Introduction to Operations Research,* 2d ed., Holden-Day, San Francisco, 1974.

Phillips, D. T., A. Ravindran, and J. J. Solberg: *Operations Research,* Wiley, New York, 1976.

Riggs, J. L., and M. S. Inoue: *Introduction to Operations Research and Management Science,* McGraw-Hill, New York, 1975.

Simmonard, M.: *Linear Programming,* W. Jewell (trans.), Prentice-Hall, Englewood Cliffs, N.J., 1966.

Taha, H. A.: *Operations Research—An Introduction,* Macmillan, New York, 1971.

Wagner, H. M.: *Principles of Management Science,* 2d ed., Prentice-Hall, Englewood Cliffs, N.J., 1975.

Whitehouse, G. E., and B. L. Wechsler: *Applied Operations Research,* Wiley, New York, 1976.

DECISION-10 MAKING METHODOLOGIES

Decision making is a frequent activity in our daily lives. Every conscious move made by a person can be viewed as the implementation of a decision. Sometimes we do nothing and may consider ourselves as having made no decision. But even then, we have clearly decided between doing something and doing nothing. In some cases, a decision may have an insignificant effect, regardless of its outcome. Most of the routine, instinctive decisions are in this category. Some decisions, on the other hand, may affect people and organizations for a long time. In fact, it is not unusual to realize the full implication of an important decision several years after it is made. Looking back today, we see, for example, that the decisions made by the Roosevelt administration during the Depression years have changed our lives and established new directions for many generations, even though their importance and long-range impacts may or may not have been fully understood when they were made.

It is easy to make a decision when complete information is available about its outcome. Unfortunately, we do not have access to such information, except on very rare occasions. The manager is expected to make decisions almost on a continuous basis with or without sufficient information. Every decision represents a risk for him and for his organization, because once a decision is made, he assumes responsibility for its outcome. After all, he is paid for his ability to make and implement decisions on behalf of the organization.

Judgment plays a key role in decision making. When faced with situations similar to previous occurrences, the engineering manager uses his experience and judgment to arrive at a decision. He compares the situation to others in terms of similarities and differences, evaluates the likelihood of the various outcomes of his actions, and decides on the action that is likely to lead to the best results. In other words, he implicitly assigns probabilities to the various results of his decisions, develops a mental model, and chooses the best alternative according to his judgment.

An important factor in decision making is the value structure of the decision maker. Given identical situations, two persons do not

298 necessarily arrive at the same decision; one may favor decisions that lead to completely different results. This behavior is explained by the differences in relative values attributed to the same outcome. These values are a function of the individual's background, beliefs, and perceptions. In more formal terms, decisions are made according to the "utilities" of the outcomes as perceived by the decision maker.

The role of an executive and how personal beliefs and philosophies affect decision making can be illustrated by examining recent experiences at the Ford Motor Company. In early 1980 this company was grappling with some awesome financial problems. It could lose approximately $1.5 billion in 1980 on top of a $1 billion loss in 1979. Part of this loss could be credited to the strong foreign car sales in the United States, but it also reflected a lack of strategic perception at the top level in company.

At the insistence of Henry Ford II, the Ford Motor Company was philosophically opposed to down-sizing its large cars until 1979. This down-sizing was started almost 6 years after General Motors Corporation established a policy of down-sizing its automobiles. Even as late as 1978, the Ford Motor Company was building large cars instead of the small, front wheel drive vehicles that were cutting into the market. Apparently Henry Ford II was convinced that most American car buyers would never accept vehicles smaller than the compact Fairmont vehicle, which was roughly equivalent to the largest cars which Ford sold overseas. "There was almost a preoccupation about keeping our larger cars around because we thought that's what the public wanted," noted Bennett E. Bidwell, vice president of Ford Motor Company.[1]

In many cases, decisions are made intuitively without any rational basis. Some of these may prove to be the right decisions after the facts become available, but this is not the rule. In a modern society, where the manager deals with increasingly complex systems and system interactions, he needs all the assistance he can get in order to arrive at the right answer at the right moment. Assistance is available in the form of analytical approaches to decision making. These approaches provide him with a rational basis to systematically evaluate the alternatives that he has. At the end, he still has to use his judgment and incorporate his value structure into his decisions. The analytical methods provide him with a framework to arrive at conclusions that are traceable and defendable according to rational decision rules.

This chapter presents a summary of methods that an engineering manager can use in making rational decisions. These methods have been developed in a body of literature called "decision theory." Before discussing the methods, we will define the terms used in decision theory.

[1] Business Week, March 31, 1980.

Action: An activity performed by the decision maker.

Strategy: A complete set of actions for the decision maker to adopt. (*Note:* This definition used in the decision theory literature is different from the definition of strategy in general-management literature.)

For example, invest $1 million in real estate, put $4 million into plant improvement. If the real estate appreciation is less than 25 percent over the next 2 years, sell it and use the proceeds for further plant improvements 2 years hence.

Alternatives: A set of strategies from which the decision maker chooses one.

Outcome (O_i): The result of alternative i if it is chosen.

Utility $U(O_i)$: The perceived value of outcome i according to the decision maker.

For example, if A_1 is the alternative chosen by the decision maker between two alternatives, he perceives the value of outcome 1 to be higher than the value of outcome 2, i.e.:

$$U(O_1) > U(O_2)$$

If the decison maker is indifferent between the two alternatives, the utilities of the two outcomes are equal, i.e.:

$$U(O_1) = U(O_2)$$

State of nature S_j: The uncontrollable conditions imposed on the decision maker by the external system.

Payoff P_{ij}: The result of alternative i under the state of nature j.

These definitions will be used in the decision methodologies discussed below.

TYPES OF DECISIONS

Decisions can be categorized in numerous ways. One major classification is according to the nature of the decision situation:

Single-criterion decisions

Multiple-criteria decisions

Single criterion decisions are easier to make. The alternatives are evaluated only in one dimension in these decisions. That decision could be the cost, the profit, or any other similar measure. In such decisions, either the objective of the decision maker has only

300   one relevant criterion or the outcomes of the various alternatives are identical with respect to all criteria except one.

Most decision situations cannot be reduced to a single-criterion case. Typically, the decision makers have to consider conflicting objectives under multiple criteria. Even in the selection of two cars, people do not decide on the basis of price alone. Styling, comfort, safety, and many other features play important roles in the decision. The multiple-criteria decisions are more complex than the single-criterion ones because it is difficult to find an alternative that dominates all others with respect to all criteria. Some alternatives are favored according to some of the criteria; some are favored according to other criteria.

An example can clarify the difference between these two types of decisions. Suppose that an aircraft manufacturer has received proposals for the landing gear assembly of a new aircraft. The engineering manager is a member of the project team, responsible for evaluating the proposals and identifying the best subcontractor. If the proposed assemblies are very similar except for their cost, the engineering manager's decision has only *one criterion*. He simply selects the subcontractor who shows the least cost in his proposal. But this is a very rare situation.

Let us look at this decision a little more closely. The criteria by which the alternatives are analyzed include the product design, the performance characteristics, the reliability of the landing gear, the track record of the subcontractor, the delivery date of the final product, as well as the cost. If the outcomes of all alternatives are identical with respect to a criterion, he can drop that criterion from consideration. In this example, the outcomes are identical for all criteria except cost. So the cost criterion remains as the only measure of effectiveness, and the least expensive alternative dominates the others in the decision. In other words, one of the alternatives is the dominant alternative, because its outcome is as good as the other alternative with respect to all criteria, and better than they with respect to one criterion. The critical assumption in this example is the similarity of the proposals. When this is not a valid assumption, the decision has to be based on multiple criteria. For example, suppose that all proposed landing gears can be used satisfactorily on concrete surface, but one of them can also be used for safe landing on sod. How much is this extra feature worth? Suppose also that the aircraft manufacturer previously had a bad experience with the subcontractor who came up with the lowest bid for the landing gear. That company was the low bidder in an earlier project, but could not deliver the hardware on time and delayed the whole project. How should this be reflected in the final decision? How about a contractor who comes up with an innovative design which is costlier but more reliable than the others? What value should be assigned to **301** that innovation?

The final decision will depend on the relative importance of each criterion as viewed by the decision maker. Thus, it will be a question of his "utilities." We will first discuss single-criterion decisions, then introduce the utility concept later in the chapter.

An important factor in all decisions is the amount of pertinent information available to the decision maker. Keep in mind that he selects the "best" alternative based on what he knows at a given point in time and what he assumes will happen during the next 3 or 4 years. What if the conditions change after the decision is made? The innovative design appears to be improving the reliability of the landing gear assembly. If the proposal is accepted, what is the likelihood that the final product will indeed perform as expected; and if it does, how much more reliable will it be? Also, if the low bidder had delayed a previous project, what is the probability that the same problem will occur? For example, suppose that the company had labor problems in the past and had not been able to deliver the hardware on time. What is the probability that it will not face a similar situation? More important, all of the subcontractors may have to deal with such problems. What is the probability that each one will deliver the final product on time? Note that the decision maker may have sufficient information to be certain about some of these issues, but for most of them he has to take a risk in his decision. Sometimes he can evaluate the risk in terms of probabilities, sometimes his decision is made in complete uncertainty. This implies another classification for decisions. We can categorize both single-criterion decisions and multicriteria decisions as follows:

1   Decisions under certainty

2   Decisions under uncertainty

3   Decisions under risk

DECISIONS UNDER CERTAINTY

When complete information is available on all alternatives and the outcomes are known with certainty, the decision maker selects the alternative which shows the best outcome. Suppose that you are enlarging the design office of your engineering firm and have received bids from several salesmen for the desks of your engineers. Two salesmen are offering identical desks at exactly the same price with exactly the same purchase conditions, but one can deliver the desks immediately while the other one will deliver in 6 months. Assuming that construction of the new office is well under way, you

302 have no reason to wait for 6 months. The decision is easily made to purchase the desks from the salesman who can deliver them right away. This is a single-criterion decision under certainty.

A modification of this problem can illustrate a multiple-criteria decision under certainty. Suppose that another design office in the city is about to be closed down. Desks from this office can be purchased at about half the price of the new desks, but some of them are too large and some too small for your purposes. You may have to rearrange your office if you buy them. Assume also that the desks that will be available 6 months later can be raised and lowered hydraulically, whereas the desks in the other engineering office have manual controls, and the desks available for immediate delivery have a fixed height that cannot be changed. Now you have three alternatives, as follows:

Decision Alternatives and Criteria

10.1

	CRITERIA				
ALTERNATIVES	1 PRICE, %	2 CONDITION	3 DELIVERY	4 ADAPTABILITY TO OFFICE	5 EASE OF USE
A_1: Immediate delivery (first manufacturer)	100	New	Now	Good	Poor (no adjustments in height)
A_2: Delayed delivery (second manufacturer)	100	New	Later	Good	Good (hydraulic control)
A_3: Used desks from another office	50	Used	Now	Bad	Medium (manual control)

Alternative A_3 has a good price but poor measures with respect to criteria 2 and 4. It is better than alternative A_1 with respect to criterion 5, but not as good as A_2. Alternative A_2 has automatic controls but late delivery. Your decision is complicated by these differences. It is necessary to compare the relative value of each criterion with respect to the others and then determine how good each alternative is within a given criterion. In other words, you have to make tradeoff decisions both among and within the five criteria before you can make a decision. These tradeoffs will involve the assignment of relative values to the criteria and to the alternatives for each criterion. Those values will be your utilities. Once you make those tradeoffs, you will be able to select the alternative that has the maximum utility for you. Of course, you might alter your decision completely if, let us say, one of the parties involved in this transaction is a minority owned business and your firm happens to be interested in developing a socially responsible image. In that case the value structure favoring a minority business is superimposed over the other considerations and could **303** very well lead you to choose that alternative.

DECISIONS UNDER UNCERTAINTY

On the other end of the decision spectrum are the decisions under uncertainty. These are the decisions whose outcomes are affected by conditions outside the decision maker's control, with the probabilities of occurrence of those conditions not known at all.

Let us consider an automobile manufacturer trying to grapple with the fuel shortage while satisfying the Environmental Protection Agency (EPA) regulations. It has already committed itself to reduce the pollutants coming out of the engine and allocated major resources for that purpose. Now it has to decide whether or not it should follow this strategy or follow a different one. A strategic move is about to be made that will affect the company for a long time to come. After long deliberations, the top management has reduced its alternatives to the following four:

A_1: Continue with the present strategy (use the current gasoline engine with continued effort on pollution control)

A_2: Improve fuel consumption of the current engine; continue with pollution control efforts

A_3: Develop diesel engine technology

A_4: Acquire a company with expertise in diesel-engine technology

Although the final decision will not be made by the engineering manager, he will play a key role in developing the company's strategy. He has to provide the basis for pros and cons of each alternative and justify his recommendations so that the top management can reach a decision.

In the analysis of this decision problem, we first have to identify the states of nature. These are the external conditions, uncontrollable by the decision maker. The company is concerned with two major factors: fuel availability and EPA regulations. Let us assume that the preliminary studies indicate that neither the fuel nor the EPA regulations will stay at their current level. The company can, therefore, safely assume that both will change, but it does not know the direction of that change. Based on this assumption, we can identify four states of nature:

S_1: Fuel available, EPA regulations relaxed

S_2: Fuel shortage, EPA regulations relaxed

S_3: Fuel available, EPA regulations strict

S_4: Fuel shortage, EPA regulations strict

Notice that S_1 is a favorable scenario for the automotive industry since there are no fuel shortages and also no strict environmental regulations. On the other hand, S_4 is unfavorable because fuel
shortage is accompanied by strict regulations. The net effect of S_2
and S_3 could be mixed because a relaxation in one factor is accompanied by a restriction in the other one. However, if the company
makes the right decision, it could benefit from each state of nature
whether it has a favorable, unfavorable, or mixed impact on the
overall industry. For example, even though S_4 is basically an unfavorable scenario, the company would stand to gain if it decides to go
with the second alternative and invests in a fuel-efficient gasoline
engine with low pollution. On the other hand, if S_1 is to be the state of
nature, that same alternative will turn out to be unnecessarily costly
and certainly not a good decision for the company. The company's
"payoff" under each state will depend on a number of other considerations, including the resource requirements for technology development, the competitive actions, the strategies of foreign-car makers,
the availability of companies to acquire, the length of development
and the risks involved in diesel technology, the technical feasibility of
improving the fuel performance of the current engine, etc.

Let us assume that after considering these and similar relevant questions, the net present value of each alternative under each
state of nature has been determined for the planning horizon of the
company. Table 10.2 shows these figures, P_{ij} in million dollars.

Looking at Table 10.2 column by column, we can make
some observations. For example, if fuel will be available and EPA
regulations relaxed, the best decision will be A_4, since all other
decisions will result in a lower payoff. Thus, if A_1 is chosen, the
company will have a "regret" of $1 million. A_2 will result in a regret
of $3 million, A_3 $2 million. Similarly, looking at the third column, if
fuel will be available but EPA will be strict, alternative A_2 will represent $1 million regret, A_3 and A_4 $4 million each.

Using this concept, a regret matrix, such as Table 10.3, can
be developed.

The regret R_{ij}^* for a particular state of nature j^* is calculated
as follows:

$$R_{ij^*} = \max (P_{ij}) - P_{ij}^*$$

Several decision principles have been developed for making
decisions under uncertainty using these two matrices. The final
decision depends on the principle chosen. Two persons could come
up with entirely different decisions using the same information if they

Payoff Matrix

STATES OF NATURE				**10.2**
ALTERNATIVES	S_1: FUEL AVAILABLE; EPA REGULATIONS RELAXED	S_2: FUEL SHORTAGE; EPA REGULATIONS RELAXED	S_3: FUEL AVAILABLE; EPA REGULATIONS STRICT	S_4: FUEL SHORTAGE; EPA REGULATIONS STRICT
A_1: Continue with present strategy	7	6	9	6
A_2: improve fuel efficiency	5	7	8	9
A_3: Develop diesel technology	6	10	5	7
A_4: Acquire company with diesel expertise	8	9	5	8

chose different decision principles. Selection of a decision principle reflects the individual's preference and risk-taking attitude.

Five decision principles will be used for this purpose. They will be illustrated using a summary of the information presented in Tables 10.2 and 10.3.

Regret Matrix

STATES OF NATURE				**10.3**
ALTERNATIVES REGULATIONS	S_1 FUEL AVAILABLE; EPA REGULATIONS RELAXED	S_2 FUEL SHORTAGE; EPA REGULATIONS RELAXED	S_3 FUEL AVAILABLE; EPA REGULATIONS STRICT	S_4 FUEL SHORTAGE; EPA REGULATIONS STRICT
A_1: Continue with present strategy	1	4	0	3
A_2: Improve fuel efficiency	3	3	1	0
A_3: Develop diesel technology	2	0	4	2
A_4: Acquire company with diesel expertise	0	1	4	1

306
For the evaluation of each alternative, we need to determine the maximum payoff, the minimum payoff, and the maximum regret that will be observed if that alternative is chosen. These values are calculated as follows:

$$(\text{Maximum payoff})_i = \max_j [P_{ij}]$$

$$(\text{Minimum payoff})_i = \min_j [P_{ij}]$$

$$(\text{Maximum regret})_i = \max_j [R_{ij}]$$

DECISION PRINCIPLES

MAXIMIN PRINCIPLE This is an ultraconservative principle with a pessimistic attitude toward the states of nature. It prepares the decision maker for the worst. The objective is to choose the alternative that will give the maximum of the minimum payoffs. The decision according to this principle is:

A_1:   Continue with the current strategy

MAXIMAX PRINCIPLE This is the optimistic principle. It considers the nature to be benevolent and suggests that, no matter what alternative is chosen, the best state of nature will occur. The objective is to get the maximum of the maximum payoffs. This principle appeals to a risk-taking decision maker. By the maximax principle, the decision is:

A_3:   Develop diesel technology

Summary of Payoff and Regret Matrices

10.4 ALTERNATIVE	MAXIMUM PAYOFF	MINIMUM PAYOFF	MAXIMUM REGRET
A_1: Continue with present strategy	9	6	4
A_2: Improve fuel efficiency	9	5	3
A_3: Develop diesel technology	10	5	4
A_4: Acquire company with diesel expertise	9	5	4

MINIMAX PRINCIPLE   The minimax principle seeks to minimize the   **307**
maximum regret. It does not concern itself directly with the payoffs
but rather deals with the loss of opportunity. It minimizes the risk of
"wrong decision" under any state of nature. According to the
minimax principle, the decision is:

A_2:   Improve fuel efficiency of present engine

LAPLACE PRINCIPLE   This principle, sometimes called the "principle of insufficient reasoning," assumes that each state of nature has
an equal probability of occurrence. The expected value of the outcome is calculated for each alternative. The alternative with the maximum average outcome is chosen.

The expected value of the outcome for alternative i is
defined as

$$E(O_i) = \sum_j p_j P_{ij} \tag{10.1}$$

where p_j = probability of the occurrence of state S_j.

In Laplace principle, $p_j = 1/j$ for all j. Therefore,

$$E(O_i) = \sum_j \frac{P_{ij}}{j}$$

$$E(O_1) = \frac{7 + 6 + 9 + 6}{4} = 7.00$$

$$E(O_2) = \frac{5 + 7 + 8 + 9}{4} = 7.25$$

$$E(O_3) = \frac{6 + 10 + 5 + 7}{4} = 7.00$$

$$E(O_4) = \frac{8 + 9 + 5 + 8}{4} = 7.50 \leftarrow \text{maximum}$$

The decision is:

A_4:   Acquire a company with diesel technology

HURWICZ PRINCIPLE   The decision maker using the Hurwicz principle first selects a coefficient of optimism. Its value varies between 0.0
and 1.0 and reflects the degree of his confidence in the occurrence of
a favorable state of nature.

The outcome of alternative i is then calculated as:

$$V(O_i) = \alpha \max_j [P_{ij}] + (1 - \alpha) \min_j [P_{ij}] \text{ for each } i$$

$\alpha = 0.0$ corresponds to the maximin principle and $\alpha = 1.0$ corresponds to the maximax principle. All other values of α represent vary-

308 ing degrees of optimism between those two extremes. For example, for $\alpha = 0.4$:

$$V(O_1) = 0.4\ (9) + 0.6\ (6) = 7.2 \quad \leftarrow$$
$$V(O_2) = 0.4\ (9) + 0.6\ (5) = 6.6$$
$$V(O_3) = 0.4\ (10) + 0.6\ (5) = 7.0$$
$$V(O_4) = 0.4\ (9) + 0.6\ (5) = 6.6$$

The decision is:

A_1: Continue with present strategy

Similarly, for $\alpha = 0.8$:

$$V(O_1) = 0.8\ (9) + 0.2\ (6) = 8.4$$
$$V(O_2) = 0.8\ (9) + 0.2\ (5) = 8.2$$
$$V(O_3) = 0.8\ (10) + 0.2\ (5) = 9.0 \quad \leftarrow$$
$$V(O_4) = 0.8\ (9) + 0.2\ (5) = 8.2$$

Now, the decision is:

A_3: Develop diesel technology

DECISIONS UNDER RISK

Decisions under risk are the decisions made with explicit recognition of the probabilities of events outside the decision maker's control. The expected value approach is used in the analysis and the information about the external conditions is incorporated into the decision as that information becomes available. We will discuss these concepts and apply them to our example.

EXPECTED VALUE OF DECISIONS   As defined in expression (10.1), the expected value of the outcome of an alternative A_i that has payoffs P_{ij} under states of nature S_j with respective probabilities p_j is

$$E(O_i) = \sum_{j=1}^{n} p_j P_{ij}$$

Expected value is not the value of an actual outcome. In fact, we may never observe that value. Assume for example, that alternative A_i has four payoffs with the following probability distribution:

S_j	S_1	S_2	S_3	S_4
P_j	(0.10)	(0.50)	(0.25)	(0.15)
P_{ij}	$500	−$100	0	$1000

The expected value is

$$E(O_i) = \sum_{j=1}^{4} p_j P_{ij}$$
$$= (0.10)(500) + (0.50)(-100) + (0.25)(0) + (0.15)(1000)$$
$$= 150$$

The only possible outcomes in this example are $500, -$100, $0, and $1000. The expected value of $150 will not even occur. It is merely the average of these outcomes weighted with their respective probabilities. It can be interpreted as the average value that will be obtained in the long run if this alternative is chosen repeatedly.

In the previous investment example, let us assume the following probabilities:

p (Fuel available, EPA regulations relaxed) $= 0.30$

p (Fuel shortage, EPA regulations relaxed) $= 0.50$

p (Fuel available, EPA regulations strict) $= 0.15$

p (Fuel shortage, EPA regulations strict) $= 0.05$

The expected value of continuing with present strategy can now be calculated.

$$E(O_1) = 0.3(7) + 0.5(6) + 0.15(9) + 0.05(6) = 6.75$$

Similarly, $E(O_2) = 6.65,$ $E(O_3) = 7.90,$ $E(O_4) = 8.05.$

According to the expected value criterion, the best decision is to acquire a company with diesel technology, because alternative 4 has the maximum expected value of $8.05 million. This does not imply that acquisition of the company is better than the other three alternatives under all states of nature. In fact, if we study the payoffs, we see that this will be a very bad decision if the state of nature happens to be S_3 (Fuel available, EPA regulations strict).

BAYESIAN DECISIONS

Bayes' theorem provides the basis for incorporating additional information into probabilistic decisions as that information becomes available. In general, we start analyzing probabilistic situations by assigning probabilities of certain events according to our earlier experience under similar conditions. These probabilities may or may not be accurate. We can improve them by obtaining additional information through experiments or surveys. After a sufficient number of these improvements, we can reach a reasonable level of accuracy for the probability distribution of the outcomes. Decisions using this approach are called "bayesian decisions" because of Bayes' theorem.

Bayes' theorem states that, when two events A and B are

310 considered, the conditional probability[2] $p(A \mid B)$ can be calculated if we know the conditional probability $p(B|A)$. Formally stated;

$$p(A \mid B) = p(B \mid A)\frac{p(A)}{p(B)}$$

or, in more general terms: if A_a, A_2, . . . ,A_n are partitions of event A such that $p(A_i) \neq 0$ for $i = 1, 2, \ldots n$, and also $p(B) \neq 0$, then

$$p(A_r|B) = \frac{p(B|A_r)p(A_r)}{\sum\limits_{i=1}^{n} p(B|A_i)p(A_i)} \qquad \text{for } r = 1, 2, \ldots, n$$

Let us explain this with an example. Suppose that a concrete structure has failed and we are interested in determining whether it was caused by a design error or the use of low-quality concrete in construction. Suppose also that, based on previous experience, we can assign the following probabilities:

p(Construction error) $= 0.7$

p(Design error) $= 0.3$

These are called the a priori probabilities. They reflect our judgment, experience, and observations of previous situations.

We can obtain additional information about the quality of the concrete by conducting a compression test on a concrete core taken from the structure. Let us suppose that such a test will be 90 percent reliable if low-strength concrete has been used and 95 percent reliable if normal-strength concrete has been used. In other words, if there was a construction error, we will detect it with 90 percent probability. Similarly, if there was no construction error, and the failure was due to a design error, the test will indicate it with a 95 percent probability.

Stated differently:

p(Low core strength | Construction error) $= 0.9$

p(Normal core strength | Construction error) $= 0.1$

and

p(Normal core strength | Design error) $= 0.95$

p(Low core strength | Design error) $= 0.05$

---

[2]Conditional probability $p(A \mid B)$ is the probability of A given B, meaning that if we know that event B has occurred, $p(A \mid B)$ represents the probability that event A will also occur. For example, suppose that 20 percent of engineers in an office have graduate degrees and that all female engineers employed in the office have at least a masters degree. If we randomly select an engineer from this office, we can say that there is a 20 percent probability that she or he has a graduate degree. Thus p(Graduate degree) $= 0.20$. However, if we know that the person we selected is female, the probability that we have selected an engineer with a graduate degree becomes 100 percent. In other words, p(Graduate degree|Female) $= 100$ percent.

With this information, we can determine the probability that the test **311** will show normal strength, $p(N)$, and low strength, $p(L)$.

$$p(L) = p(L|C)p(C) + p(L|D)p(D)$$
$$p(L) = 0.9(0.7) + 0.05(0.3) = 0.645$$

and

$$p(N) = 1 - 0.645 = 0.355$$

After developing these conditional probabilities, suppose we conducted the test and it showed normal strength. We can now calculate the probabilities of design and construction errors as follows:

$$p(C|N) = \frac{p(N|C)p(C)}{p(N)}$$

$$= \frac{(0.1)\,0.7}{0.355}$$

$$= 19.7 \text{ percent}$$

Therefore, $p(D|N) = 100 - 19.7 = 80.3$ percent.

Similarly, if the test shows low strength:

$$p(C|L) = \frac{p(L|C)p(C)}{p(L)}$$

$$= \frac{(0.9)0.7}{0.645}$$

$$= 97.7 \text{ percent}$$

and $p(D|L) = 100 - 97.7 = 2.3$ percent.

In summary, by performing the concrete test, we have improved our knowledge of each type of error and obtained the a posteriori probabilities as shown in Table 10.5.

| | | A Priori and A Posteriori Probabilities | |
| | | A POSTERIORI PROBABILITIES, % | **10.5** |
TYPE OF ERROR	A PRIORI PROBABILITIES, %	IF TEST SHOWS LOW STRENGTH	IF TEST SHOWS NORMAL STRENGTH
Design error	30.0	2.3	80.3
Construction error	70.0	97.7	19.7
Total	100.0	100.0	100.0

312
The additional information obtained from the compression test has thus provided us with the capability of making a much more precise statement about the failure and arriving at a better decision.

We will use the bayesian decision concept when we discuss the value of information in the next section.

VALUE OF INFORMATION If the state of nature is known with certainty, the decision, of course, is to choose the best alternative for that particular state. This requires "perfect information" prior to the occurrence of the state. For example, if the automobile manufacturer is sure that there will be plenty of fuel and relaxed EPA regulations, looking at the first column of Table 10.1 it immediately decides to acquire a company with diesel expertise. Its decisions are similarly simple for all states of nature, as shown in Table 10.6.

The concept of perfect information is equivalent to the prediction of future events with certainty. We have never been able to make such predictions in nontrivial situations; we never will be. However, even though there can be no perfect information, we can use this concept to determine the value of information to the decision maker.

Suppose for a moment that perfect information is indeed available, at a cost. How much should we be willing to pay for it?

Before we recommend the purchase of such information, we should determine the expected value of our decision assuming that we know exactly what the external conditions will be. Note that no matter how accurate our information is, we cannot change the probabilities of the states of nature. All we can do is make better decisions using the additional information. Since the probabilities will not be affected, we can conclude that in our example there is a 0.3 probability that the information source will indicate S_1, a 0.5 probability that it will indicate S_2, etc.[3]

Once that information is available to us, we will make the best decision corresponding to the state of nature indicated. Therefore, the expected value of our decision with perfect information will be

$$(0.3) (8) + (0.5) (10) + (0.15) (9) + (0.05) (9) = \$9.20 \text{ million}$$

Remember that the expected value of the best decision without information was $8.05 million. The value of "perfect information" is the difference between these two, i.e.:

$$9.20 - 8.05 = \$1.15 \text{ million}$$

This figure shows that additional information, in this case, no matter how reliable, is not worth more than $1.15 million for us. In reality, of

---

[3]These are the probabilities of the states of nature that we assumed when we calculated the expected value of the decision without perfect information.

Decisions with Perfect Information

10.6

STATE OF NATURE	MAXIMUM PAYOFF	ALTERNATIVE TO CHOOSE
S_1: Fuel available; EPA regulations relaxed	8	Acquire company with diesel expertise
S_2: Fuel shortage; EPA regulations strict	10	Develop diesel technology
S_3: Fuel available; EPA regulations relaxed	9	Continue with present strategy
S_4: Fuel shortage; EPA regulations relaxed	9	Improve fuel efficiency

course, its value is considerably lower, because it will never be perfect information.

After establishing the upper limit for the value of information, let us look at the realistic case of partial information. Let us assume, for example, that we are considering using the services of a management consulting firm that specializes in the automotive industry. Its forecast will indicate if the combined fuel and EPA conditions will be favorable, unfavorable, or mixed for the industry. Suppose also that based on previous performance this firm has developed the following probabilities as a measure of reliability for its forecasts:

If the state of nature will be S_1 (Fuel available, EPA regulations relaxed), the forecast is likely to be:

F = Favorable, with a probability of 0.4

M = Mixed, with a probability of 0.4

U = Unfavorable, with a probability of 0.2

In other words:

$$p(F \mid S_1) = 0.4$$
$$p(M \mid S_1) = 0.4$$
$$p(U \mid S_1) = 0.2$$

Let us assume that all conditional probabilities are shown in Table 10.7.

Using the probabilities of the four states of nature, i.e.,

$$p_1 = p(S_1) = 0.30$$
$$p_2 = p(S_2) = 0.50$$

Conditional Probabilities for the Forecasts

10.7

IF THE STATE OF NATURE IS	ECONOMIC FORECAST			Σ
	F = FAVORABLE	M = MIXED	U = UNFAVORABLE	
S_1	0.4	0.4	0.2	1.0
S_2	0.3	0.6	0.1	1.0
S_3	0.2	0.3	0.5	1.0
S_4	0.0	0.4	0.6	1.0

$$p_3 = p(S_3) = 0.15$$

$$p_4 = p(S_4) = 0.05$$

the probability of each forecast is calculated.

$$p(F) = \sum_{j=1}^{4} p(F|S_j)p(S_j)$$
$$= 0.4(0.3) + 0.3(0.5) + 0.2(0.15) + 0.0(0.05) + 0.30$$

Similarly,

$$P(M) = 0.49 \quad \text{and} \quad P(U) = 0.21$$

We can also calculate the conditional probabilities of the four states of nature for each of the three economic forecasts.

$$p(S_j \mid F) = p(F \mid S_j) \frac{p(S_j)}{p(F)} \qquad (10.2)$$

$$p(S_j \mid M) = p(M \mid S_j) \frac{p(S_j)}{p(M)} \qquad (10.3)$$

$$p(S_j \mid U) = p(U \mid S_j) \frac{p(S_j)}{p(U)} \qquad (10.4)$$

The conditional probabilities $p(F \mid S_j)$, $p(M \mid S_j)$, and $p(U \mid S_j)$ are shown in Table 10.7. The probabilities of the four states of nature $p(S_j)$ are the a priori probabilities, i.e., $p(S_1) = 0.3$, $p(S_2) = 0.5$, $p(S_3) = 0.15$, and $p(S_4) = 0.05$. These values and the probabilities of the three forecasts are combined to calculate all conditional probabilities using the Bayesian relationships of expressions (10.2) through (10.4). Table 10.8 shows these probabilities.

Let us assume that we agreed to purchase the service of the forecasting firm and it indicated that the overall conditions will be favorable. Based on this information, the expected value of the outcome of each of the four alternative decisions is determined as follows:

$$E(O_i|F) = \sum_{j=1}^{4} p(S_j|F)P_{ij}$$

Conditional Probabilities of the States of Nature
for Each Forecast

10.8

STATE OF NATURE PROBABILITY	ECONOMIC FORECAST		
	F = FAVORABLE	M = MIXED	U = UNFAVORABLE
$p(S_1)$	0.40	0.25	0.28
$p(S_2)$	0.50	0.62	0.23
$p(S_3)$	0.10	0.09	0.35
$p(S_4)$	0.00	0.04	0.14
Total	1.00	1.00	1.00

The conditional probabilities $p(S_j|F)$ are in Table 10.8 and the payoffs, P_{ij} are shown in Table 10.2, the payoff matrix.

$$E(O_1 \mid F) = (0.4)\,(7) + (0.5)\,(6) + (0.1)\,(9) + (0.0)\,(6) = 6.70$$

$$E(O_2 \mid F) = (0.4)\,(5) + (0.5)\,(7) + (0.1)\,(8) + (0.0)\,(9) = 6.30$$

$$E(O_3 \mid F) = (0.4)\,(6) + (0.5)\,(10) + (0.1)\,(5) + (0.0)\,(7) = 7.90$$

$$E(O_4 \mid F) = (0.4)\,(8) + (0.5)\,(9) + (0.1)\,(5) + (0.0)\,(8) = 8.20$$

Expected values with other economic forecasts are also calculated in a similar way as shown in Table 10.9.

As indicated in Table 10.9, if the forecast is favorable, the maximum expected value is obtained using alternative 4. If the forecast is mixed, the best alternative is A_3; and if the forecast is unfavorable, the best alternative is A_1.

Expected Values of the Alternatives after
Economic Forecasts

10.9

ALTERNATIVES	ECONOMIC FORECAST		
	F	M	U
A_1: Continue with present strategy	6.70	6.52	7.33
A_2: Improve fuel efficiency	6.30	6.67	7.07
A_3: Develop diesel technology	7.90	8.43	6.71
A_4: Acquire company with diesel expertise	8.20	8.35	7.18
Best decision	A_4	A_3	A_1

Since we have the estimates for the probabilities of the three forecasts, we can now calculate the expected value of the decision with this additional information:

$$(0.3) (8.20) + (0.49) (8.43) + (0.21) (7.33) = 8.13$$

The value of information in this case is $8.13 - 8.05 = \$0.08$ million. Thus, if the forecast is offered at a price less than \$80,000, the company can justify purchasing that service according to the expected-value criterion.

In this example, the calculations for the value of information followed the bayesian decision approach by evaluating the sequence of decisions in a systematic way. Such decisions can also be represented pictorially as the branches of a symbolic tree called the "decision tree."

THE DECISION TREE

A decision tree is a graphical representation of the decision process. Sequential decisions are drawn in the form of the branches of a tree, stemming from an initial decision point and extending all the way to the final outcomes. Each path through the branches of the tree represents a separate series of decisions and probabilistic events. The decisions are evaluated by calculating the expected value of each path. The analysis starts at the end points of the tree and continues by "rolling back" the decision tree to the starting point.

There are five components of a typical decision tree:

1 Decision points: Points representing the decisions to be made, the alternatives to be chosen. Decision points are shown as square boxes in this book.

2 Chance points: Points where a probabilistic event occurs. These points are shown as circles in this book.

3 Branches: The lines connecting the decision points and the chance points in a sequential way. The branches may represent alternative decisions or states of nature.

4 Probabilities: The probabilities of the chance-dependent events are shown on the branches representing those events. In most cases, these are conditional.

5 Payoff: The payoff of each alternative is placed at the tip of the branch that represents the sequence of decisions and events included in that alternative.

The automobile manufacturer's decision problem discussed above is depicted as a decision tree in Fig. 10.1.

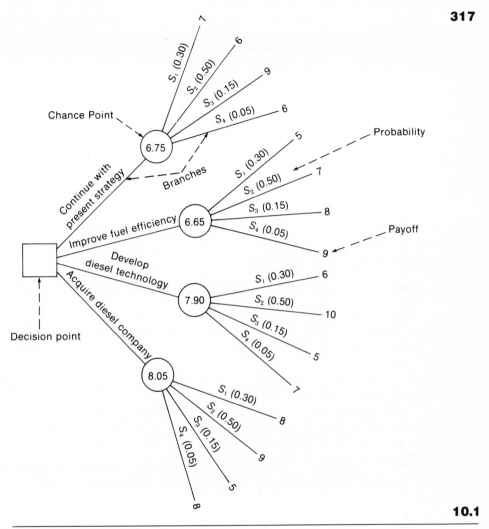

Decision tree.

When the tree is rolled back, the expected value of each alternative is calculated as follows:

$$E(\text{Present strategy}) = (0.3)(7) + (0.5)(6) + (0.15)(9) + (0.05)(6) = 6.75$$

$$E(\text{Improved fuel efficiency}) = 6.65$$

$$E(\text{Development of diesel technology}) = 7.90$$

$$E(\text{Acquisition of diesel firm}) = 8.05$$

These, of course, are equal to the expected values previously calculated for each alternative. The decision tree is not necessary to

318   make these calculations. It is just a convenient model representing the decision process for a better understanding of the relationships among the decision elements.

Calculations for the value of information can be performed on a slightly modified decision tree. If we want to calculate the value of information, we typically start the tree with a chance point, rather than a decision point. This is because we are faced with a probabilistic event concerning the nature of the information that we receive even before selecting an alternative.

The decision tree shown in Fig. 10.2 depicts the conditions described in the example with economic forecast. Note that the whole decision-tree structure of Fig. 10.1 is repeated three times in Fig. 10.2, once for each forecast. Regardless of what the forecast is, the decision maker always has the same four alternatives. The states of nature or their probabilities cannot be controlled by the decision maker. No matter what decision is made, the states of nature will be S_1 with a probability of 0.3, S_2 with 0.5, S_3 with 0.15, and S_4 with 0.05. However, in Fig. 10.2, the decision maker will see the forecast before selecting an alternative. This additional information will provide a better insight into the probabilities of the states of nature. The relevant probabilities are the conditional probabilities in this case. Even though the overall probabilities do not change, we know from previous experience that, for example, if the forecast is favorable, we can safely assume that the probability of a simultaneous fuel shortage and strict EPA regulations is negligible; and the probability of fuel availability together with relaxed EPA regulations is as high as 0.4. These are the conditional probabilities shown in Table 10.8.

The expected values at the chance points can now be calculated. For example,

$$E(a) = E(O_1|F) = \sum_{j=1}^{4} p(S_j|F)p_{1j}$$

$$= (0.4)(7) + (0.5)(6) + (0.1)(9) + (0.0)(6)$$

$$= 6.70$$

When these calculations are performed for points a through 1, the maximum expected value for each economic forecast is determined as follows:

$E(d) = E(O_4|F) = 8.20$ Maximum when the forecast is "favorable"

$E(g) = E(O_3|M) = 8.43$ Maximum when the forecast is "mixed"

$E(i) = E(O_1|U) = 7.33$ Maximum when the forecast is "unfavorable"

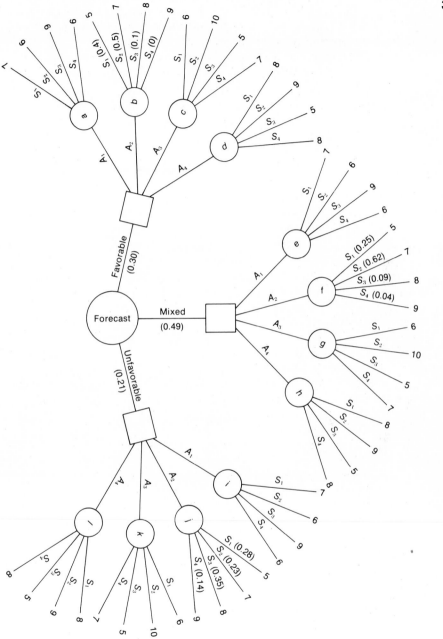

10.2

Decision tree to calculate the value of information.

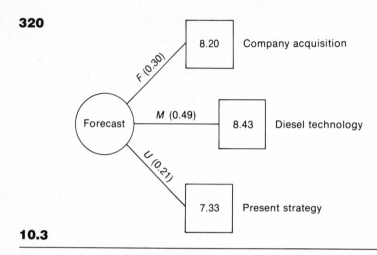

10.3

Therefore based on the expected value criterion, the decisions are:

Acquire company with diesel expertise if forecast is favorable

Develop diesel technology if forecast is mixed

Continue with present strategy if forecast is unfavorable

The decision tree is now reduced to that shown in Fig. 10.3. Rolling it back further, the expected value of the decision with forecast is

$$(0.30)\ (8.20) + (0.49)\ (8.43) + (0.21)\ (7.33) = 8.13$$

The value of forecast is the difference between 8.13 and 8.05, as calculated before.

The expected-value criterion used in the decision-tree applications assumes that the value assigned to monetary payoff is directly proportional to the payoff, regardless of the risks involved. It implies that people are neutral toward risk. This is not always true. When there are high risks in some outcomes, risk averters avoid them even when the payoff is high. On the other hand, risk takers prefer such outcomes to low-risk–low-reward alternatives. Thus, the value assigned to a payoff depends not only on the payoff itself, but also on the value structure of the decision maker.

UTILITY THEORY   The utility theory developed by Von Neuman and Morgenstern[1] assumes rational behavior of people in selecting among alternatives but does not assume risk neutrality of decision makers.

[1]J. Von Neumann and O. Morgenstern, *Theory of Games and Economic Behavior*, rev. ed., Princeton: Princeton, N.J., 1953.

We can develop a curve, known as the "utility function," for **321** any decision maker by asking him a series of questions about the value he assigns to various levels of reward. The relative values assigned to the rewards are measured by utiles and plotted on the y axis. The rewards are plotted on the x axis.

Let us assume that we will develop the utility curve for a decision maker in the 0- to $10-million range. To start the process, we can define the end points of the curve by setting $U(0) = 0.0$ and $U(10 \text{ million}) = 1.0$. The next step is to present the decision maker with a series of gambling situations, such as:

Do you prefer $5 million with certainty, or a gamble involving a 50 percent probability of $10 million and a 50 percent probability of $0?

If he is a risk averter, he is likely to take the $5 million. On the other hand, a risk taker could very well prefer the gamble. If he takes the $5-million option, it indicates that

$$U \text{ (5 million)} > (0.5)U(0) + (0.5)U(10 \text{ million})$$

or

$$U \text{ (5 million)} > 0.5 \ (0) + 0.5 \ (1)$$

$$U \text{ (5 million)} > 0.5$$

We then start changing the odds until he reaches the point of indifference. For example, he may be indecisive between $5 million with certainty and the gamble involving 70 percent probability of $10 million and 30 percent probability of $0. At this point, we can determine his utility for $5 million as follows:

$$U \text{ (5 million)} = 0.3 \ (0) + 0.7 \ (1) = 0.7$$

Now we have three points on the curve, and we can use any two of them to obtain the other points. For example, we present him with a situation where $7 million is available to him with certainty, but he could also take a gamble that gives him $10 million with probability p and $5 million with probability $1 - p$. In other words, we ask him to make the following comparison:

$$U \text{ (7 million) versus } (p)U(10) + (1 - p) \ U(5 \text{ million})$$

or

$$U \text{ (7 million) versus } p(1) + (1 - p)(0.7)$$

that is,

$$U \text{ (7 million) versus } (p + 0.7 - 0.7p) = 0.7 + 0.3p$$

By varying p, we again reach the point of indifference and determine the utility of $7 million in relative units. This method is called the

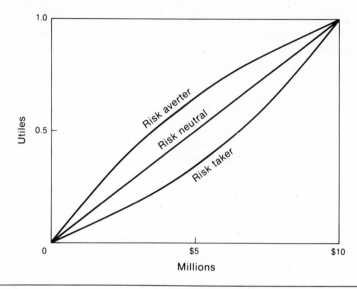

10.4

Utility functions.

"standard gamble." After obtaining several points similar to these, we combine them by a smooth curve and use the utiles instead of the dollar values in calculations. When utility theory is applied to a decision tree, the values at branch ends are expressed as the utilities of the payoffs rather than the payoffs themselves. The rest of the calculations are the same as before.

The curvature of the utility function can be interpreted as the indication of the decision maker's degree of willingness to take risks. A concave downward curve shows risk aversion, a concave upward curve shows willingness to gamble, and a straight line represents risk neutrality, as shown in Fig. 10.4.

Once the utility function is developed for a decision maker, its scale can be changed by a linear transformation in the form:

$$U_j^{\text{new}} = a + bU_j^{\text{old}}$$

where U_j^{old} = value of the original function at point j
  U_j^{new} = value of the transformed function at the same point
  a and b = constants.

This transformation is possible because of the interval scale[5] used in the development of utility functions.

The utility concept is particularly useful when nonmonetary outcomes are considered. Since most decisions involve multidimensional issues, the tradeoffs among the various criteria become critical

---

[5]Interval scale has no absolute zero, but it has a consistent distance between any pair of measurements. It permits linear transformations. Temperature is an example of this scale of measurement. Celsius degrees can be converted to Fahrenheit by the transformation $F° = a + b(C°)$, where $a = 32$ and $b = 9/5$.

in selecting an alternative. For example, market share, return on investment, and social responsibility are three criteria that need to be evaluated simultaneously in many decisions. They all use different units of measurement. In many situations, an improvement along one dimension leads to a sacrifice along the others. The decision maker has to decide how much he is willing to reduce his profits for a unit increase in market share, or for a measurable improvement in his organization's public image in terms of social responsibility. These tradeoffs can be made if all the criteria can be expressed and evaluated by the same units of measurement. If utilities are used instead of dollars, market share, and image indices, a common measure can be developed for evaluations and tradeoff decisions. This is the general approach used in multiple-criteria decision problems.

DECISIONS UNDER MULTIPLE CRITERIA

When decisions are made under multiple criteria, analysis is based on a comparison of the value of a unit change along each criterion. One of the common methods used for this purpose is an additive model with relative weights attached to the criteria. The assumption for such a model is that the combined utility function of the multivalued outcomes is additive. This means that we can state the combined utility function as

$$U(A_i) = U(O_{i1}, O_{i2}, \ldots O_{in})$$
$$= w_1 u_1(O_{i1}) + w_2 u_2(O_{i2}) + \ldots + w_n u_n(O_{in})$$

In this expression, the outcome of alternative i is measured along n criteria. Suppose that the automobile manufacturer in our example

Alternatives-Criteria Matrix for Multiple-Criteria Decisions

10.10

ALTERNATIVES	CRITERIA		
	1. NET PRESENT VALUE (NPV)	2. MARKET SHARE	3. SOCIAL RESPONSIBILITY
A_1: Continue with existing strategy	$O_{11} = \$6M$	$O_{12} = 15\%$	$O_{13} = $ Medium
A_2: Improve fuel efficiency	$O_{21} = \$7M$	$O_{22} = 25\%$	$O_{23} = $ Good
A_3: Develop diesel technology	$O_{31} = \$10M$	$O_{32} = 20\%$	$O_{33} = $ Medium
A_4: Acquire company with diesel expertise	$O_{41} = \$9M$	$O_{42} = 30\%$	$O_{43} = $ Poor

324   wants to evaluate the four alternatives with respect not only to their net present value, but also to the market share and social responsibility. Let us assume that the following values have been obtained for the outcomes under S_2 (fuel shortage and relaxed EPA rules).

The utility functions for profits and market share can be developed using the standard gamble approach. The same procedure is also applicable to social responsibility, even though the utility function in that case is discrete rather than continuous. Let us assume that the utility functions for the three criteria have been developed as shown in Figs. 10.5 to 10.7. Note that in these functions the utility of the best outcome is considered to be equal to 1.0 and the utility of the worst outcome is given a value of zero. This is a convenient scale for comparing alternatives.

Using the utility functions depicted in Figs. 10.5 through 10.7, we can evaluate each alternative as a linear additive model. This form of model is appropriate only if the criteria are preferentially independent, i.e., only if the decision maker's judgment concerning the values of one criterion is not affected by the values of the other criteria. This does not necessarily imply technical independence. As

Utility function for NPV of profits.

10.5

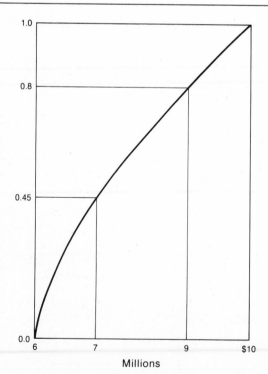

Millions

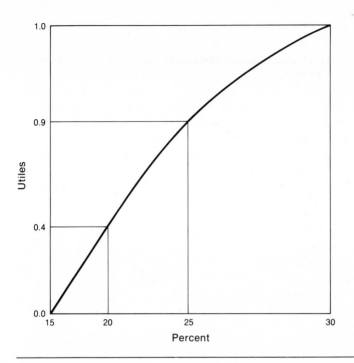

Utiles

Percent

10.6

Utility function for market shares.

long as the decision maker feels that the criteria are independent decision elements, we can justify an additive model.[6]

$$U(A_1) = w_1u(\$6 \text{ million}) + w_2u(15\%) + w_3u(\text{Medium})$$
$$= w_1(0) + w_2(0.0) + w_3(0.6)$$

[6]Description of tests for preferential independence can be found in E. S. Buffa and J. S. Dyer, *Management Science/Operations Research*, Wiley, New York, 1977.

Utility scale for social responsibility.

10.7

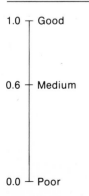

1.0 — Good

0.6 — Medium

0.0 — Poor

326   Similarly,

$$U(A_2) = w_1(0.45) + w_2(0.9) + w_3(1.0)$$
$$U(A_3) = w_1(1.0) + w_2(0.4) + w_3(0.6)$$
$$U(A_4) = w_1(0.8) + w_2(1.0) + w_3(0.0)$$

Each of the above expressions can be reduced to a single numerical value after the weights w_1, w_2, and w_3 are determined. The w's are the relative weights reflecting the importance of each criterion going from its worst value to the best value.

ASSESSMENT OF PRIORITY WEIGHTS

The weights in the additive utility function can be obtained by pairwise comparisons of the criteria. The decision maker is asked to compare the relative importance of going from the worst to the best value of two criteria. He allocates a total of 100 points to the two criteria reflecting his judgment of the relative importance of each. For n criteria, $n(n-1)/2$ such comparisons are made. The pairs are randomized before being presented to the decision maker. Because of the allocation of a constant (100 points) to the pair in the comparisons, this method is sometimes called the "constant sum method".[7] In the example, there will be $3(2)/2 = 3$ comparisons to obtain the relative weights. Let us assume that the decision maker makes these comparisons and provides the values shown below:

Criterion 2   (60)

versus

Criterion 1   (40)

Criterion 3   (30)

versus

Criterion 2   (70)

Criterion 1   (60)

versus

Criterion 3   (40)

[7] J. P. Guilford, *Psychometric Methods*, 2d ed., McGraw-Hill, New York, 1954; Dundar F. Kocaoglu, "Hierarchical Decision Model under Multiple Criteria," paper presented at TIMS/ORSA meeting, Washington, D. C. May 5, 1980.

When the decision maker makes these allocations, he implicitly as-   **327**
signs a value to each criterion. Those values can be recovered as
follows:

1   Form a matrix using the points allocated to criteria in pairwise comparisons:

Criterion

Criterion   1   2   3

$$
\begin{array}{c}
1 \\
2 \\
3
\end{array}
\begin{bmatrix}
\cdots & 60 & 40 \\
40 & \cdots & 30 \\
60 & 70 & \cdots
\end{bmatrix}
$$

Matrix A

In this matrix a_{ij} represents the points assigned to j when it is
compared with i. Thus $a_{12} = 60$ and $a_{21} = 40$ mean that when
criterion 2 was compared with criterion 1, they were assigned 60 and 40 points, respectively.

2   Form a second matrix (matrix B) by calculating
the ratios of reciprocal cells of matrix A—i.e., $b_{ij} = a_{ij}/a_{ji}$ for
all i, j except for $i = j$. $b_{ii} = 1.00$.

$$
\begin{array}{c}
1 \\
2 \\
3
\end{array}
\begin{bmatrix}
1.0 & 3/2 & 2/3 \\
2/3 & 1.0 & 3/7 \\
3/2 & 7/3 & 1.0
\end{bmatrix}
$$

Matrix B

Note that each row of matrix B contains the ratios of the
judgmental values assigned by the decision maker to the
various criteria as shown in Table 10.11.

3   Divide each element of matrix B by the element in
the same row of the next column. Do this for columns 1
through $n - 1$, and form matrix C.

Cell Values in Matrix B

	1	2	3	**10.11**
1	w_1/w_1	w_2/w_1	w_3/w_1	
2	w_1/w_2	w_2/w_2	w_3/w_2	
3	w_1/w_3	w_2/w_3	w_3/w_3	

$$c_{ij} = \frac{b_{ij}}{b_{i,j+1}} \qquad \begin{array}{l} for \ i = 1, \ldots, n \\ j = 1, \ldots, n-1 \end{array}$$

$$\begin{array}{c} \begin{array}{cc} w_1/w_2 & w_2/w_3 \end{array} \\ \begin{array}{c} 1 \\ 2 \\ 3 \end{array} \begin{bmatrix} 2/3 & 9/4 \\ 2/3 & 7/3 \\ 9/14 & 7/3 \end{bmatrix} \end{array}$$

Matrix C

4   Calculate the mean of the column values:

$$\frac{w_1}{w_2} = \frac{1}{3}\left(\frac{2}{3} + \frac{2}{3} + \frac{9}{14}\right) = \frac{83}{126}$$

$$\frac{w_2}{w_3} = \frac{1}{3}\left(\frac{9}{4} + \frac{7}{3} + \frac{7}{3}\right) = \frac{83}{36}$$

5   Assign the value of unity to w_3; calculate w_2 and w_1, using the mean ratios obtained in step 4.

$$w_2 = \frac{w_2}{w_3} \, w_3 = \frac{83}{36} \times 1 = 2.31$$

$$w_1 = \frac{w_1}{w_2} \frac{w_2}{w_3} \, w_3 = \frac{83}{36} \times \frac{83}{126} \times 1 = 1.52$$

6   Normalize the weights obtained in step 5.

$$w_i^{\text{normalized}} = \frac{w_i}{\sum\limits_{i=1}^{n} w_i}$$

Thus,

$$w_1 = \frac{1.52}{1.0 + 2.31 + 1.52} = 0.31$$

$$w_2 = \frac{2.31}{1.0 + 2.31 + 1.52} = 0.48$$

$$w_3 = \frac{1.0}{1.0 + 2.31 + 1.52} = \underline{0.21}$$

$$\sum = 1.00$$

7   Repeat steps 3 to 6 for all possible orientations of the columns of matrix B. Note that there are $n!$ column ori-

entations for n criteria. In our example, where $n = 3$, there are $3! = 6$ orientations listed below:

$$1 - 2 - 3$$
$$1 - 3 - 2$$
$$2 - 1 - 3$$
$$2 - 3 - 1$$
$$3 - 1 - 2$$
$$3 - 2 - 1$$

The criteria weights for the first orientation were already determined. Calculations for the other orientations are also similar.

COLUMN ORIENTATION	w_1	w_2	w_3
$1 - 2 - 3$	0.31	0.48	0.21
$1 - 3 - 2$	0.31	0.48	0.21
$2 - 1 - 3$	0.31	0.48	0.21
$2 - 3 - 1$	0.31	0.48	0.21
$3 - 1 - 2$	0.31	0.48	0.21
$3 - 2 - 1$	0.31	0.48	0.21
Mean =	0.31	0.48	0.21

If the decision maker could assign consistent relative weights to the criteria and allocate the 100 points in pairwise comparisons with consistency, it would be easy to recover the criteria weights by using any row of matrix B. However, such an assumption cannot be made. People cannot consistently follow the rules of rationality in decisions. In many situations, we make the following observations:

$$A > B \quad (A \text{ preferred to } B)$$
$$B > C \quad (B \text{ preferred to } C)$$
$$C > A \quad (C \text{ preferred to } A)$$

In order to be consistent, the third relationship would have to be $A > C$, but in reality, we may not feel that way. This is called intransitivity. For example, when a weapon system is being considered, we could feel that

Technical superiority is more important than cost.

Cost is more important than political acceptability.

But

> Political acceptability is more important than technical superiority.

This is a fact that we cannot change. Therefore, we cannot assume that people will always be consistent in their judgments when they compare two items. Even if a decision maker is consistent in the ranking of alternatives, it is unlikely that he will consistently assign the same relative weight to an alternative when he compares it with all the others one by one. In other words, even if he feels that $A > B$, $B > C$, and $A > C$, he cannot be expected to give consistent proportional values between A and B, B and C, and A and C when he makes those comparisons. If he were comparing physical properties such as weights or distances, he could maintain consistency in all comparisons. One yard is always equal to three feet regardless of whether it is compared with an inch or a mile. When subjective measures are used in comparisons, the values assigned to the elements tend to fluctuate in the decision maker's mind. Because of these fluctuations, the w_i/w_j values in a column of matrix C are not constant for all rows of that column. Steps 4 and 7 smooth out the effects of inconsistent evaluations. The final result is the vector of average priority values assigned to the criteria by the decision maker.

The variance of w_i values obtained in $n!$ column orientations in step 7 is a measure of inconsistency of the subjective values used in pairwise comparisons. In our example, the assignments shown in matrix A are consistent enough not to show any variation in the w_i obtained in step 7, but in other situations this may not be the case.

Using the w_i as the criteria weights in the multiple-criteria decision example, we can now evaluate each of the four alternative strategies.

$$U(A_1) = 0.31 \, (0.0) + 0.48 \, (0.0) + 0.21 \, (0.6) = 0.13$$

$$U(A_2) = 0.31 \, (0.45) + 0.48 \, (0.9) + 0.21 \, (1.0) = 0.78$$

$$U(A_3) = 0.31 \, (1.0) + 0.48 \, (0.4) + 0.21 \, (0.6) = 0.63$$

$$U(A_4) = 0.31 \, (0.8) + 0.48 \, (1.0) + 0.21 \, (0.0) = 0.73$$

Alternative 2 (improving fuel efficiency) maximizes the expected utility. It is the alternative to be chosen when we use this approach.

The approach described above for priority weights can also be used to assess subjective probabilities and to solve decision problems involving multilevel decisions.

SUBJECTIVE PROBABILITIES

Earlier in this chapter, we mentioned that the probabilities used in our decisions reflect the degree of our belief in the occurrence of one-

time events. Usually such probabilities cannot be obtained by ob- **331**
serving the frequency of those events in repetitive experiments. We
need subjective methods to develop them.

Probabilities can be either discrete or continuous. Discrete
probabilities are used when discrete events are considered. The four
states of nature were presented as four discrete events in our example in this chapter. The probabilities associated with those states
were discrete. Continuous probabilities are used when the event is
defined over an interval, rather than at a discrete point.

There are three characteristics of probabilities:

1   Probabilities take on values between zero and
one.

2   Probabilities of the outcomes spanning the total
events sum to one, i.e.:

$$\sum_i p_i = 1.0 \qquad \text{for discrete probabilities}$$

and

$$\int_{-\infty}^{\infty} f(x)\,dx = 1.0 \qquad \text{for continuous probabilities}[8]$$

3   Probabilities are measured in ratio scale. In other
words, there is an absolute zero in probability measurements, and if $p(x_1) = a \cdot p(x_2)$, we conclude that the likelihood
of occurrence of x_1 is a times that of x_2.

The values obtained by the method presented above for priority
measurements satisfy these three conditions. We describe the states
of nature to the decision maker and ask him to distribute 100 points
to the two elements of each pair of states. He makes the allocation in
such a way that it represents his belief for each state to occur in comparison with the other state. In other words, he expresses his subjective judgment for the probability of each state of nature to occur
when the universe consists of only two possible states at a given time.
After the computations are completed, the subjective probabilities
are obtained in a way similar to that described for obtaining relative
weights in the previous paragraph.

Let us consider the four states of nature, and assume
that the following values have been obtained in the comparisons of
randomized pairs:[9]

---

[8]$f(x)$ is the commonly used notation to denote the continuous probability function associated with variable x.
[9]Zero is not used in pairwise comparisons in this method because of computational difficulties. If the
decision maker wants to allocate 100 versus 0, he should allocate 99 versus 1.

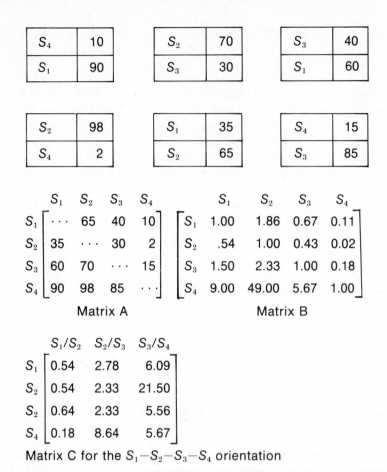

S_4	10
S_1	90

S_2	70
S_3	30

S_3	40
S_1	60

S_2	98
S_4	2

S_1	35
S_2	65

S_4	15
S_3	85

	S_1	S_2	S_3	S_4
S_1	$\cdots$	65	40	10
S_2	35	$\cdots$	30	2
S_3	60	70	$\cdots$	15
S_4	90	98	85	$\cdots$

Matrix A

	S_1	S_2	S_3	S_4
S_1	1.00	1.86	0.67	0.11
S_2	.54	1.00	0.43	0.02
S_3	1.50	2.33	1.00	0.18
S_4	9.00	49.00	5.67	1.00

Matrix B

	S_1/S_2	S_2/S_3	S_3/S_4
S_1	0.54	2.78	6.09
S_2	0.54	2.33	21.50
S_2	0.64	2.33	5.56
S_4	0.18	8.64	5.67

Matrix C for the $S_1-S_2-S_3-S_4$ orientation

There are $4! = 24$ column orientations in this example. When they are all evaluated, the following average values are obtained for the subjective probabilities:

	Mean
$p(S_1)$	0.24
$p(S_2)$	0.57
$p(S_3)$	0.17
$p(S_4)$	0.02
$\Sigma = 1.00$	

Once the above distribution is obtained, the subjective probabilities can be used in conjunction with the outcome of each state of nature in the expected value calculations.

HIERARCHICAL DECISIONS   **333**

Operational decisions are made in support of higher-level goals and objectives. When the objectives are fulfilled, the final results of the operational decisions are transformed into benefits. The hierarchical nature of this process is easy to observe but difficult to quantify. We can make the measurements easier if we divide the space between the top and bottom of the decision hierarchy into intermediate levels. The number of levels depends on the logical sequence of the decisions involved. If too many levels are identified, we have to take too many measurements; if too few levels are used, we run into difficulties because of excessive aggregations.

In most decision problems, the hierarchical relations lend themselves to three major levels:

1   Impact level

2   Target level

3   Operational level

Impact level contains the objectives and the benefits. This is frequently the starting point to trigger the decision process. We do this by setting broad objectives and defining the benefits expected from our activities. For example, development of a pollution-free, fuel-efficient engine is an objective. When fulfilled, it leads to such benefits as clean environment, energy independence, market leadership, etc. When we use these broad terms for objectives and benefits, we cannot define a precise measure of effectiveness for them. This leads us to the *target level*. Once we identify measurable goals by disaggregating the objectives, we can start developing a basis for measurements. For example, we can set 40 miles per gallon as a goal, a 50 percent reduction in pollutants as another goal. Both of these contribute to the fulfillment of the pollution-free, fuel-efficient engine objective and both are measurable. These goals are achieved by the actions and strategies at the operational level. One strategic action could involve a new design; another one could result in the use of lightweight materials in the engine block; another action could lead to an energy recovery device in the engine. These are all operational level decisions contributing to the goals at the next level of the hierarchy. Note that each operational decision contributes to several, maybe all, objectives and benefits. The overall decision process can be viewed as a network of relationships among the decisions. The nodes and arcs of the network can be evaluated through sequential applications of pairwise comparisons described in the previous section. A typical decision hierarchy is depicted in network form in Fig. 10.8.

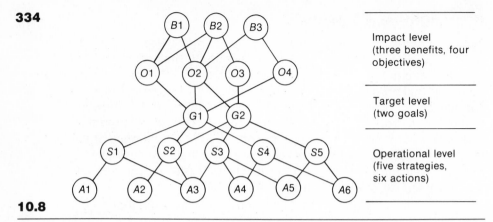

10.8

A typical decision hierarchy.

The evaluations in such a hierarchy require the assignment of a numerical value to each branch of the network shown in Fig. 10.8. The values are assigned in such a way as to represent the relative contribution of an element at one level to an element at the next level. When this measurement process is completed, an evaluative model emerges for the relative measure of effectiveness of each element at the bottom of the decision hierarchy in terms of the elements at the top.

An example of the hierarchical decision model is worked out in the Appendix.[10]

SUMMARY

Engineering managers make and implement decisions as they manage the material, financial, and human resources of the organization. All these decisions involve varying degrees of risk and uncertainty because of the probabilistic nature of the events and their outcomes.

Decisions can be classified as single-criterion decisions and multiple-criteria decisions according to the nature of the situation. They can further be broken into three categories:

1   Decisions under certainty

2   Decisions under uncertainty

[10]The mathematical details of the hierarchical decision model can be found in D. F. Kocaoglu, "A Systems Approach to the Resource Allocation Process in Police Patrol," Ph.D. dissertation, University of Pittsburgh, 1976.

3   Decisions under risk   **335**

The decision maker's utility plays an important role in making a selection from the available alternatives. *Utility* is the relative value that the decision maker assigns to the various outcomes. It is a subjective concept reflecting the individual's personal value structure. The utility concept facilitates the analysis of multicriteria decisions where the criteria have different measures along different dimensions.

Decision trees can be constructed to depict the decision process as a logical sequence of steps consisting of decision points and probabilistic event points. We can evaluate the various decision alternatives by "rolling back" the decision tree and comparing the expected values of the branches. The decision tree concept is based on bayesian analysis.

Decision analysis involves the measurement and evaluation of subjective factors to a great extent. Utility theory offers a methodology for such measurements under the assumptions of rationality. Another approach is the use of pairwise comparisons in the development of hierarchical relationships for complex decisions. The hierarchical decision model can be used for the assessment of subjective probabilities and priority weights, and the analysis of level decisions.

DISCUSSION QUESTIONS

1   A manager is paid for his ability to make and implement decisions on behalf of the organization. Comment on this statement.

2   Good decisions do not always lead to good outcomes. Give an example of a bad outcome that may result from a good decision, and a good outcome that may result from a bad decision.

3   Can a decision maker have "perfect information"? Describe situations where he comes close to it.

4   Why is the Laplace principle called the "principle of insufficient reasoning"?

5   Although the probability of a major failure is very low in nuclear power plants, such an outcome could be disastrous. On the other hand, ignoring nuclear power could mean the loss of a significant source of energy and lead to potentially disastrous outcomes for the industrial system. Discuss what role the engineers can play in bringing these tradeoffs to the public's attention.

6   The primaries and the political debates among the presidential candidates facilitate a bayesian decision process for the voters in election years. Explain how.

7   Discuss the advantages of the decision tree over a payoff table in sequential decisions.

336

8 Find a partner and develop his (her) utility function between 0 and $50,000. Is your partner a risk taker or risk averter?

9 Discuss what you might expect if you attempt to develop the utility function of your partner for outcomes between 0 and 50 million dollars.

10 The following pairwise comparisons have been made among three elements.

A: 90	A: 10	B: 90
B: 10	C: 90	C: 10

Calculate the relative weights of A, B, and C by the constant sum method. Comment on the results. Do you accept them? Why or why not?

11 As a successful engineer you expect to get a promotion within a year in your company. You have also been contacted by two other companies. One is located across the country and has offered you a job two levels above your current position and a 20 percent raise in salary. The other company has offered a 3-year contract in an overseas project at double your current salary. Discuss in detail how you can develop a multicriteria decision model for your decision.

12 Probability of a future event can be estimated in two ways:

(a) *Objectively*, by developing a frequency distribution from observations on previous occurrences of similar events.

(b) *Subjectively*, by quantifying the judgment of decision makers to reflect the degree of their belief in the occurrence of the event.

Make a list of real-life examples for which each of these methods is particularly useful. Discuss the appropriateness of your selection of the method in each case.

KEY QUESTIONS FOR THE ENGINEERING MANAGER

1 What key decisions have been made in the past in this organization which were influenced by the value system of the decision maker?

2 What quick and intuitive decisions have I recently made? What have been the outcomes of these decisions?

3 If I perceived that my general manager was becoming too involved in the details of engineering decisions, what would I do? Would I be comfortable bringing this to his attention?

4 What are the differences and similarities between engineering-oriented and management-oriented decisions in this organization?

5 What assistance do the engineers working with me and for me

require in developing analytical approaches to management decision making?

337

6 Can I analyze recent engineering-management decisions in this organization through a rational decision methodology?

7 What were some of the conflicting objectives involved in the key decisions made in the past few years in my department/division?

8 What information is lacking in the current decisions facing this organization?

9 What sort of a decision process is involved in this organization to evaluate the external events outside our control?

10 In which recent situations could a decision tree have helped me make a better decision?

11 Has the lack of timely decision making created any recent problems in this organization?

12 Are the managers in this organization evaluated on the basis of the quality of their decision or merely on the outcomes?

BIBLIOGRAPHY

Brown, R. V., A. S. Kahr, and C. Peterson: *Decision Analysis for the Manager,* Holt, New York, 1974.

Chernoff, H., and L. Moses: *Elementary Decision Theory,* Wiley, New York, 1959.

Easton, A.: *Complex Managerial Decisions Involving Multiple Objectives,* Wiley, New York, 1973.

Fishburn, P. C.: *Decision and Value Theory,* Wiley, New York, 1964.

——: "Utility Theory," *Management Science,* vol. 14, 1968, p. 335.

Guilford, J. P.: *Psychometric Methods,* 2d ed., McGraw-Hill, New York, 1954.

Gullison, H.: "Measurement of Subjective Values," *Psychometrica,* vol. 59, 1951.

Hammond, J. S.: "Better Decisions with Preference Theory," *Harvard Business Review,* vol. 45, November-December 1967, p. 123.

Keeney, R. L., and H. Raiffa: *Decisions with Multiple Objectives,* Wiley, New York, 1976.

Kocaoglu, D. F., "A Systems Approach to the Resource Allocation Process in Police Patrol," Ph.D. dissertation, University of Pittsburgh, 1976.

——, "Hierarchical Decision Model under Multiple Criteria," paper presented at TIMS/ORSA meeting, Washington, D. C., May 5, 1980.

338 Luce, R. D., and H. Raiffa: *Games and Decisions,* Wiley, New York, 1958.

Magee, J. R.: "Decision Trees for Decision Making," *Harvard Business Review,* vol. 42, July-August 1964, p. 126.

Magee, J. F.: "How to Use Decision Trees in Capital Investment," *Harvard Business Review,* vol. 42, September-October 1964, p. 79.

Page, A. N.: *Utility Theory: A Book of Readings,* Wiley, New York, 1968.

Pratt, J., H. Raiffa, and R. Schlaifer: *Introduction to Statistical Decision Theory,* McGraw-Hill, New York, 1965.

Raiffa, H.: *Decision Analysis,* Addison-Wesley, Reading, Mass., 1968.

————, and R. Schlaifer: *Applied Statistical Decision Theory,* Harvard Graduate School of Business Administration, Division of Research, Boston, 1961.

Schlaifer, R.: *Analysis of Decisions under Uncertainty,* McGraw-Hill, New York, 1969.

————: *Probability and Statistics for Business Decisions,* McGraw-Hill, New York, 1959.

Swalm, R.O.: "Utility Theory—Insights into Risk Taking," *Harvard Business Review,* vol. 44, November-December 1966, p. 123.

Von Neumann, J., and O. Morgenstern: *Theory of Games and Economic Behavior,* rev. ed., Princeton, Princeton, N.J., 1953.

Winkler, R. L.: *An Introduction to Bayesian Inference and Decision,* Holt, New York, 1972.

PROJECT **11**
SELECTION,
CONTROL AND
EVALUATION

Engineering projects are building blocks of an organization's strategy to achieve overall objectives. Some engineering projects are successful in achieving those objectives; some never come to fruition, perhaps having proved technically infeasible, too costly to develop, or not adaptable to market demand. This chapter deals with the selection of new projects, the scheduling and control of ongoing projects, and the evaluation of completed projects. Both technologists and those who have responsibility for the management of projects should find the chapter useful. The methods used for project evaluations range from informal discussions to highly structured approaches and models. In this chapter, we discuss the methods that follow a systematic procedure and utilize both subjective and objective data for the evaluation. The methods are presented in sufficient detail so that engineers and engineering managers can use them in their current and future projects.

The engineering manager is responsible for evaluating his projects at all stages of the project life cycle.[1] More specifically, he conducts three types of evaluation:

1   Preproject evaluation for the *selection* of the best project from among the prospective projects

2   Ongoing evaluation for the *control* of a project in progress

3   Postproject evaluation for the *assessment* of the success and efficacy of a completed project

---

[1]See Chap. 3 for a discussion of the project life-cycle concept.

340 In preproject evaluation, the projects are compared with respect to relevant criteria defined for project selection purposes. There is a plethora of approaches that managers may choose to use in selecting projects. A few years ago, Karger and Murdick[2] surveyed several major industrial firms to determine their methods for identifying and developing new products. The responding companies indicated that they used three major dimensions in their evaluations: technical evaluation, market evaluation, and economic evaluation.

Technical evaluation assesses the technical feasibility of developing the project within the time limits established by the company. *Marketing evaluation* addresses the question of whether the project output is compatible with the company's product lines and market strategy. *Economic evaluation,* which is usually subsequent to the first two evaluations, examines the project from the perspective of the company's overall strategic objectives and examines whether investments to support the project are justified by the expected returns.

After the decision is made for the selection of a project, the engineering manager becomes responsible for the implementation of that decision. He works with a team of professionals to complete the project within the time, cost, and technical requirements. In the course of the project, information feedback is generated for the evaluation of the project performance. This process in an ongoing project has been called a *normative* evaluation.[3] Project budgets, schedules, technical parameters and control networks are used for such a purpose.

As the project winds down, a different kind of evaluation is necessary to assess how the work has progressed and what has been achieved by the project. This postproject assessment is called *summative* evaluation by Scriven.[4] It serves two purposes. First, it measures the efficacy of a completed project. Second, the lessons learned in the postproject evaluation provide good insights for the management of future projects.

PREPROJECT EVALUATION

One of the key decisions for an engineering manager is the selection of projects to undertake. The freedom that he has in this decision is usually conditioned by the strategic plans of the organization. Because of the diversity of markets—often geographically dispersed—and the extent of competitive technology in products, the

---

[2]D. W. Karger and R. G. Murdick, *New Product Venture Management,* Gordon and Breach, New York, 1972.

[3]C. H. Weiss, *Evaluation Research,* Prentice-Hall, Englewood Cliffs, N.J., 1972.

[4]M. Scriven, "The Methodology of Evaluation," in *Perspectives of Curriculum Evaluation,* R. W. Tyler, R. M. Gagne, and M. Scriven (eds.), AERA Monograph Series on Curriculum Evaluation, no. 1, Rand McNally, Chicago, 1967, pp. 39–83.

engineering manager is constrained to a technological strategy en- **341**
dorsed by the organization's senior management. For example, the
huge $14 billion Phillips company based in Holland does not launch a
new development unless it is suitable for most, if not all, of its major
markets. To survive, the development projects that are undertaken by
the company must have high quality, be suitable for production, and
be supportable by field engineering.

Within the framework of corporate strategy, the engineering manager needs a rational methodology for the selection of engineering projects. He is responsible for acquainting his professionals
with the various methods available for project selection and encouraging the rigorous use of such methods. Members of a project team
can be used efficiently as a "decision body" for this purpose. Such a
group effort facilitates a dialog in the team that helps to answer
such questions as: What project should be selected for funding?
Why has a particular project been selected? How does the project fit
into the strategic purpose of the organization?

In the ensuing discussion, we will speak of the project selection decision maker as if he were the engineering manager. Of course,
the ultimate decision does rest with this manager, assuming such
decision is compatible with the strategic direction desired by the
chief executive of the organization.

A wide variety of project selection methods has been developed for engineering projects, R&D projects, marketing strategies,
and large-scale programs. A summary will be given here to familiarize
the reader with the methods that are available.

We can classify the methods in order of increased sophistication as follows:

> Rank-ordering methods:
>> Simple rank ordering
>> Q-sorting
>> Weighted ordering
>> Successive comparisons
>
> Scoring models:
>> Simple scoring model
>> Probabilistic scoring model
>
> Utility models
>> Simple utility model
>> Probabilistic utility model
>
> Hierarchical models

RANK-ORDERING METHODS

Rank-ordering methods are used for listing projects according to
their overall desirability. The engineering manager and his profes-

342 sionals look at all projects and use their judgment to rank the projects from the most desirable to the least desirable.

SIMPLE RANK ORDERING   In this method, a square matrix is developed, with rows and columns corresponding to the projects under consideration. The engineering manager compares the project in a column with the projects in the rows one at a time. If he prefers the project in column j to the project in row i, he places a plus sign in cell ij. When the matrix is completed, the number of pluses in the columns are counted. If the comparisons are consistent, the pluses in a column equal the minuses in the corresponding row and their sum determines the ranking of the projects.

For example, if there are n projects, P_1, P_2,..., P_n, the decision maker starts comparing P_1 in the first column with the other projects. If P_1 is preferred to P_2, he places a plus in cell (2, 1). If he prefers P_3 to P_1, he assigns a minus to cell (3, 1), etc. After comparing P_1 with all the other projects and putting the pluses and minuses into the cells of the first column, he continues with the second column and compares P_2 with all the other projects. Of course, if he is consistent, he will now say that P_2 is *not* preferred to P_1, because he had decided P_1 was better than P_2 in the first comparisons. Thus, he puts a minus in cell (1, 2).

When the comparisons are completed, the number of pluses in the columns will vary between zero and $n - 1$. The number of pluses under a project indicates the number of other projects to which that particular project is preferred. The project with $n - 1$ pluses will have the highest rank; the others will follow it.

Let us assume that five prospective projects have been compared with each other and the following matrix has been developed:

P_i \ P_j	P_1	P_2	P_3	P_4	P_5	Number of minuses
P_1	✕	−	+	+	−	2
P_2	+	✕	+	+	−	1
P_3	−	−	✕	−	−	4
P_4	−	−	+	✕	−	3
P_5	+	+	+	+	✕	0
Number of pluses:	2	1	4	3	0	

This matrix shows that consistent comparisons have been made. The number of pluses in each column is the same as the number of minuses in the corresponding row, and they range from zero to 4. The projects are ranked as P_3, P_4, P_1, P_2, P_5 in order of decreasing desirability.

We cannot expect that the comparisons will always be **343** consistent. There can be two types of inconsistencies in comparisons.

1   When the decision maker compares P_1 with P_2, he puts a $+$ in cell (2, 1). Later, when he compares P_2 with P_1, he again puts a $+$ in cell (1, 2).

2   When three or more projects are ranked in such a way that P_1 is preferred to P_2, P_2 is preferred to P_3, and P_3 is preferred to P_1, implying a three-way "circular preference profile" as follows:

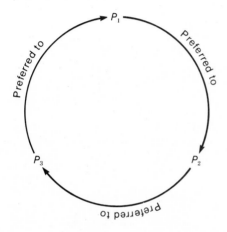

This is called "intransitivity" in decision theory terminology.

When inconsistencies occur, multiple columns have the same number of pluses and they may differ from the number of minuses in the corresponding rows. If this occurs, we have to determine where the inconsistencies are and make new comparisons for the affected projects.

This method has some obvious advantages. First, it is easy to use. Second, it forces the decision maker to make explicit choices in a systematic way. Third, the inconsistencies inherent in the thinking process can be identified by going over the matrix a few times. On the other hand, it has a major disadvantage: the method indicates that a project is more desirable than another one, but it does not provide any information as to how desirable. This disadvantage can be alleviated to some extent by weighting the projects after the ranking is obtained. A weighting scheme is described under the weighted-ordering procedure later in this chapter.

Q-SORTING   The Q-sorting method is also used for rank-ordering projects. It uses a very simple procedure, and the results are usually

344 obtained in a shorter time than the simple rank-ordering method described above. The project titles are written on small cards and given to the decision maker. He divides them into two groups, one group containing projects preferable to the other group. He then evaluates the projects in each group and divides them into two more groups, using the same approach. He keeps doing this until all projects are ranked, from the most desirable to the least desirable.

Let us suppose there are ten prospective projects and the engineering manager feels that P_3, P_6, P_8, and P_9 are superior to the rest. He forms two clusters as follows:

Step 1

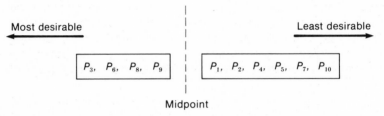

He then divides each group into two subgroups:

Step 2

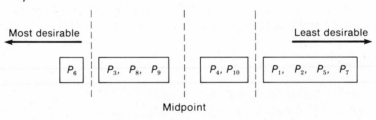

At this stage, P_6 has already been chosen as the best project, since there is only one item in that cluster.

Continuing with the same process, assume that the subsequent clusters are formed as follows:

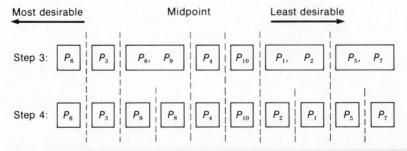

After step 4, the manager might wish to compare the two projects adjacent to the midpoint. Since the midpoint was established when the complete set of projects was initially divided into two groups, the difference between P_8 and P_4 could be so small that he may want to switch them. If no such change is made, the projects are rank-ordered as grouped in step 4.

Q-sorting has the same advantages and disadvantages as simple rank ordering.

WEIGHTED ORDERING   The two methods described above rank the projects in the order of preference. When we see such a list, we know that a project is better than all other projects that are ranked lower than itself. We do not know, however, how much better that project is in comparison with another project. That relative measure can be obtained by applying a weight factor to the projects after their rank ordering is complete. The weights are determined by comparing each project with the next lower-ranked project and dividing a total of 100 points between them to indicate the relative level of preference for each.

Suppose the ten projects rank ordered by Q-sorting in the previous section are to be weighted. Starting with P_6 (the highest ranked project) and P_3 (the next lower ranked project), nine pairs are formed and the total of 100 points is allocated to the two projects in each pair such that if a project is considered to be about twice as good as the next one, they are assigned 67 and 33 points respectively. If one is three times better than the next, then 75 and 25 points are assigned.

Assume the following comparisons are made:

| P_6: 55 | P_3: 60 | P_9: 70 | P_8: 51 | P_4: 75 |
| P_3: 45 | P_9: 40 | P_8: 30 | P_4: 49 | P_{10}:25 |

| P_{10}: 60 | P_2: 65 | P_1: 55 | P_5: 52 |
| P_2: 40 | P_1: 35 | P_5: 45 | P_7: 48 |

Letting P_7 have a value of 1.0, the other projects are assigned respective values as follows:

$$P_5 = \tfrac{52}{48} = 1.08$$
$$P_1 = 1.08\tfrac{55}{45} = 1.32$$
$$P_2 = 1.32\tfrac{65}{35} = 2.46$$
$$P_{10} = 2.46\tfrac{60}{40} = 3.69$$
$$P_4 = 3.69\tfrac{75}{25} = 11.07$$
$$P_8 = 11.07\tfrac{51}{49} = 11.52$$

$$P_9 = 11.52\tfrac{70}{30} = 26.87$$

$$P_3 = 26.87\tfrac{60}{40} = 40.31$$

$$P_6 = 40.31\tfrac{55}{45} = 49.27$$

In this evaluation, the most desirable project, P_6, is viewed as about fifty times better than the least desirable alternative, P_7.

SUCCESSIVE COMPARISONS   This is another weighting procedure, developed by Churchman, Ackoff, and Arnoff.[5] It can be used for assigning weights to projects when the combined value of two or more projects is equal to the sum of their individual values as perceived by the decision maker. The method uses the following procedure:

1   Rank-order the projects by any method, such as rank ordering or Q-sorting. Let us say that we have obtained the following order:

$$P_1, P_3, P_2, P_4, P_5$$

2   Assign a value of 1.0 to the top-ranked project and successively lower values to the others.

$$V_1 = 1.0$$

$$V_3 = 0.8$$

$$V_2 = 0.6$$

$$V_4 = 0.4$$

$$V_5 = 0.2$$

3   Make the comparison:

$$P_1 \text{ versus } (P_3 + P_2 + P_4 + P_5)$$

If the combined projects are preferred to P_1, drop P_5 from the combination and make another comparison until P_1 is preferred to the sum of the others. For example, let us say that the decision maker feels that P_1 is preferred to P_3 and P_2 taken together but considered not as attractive as the combination of P_3, P_2, and P_4. This implies:

$$(V_3 + V_2) < V_1 < (V_3 + V_2 + V_4)$$

$$(0.8 + 0.6) < V_1 < (0.8 + 0.6 + 0.4)$$

$$1.4 < V_1 < 1.8$$

Revise the original value of P_1 by setting it between 1.4 and 1.8. Let us say $V_1 = 1.6$.

---

[5]C. West Churchman, Russell L. Ackoff, and E. Leonard Arnoff, *Introduction to Operations Research*, Wiley, New York, 1957.

4   Repeat step 3 for the comparison of P_3 versus the combination of all projects ranked lower than P_3. Suppose P_3 is preferred to $P_2 + P_4$, but there is an indifference between P_3 and $P_2 + P_4 + P_5$. Thus,

$$(V_2 + V_4) < V_3 = (V_2 + V_4 + V_5)$$
$$(0.6 + 0.4) < V_3 = (0.6 + 0.4 + 0.2)$$
$$1.0 < V_3 = 1.2$$

Assign a value of 1.2 to V_3.

5   Repeat the above step for a comparison of P_2 versus $P_4 + P_5$. Suppose we find that

$$V_4 < V_2 < (V_4 + V_5)$$
$$0.4 < V_2 < (0.4 + 0.2)$$
$$0.4 < V_2 < 0.6$$

Let $V_2 = 0.5$ and keep $V_4 = 0.4$ and $V_5 = 0.2$.

6   Using the new values, go backwards to check the inequalities developed above. Revise project values as necessary. In other words, start the backward comparisons with the following values:

$$V_1 = 1.6$$
$$V_3 = 1.2$$
$$V_2 = 0.5$$
$$V_4 = 0.4$$
$$V_5 = 0.2$$

For step 5:

Check $V_4 \overset{?}{<} V_2 \overset{?}{<} (V_4 + V_5)$

$$0.4 < 0.5 < (0.4 + 0.2) \qquad \text{OK}$$

For step 4:

Check $(V_2 + V_4) \overset{?}{<} V_3 \overset{?}{<} (V_2 + V_4 + V_5)$

$$(0.5 + 0.4) < 1.2 \not< (0.5 + 0.4 + 0.2)$$

Let $V_3 = 1.1$

For step 3 (using the revised value of V_3):

$$(V_3 + V_2) \overset{?}{<} V_1 \overset{?}{<} (V_3 + V_2 + V_4)$$
$$(1.1 + 0.5) \not< 1.6 < (1.1 + 0.5 + 0.4)$$

Let $V_1 = 1.8$

7   Make the forward and backward comparisons as many times as needed. Note that in this example consistent values have already been reached. The final weights assigned to the projects are as follows:

$$V_1 = 1.8$$

$$V_3 = 1.1$$

$$V_2 = 0.5$$

$$V_4 = 0.4$$

$$V_5 = 0.2$$

8   For uniformity of comparisons, normalize these values by dividing each one by the sum of values, i.e.,

$$V_i = \frac{V_i}{\sum\limits_{i=1}^{n} V_i} \qquad \text{for } n \text{ projects}$$

In this case,

$$\sum\limits_{i=1}^{5} V_i = 4.0$$

Therefore, the normalized values are

$$V_1 = 1.8/4.0 = 0.45$$

$$V_2 = 1.1/4.0 = 0.275$$

$$V_3 = 0.5/4.0 = 0.125$$

$$V_4 = 0.4/4.0 = 0.10$$

$$V_5 = 0.2/4.0 = \underline{0.05}$$

$$\text{Total} = 1.00$$

SCORING MODELS

The weighting concept described in the previous models can also be extended to the evaluation of projects under multiple criteria. This is the approach used in scoring models.

SIMPLE SCORING MODEL   The preceding methods—simple rank ordering, Q-sorting, weighted ordering, and successive comparisons—require a perception of the *overall attractiveness* of projects. To give a precise definition for overall attractiveness is very difficult. A more realistic method is the evaluation of each project with respect to a number of criteria simultaneously. This is done as follows:

1   Identify relevant criteria. For most project selection situations, five to ten criteria are sufficient.

2   Define a range for each criterion. The range could be from "very good" to "very bad" for qualitative criteria, and from the highest measurable value to the lowest for quantitative criteria, such as return on investment (ROI).

3   Divide the range of each criterion into intervals. The recommended number of intervals is between five and nine.

4   Rank the criteria in the order of their relative importance and assign weights to them.

5   Assign a numerical value to each interval.

6   Evaluate each project, one at a time, according to the criteria by indicating the interval to which it belongs.

7   Calculate the product of the criterion weights and the criterion intervals. The sum of these products is the project score.

8   Repeat steps 6 and 7 for each project.

9   Rank the projects according to their scores.

The definition of relevant criteria is very critical in scoring models. A dialog among the decision makers and the major clientele groups is often necessary in the first two steps above. The criteria should be clearly defined and easy to evaluate. They should lend themselves to meaningful measurements, either subjectively or objectively. If the scoring model is to be used for evaluating a project proposed by subcontractors, such as the landing gear assembly mentioned in Chap. 10, the relevant criteria could include durability, maintainability, design characteristics, cost, completion time, subcontractor's track record, etc. If we are selecting a project to develop a product in line with corporate objectives, the criteria would include growth potential, profitability, competitive edge, etc.

Use of the simple scoring model is illustrated in Table 11.1. The six criteria listed in the table are all subjective. In addition to these, readily quantifiable criteria can also be included in the list. Depending on the nature of projects under consideration, such relevant criteria may include ROI, projected market share, and net present value.

An important consideration in this model is the precise definition of intervals, representing the levels at which the criteria are satisfied by each project. For example, the "very good" level for the "use of available personnel" may be used for a project that fully utilizes the engineers and draftsmen currently on the company payroll, while a "very poor" level for the same criterion may be assigned to a project that cannot use any of the technical expertise

Simple Scoring Model for a Project

11.1		CRITERION INTERVAL VALUES(C)					
		1.0	0.8	0.6	0.4	0.2	
CRITERION	CRITERION WEIGHT(W)	VERY GOOD	GOOD	MEDIUM	POOR	VERY POOR	RATING $= W \times C$
Use of available personnel	0.15		X				$(0.15)(0.8) = 0.12$
Growth potential	0.20			X			$(0.20)(0.6) = 0.12$
Profitability	0.30		X				$(0.30)(0.8) = 0.24$
Competitive advantage	0.05				X		$(0.05)(0.4) = 0.02$
Consistence with technical competence	0.20		X				$(0.20)(0.8) = 0.16$
Compatibility with existing projects	0.10	X					$(0.10)(1.0) = 0.10$
Sum:	1.00					Project score:	0.76

available in the company and requires that a project team be formed entirely from outside sources. These levels should be clearly defined for the relevant range of each criterion and should be communicated to the decision makers before the model is developed.

After the project scores are calculated for all projects, the subjective probability for the success of each project is determined. Let us assume that the project shown in Table 11.1 is assigned a 0.85 probability that it will be successful. The expected project score is $(0.85)(0.76) = 0.65$. This is the relative value of a specific project. It is compared with the expected scores of the other prospective projects in order to rank them in the order of their attractiveness.

Notice that, in this example, "profitability" is viewed as being twice as important as the "use of available personnel" and three times as important as "compatibility with existing projects." Also, the relative values of the criterion levels imply that "good" is twice as attractive as "poor" and four times better than the "very poor" category on a scale from 0 to 1.

The results of such a model would indicate that an expected score of 0.56 is better than one of 0.30, but it would not necessarily mean that 0.56 is better than 0.54. When the scores of two projects are so close to each other, we would have good reason to believe that both are superior to one with the score of 0.3, and we would analyze the two candidates more closely. In other words, the scoring model is a good approach to help us introduce a systematic method in judging the intangible characteristics of projects, but it is not a substitute for judgment. Rather, it facilitates the exercise of judgment.

PROBABILISTIC SCORING MODEL The "probability of success" **351** concept used in the simple scoring model incorporates risk into the project selection decision in a very limited sense. It assumes that the project will have a probability p to give us the profile shown in Table 11.1 and a probability of $1 - p$ that all criteria will be at an unacceptable level with a relative value of zero. This concept is extended to a more precise definition of risk in the probabilistic scoring model shown in Table 11.2.

In Table 11.2, the criterion weights and the criterion-level values are the same as those in the previous example. However, the evaluation of the project is done not by merely indicating the level for each criterion, but by assigning probabilities for each level of attainment. The probabilities add to 1.00 for each criterion.

The expected rating for a criterion is calculated by summing the products of interval values and achievement probabilities and multiplying the result by the criterion weight. The calculation for the first criterion is:

$$0.15[(0.3)(1.0) + (0.5)(0.8) + (0.2)(0.6) + (0.0)(0.4) + (0.0)(0.2)] = 0.12$$

After the ratings are calculated for all criteria, the expected project score is determined as the sum of ratings. This score is then compared with the expected scores of the other projects.

Probabilistic Scoring Model for a Project

CRITERION	CRITERION WEIGHT	CRITERION INTERVAL VALUES					EXPECTED RATING
		1.0 VERY GOOD	0.8 GOOD	0.6 MEDIUM	0.4 POOR	0.2 VERY POOR	
Use of available personnel	0.15	0.3	0.5	0.2	——	——	0.12
Growth potential	0.20	0.1	0.1	0.4	0.2	0.2	0.11
Profitability	0.30	0.3	0.4	0.1	0.1	0.1	0.22
Competitive advantage	0.05	——	0.1	0.2	0.5	0.2	0.02
Consistence with technical competence	0.20	0.1	0.6	0.2	0.1	——	0.15
Compatibility with existing projects	0.10	0.5	0.3	0.2	——	——	0.09
Sum:	1.00				Expected project score:		0.71

11.2

Relative Values Based on Criterion Distribution

11.3 CRITERION INTERVAL	RELATIVE VALUE
Below $\mu - 1.75\sigma$	1
$\mu - 1.75\sigma$ to $\mu - 1.25\sigma$	2
$\mu - 1.25\sigma$ to $\mu - 0.75\sigma$	3
$\mu - 0.75\sigma$ to $\mu - 0.25\sigma$	4
$\mu - 0.25\sigma$ to $\mu + 0.25\sigma$	5
$\mu + 0.25\sigma$ to $\mu + 0.75\sigma$	6
$\mu + 0.75\sigma$ to $\mu + 1.25\sigma$	7
$\mu + 1.25\sigma$ to $\mu + 1.75\sigma$	8
Above $\mu + 1.75\sigma$	9

The scoring model approach is based on the assignment of subjective values to the various levels of criteria. This assignment is done by dividing the measure of each criterion into several intervals and defining a numerical value for each interval. In our example the criteria were divided into five intervals and the relative values between 0.2 and 1.0 were assigned to them. Other schemes can also be used. For example, Moore and Baker[6] recommend that the measure of each criterion be divided into nine equal intervals and a value of 1 be given to the worst and 9 to the best interval. They also recommend that the width of each interval be half of the standard deviation of the distribution of the criterion measure. If the mean of the distribution of the criterion measure (for example, the net present values of prospective projects) is denoted by μ and the standard deviation by σ, the criterion-level values are defined as in Table 11-3.

When the scoring model is used in project selection, it is important to evaluate all projects by the same model in order to assure consistent decisions.

UTILITY MODELS

The critical assumption behind the scoring models is the implication that the relative values of criterion intervals are proportional to the numerical values assigned to them. However, as we discussed in Chap. 10, the values assigned to outcomes by the decision makers are defined by their utility functions and do not necessarily follow a linear relationship with the outcomes. Following this argument, we can extend the scoring models to utility models.

[6]J. R. Moore, and N. R. Baker, "An Analytical Approach to Scoring Model Design: An Application to Research and Development Project Selection," Graduate School of Business, Stanford University, Stanford, Calif.

SIMPLE UTILITY MODELS   In simple utility models we first develop
a utility function for each criterion by assigning 0 utile to the
minimum level and 1 utile to the maximum. These functions take
various forms, as discussed in Chap. 10. After they are developed,
we establish the criterion weights by comparing the relative value of
going from the worst to the best level in one criterion versus another
criterion. This can be done by using the constant-sum method
presented in Chap. 10.

After that, we determine the utility of each project along
each criterion and multiply it by the relative weight of that
criterion. The sum of these products gives us the utility for the project. If we also multiply it by the probability of success, we obtain the
expected utility of the project. This way the projects are ranked not
according to the expected scores, but according to their expected
utilities.

This approach assumes that the project will fail with a
probability of $1 - p$ and the utility of failure will be equal to zero. In
reality, a project could have various levels of success. If we determine the probability distribution of a project's outcome for each criterion, we can combine the probabilities and the corresponding
utilities to obtain a more realistic expected value for the project's
utility. This is the approach taken in probabilistic utility models.

PROBABILISTIC UTILITY MODELS   After the criterion weights and
utility functions are developed, let us assume that we have also developed subjective probability distributions for the range of outcomes
that can be obtained from a project along each criterion. For example, suppose that two of the relevant criteria are the "net present
value" and the "growth potential" of the project, and these criteria
exhibit the following distributions for a certain project (Fig. 11.1).

Probability distributions for project criteria.

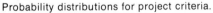

11.1

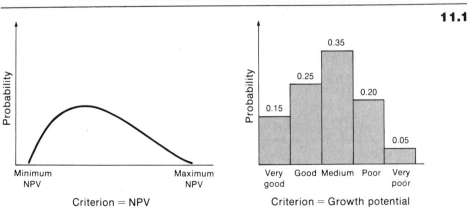

Criterion = NPV   Criterion = Growth potential

354   Since we are assuming that we have developed the utility functions for these criteria, we can determine the expected utility of the project as follows.

Let E_k = expected utility of project k

W_i = relative weight of criterion i

U_i = utility function for criterion i

V_{ijk} = level j of the outcome of project k for criterion i

P_{ijk} = probability that the level V_{ijk} will be achieved

The expected utility of project k is:

$$E_k = \sum_i^1 W_i \sum_j^3 U_i(V_{ijk})(P_{ijk})$$

In the above expression, j represents the intervals into which the criteria are divided. For discrete criteria, these are the criterion levels as defined earlier. For example, if the first criterion ($i = 1$) is "growth potential," then the first interval ($j = 1$) is "very good," the second interval is "good," and so on. Hence, V_{123} denotes that project 3 has good growth potential. Going back to the utility function for that criterion, we can easily determine the utility of being in the second interval. This will be the value represented by $U_1(V_{123})$. We then multiply it by the associated probability p_{123} and repeat these calculations for all intervals of all criteria. The probabilities are obtained from distributions similar to Fig. 11.1. Assuming that Fig. 11.1 has been developed for project 3, the value of p_{123} is 0.25. In case of continuous criteria such as the net present value, a large number of intervals can be used for the purpose of numerical integration by the computer. In that case, V_{ijk} corresponds to a small interval such as $150,000 to $155,000, so that it can still be associated with a finite probability.

So far, our discussion has been limited to the methods of ranking the projects according to their overall attractiveness or evaluating them with respect to a set of predefined criteria. Another approach to project selection is the programmatic approach described in the next section.

HIERARCHICAL MODELS
Projects can be viewed at three major hierarchical levels, as mentioned at the end of Chap. 10:

1   Impact level, which deals with issues related to the strategic purpose of the organization

2   Target level, which defines the project goals that are to be achieved

3   Operational level, which refers to the projects, **355**
strategies, and actions that are under consideration

The decision elements at each of these levels can be developed from
organizational policy statements and strategic and operational objectives. At the impact level, we have to ask ourselves what we will ultimately get out of the project effort and determine if the final results
will have a positive impact on the strategic objectives of the organization. These objectives are typically established at the corporate level
in major corporations, or at the general manager level in smaller organizational units. Some examples of such objectives are listed
below:

Market leadership

Technical superiority

Profit maximization

Social responsibility

Long-term growth

Long-term stability

Broad markets with balanced products

Strengthening the cash position

Strategic survival

These are broad, timeless statements that establish the general direction of the organization and become a framework for the decision
makers throughout the organization.
        At the lower levels of the decision hierarchy, the organiza-
tional units establish specific goals in support of the strategic objectives. Typically, these are developed for the departments and divisions in major corporations. The following list includes some examples of such goal statements:

Improve product reliability by 10 percent

Develop new technologies

Improve productivity by 30 percent over the next 4 years

Penetrate into new markets

Diversify into new products

Reduce the air pollution caused by the plant to half of the
current level

The goals established for a department or division are such that,

356   when they are achieved, each one contributes to one or more of the organizational strategic objectives. The relative contributions are not necessarily equal to each other. Some goals may have a significant impact on one objective and a negligible impact on another one. By quantifying these contributions, we can develop a relative measure of each goal's contribution to each objective.

The engineering manager operates as a decision maker in a domain constrained by these higher level goals and objectives. He can determine the relative value of each proposed project, but it has very little meaning unless it contributes to the strategic purposes of the organization. Viewed from this perspective, the project selection process becomes a much more important activity than a simple decision at the project level. When the new projects could affect the strategic direction of an organization, it is important that their selection reflect the goals, objectives, strategies, and value structures implied in the decision hierarchy of the organization. The hierarchical model is used for this purpose. It is developed systematically by obtaining relative measures among the levels of the decision hierarchy and also within the elements of each level. The concept is based on the idea that in a decision system, everything is related to everything else. The system interactions and interrelationships are evaluated by using measurable values or by quantifying the decision makers' judgments. When the model is complete, it provides information not only about the relative value of a project but also about the relative impact of each project on each strategy, goal, and objective of the organization.

Referring to Fig. 10.8 and considering five levels similar to those shown in the figure, we find that the basic model takes the following form:

$$V_i = W_m \sum_l \sum_k \sum_j P_{ij} S_{jk} G_{kl} O_{lm}$$

where V_i = relative value of project i

W_m = relative weight of benefit m

P_{ij} = relative contribution of project i to strategy j

S_{jk} = relative contribution of strategy j to goal k

G_{kl} = relative contribution of goal k to objective l

O_{lm} = relative impact of objective l on benefit m

The relative measures used in the model are P_{ij}, S_{jk}, G_{kl}, and O_{lm}. As we combine them for the final result, we obtain numerical values for the arcs and nodes of the network shown in Fig. 10.8. Those values represent the impact relationships among decision elements at all levels of the hierarchy.

An example of the hierarchical model is presented in the Appendix.

ONGOING PROJECT EVALUATION

During the course of a project's life, it should be reviewed periodically by the engineering manager and members of the project team. For example, one of the authors is a consultant to a product development effort in a high-technology, fast-moving market situation. This project is reviewed every 2 weeks by the project team; the responsible engineering manager is present at these reviews along with the organization's financial, production, and marketing managers. The president of the company attends as many review meetings as his schedule permits. During a recent review meeting it was discovered that a proposed step-up in the technical configuration of the product was incompatible with existing manufacturing know-how. A parallel development in manufacturing technology was immediately initiated to support product technology. Had the review meetings not been held on a regular basis, the information on the lag in production technology might have been missed or delayed; the result could have been a delay in getting the product to market, a delay which would have worked to the advantage of the competition.

Project evaluation through an ongoing review process gives continuous feedback on how the project is progressing. There is no substitute for such a process of getting the project team together to discuss how things are going along the following lines:

What is going wrong?

What is going right?

What problems (and opportunities) appear to be emerging?

How is our competition doing?

Where is the project with respect to schedule, cost, and technical objectives?

Does the project continue to fit into an overall company strategy?

Questions of this nature can serve to motivate discussions within the project team and help to get the status of the project out for all to see.

In addition to the review meetings, a formalized approach to ongoing evaluation also includes the project scheduling and control activities. Project control is an important part of project management responsibility and includes:

Establishment of an anticipated date of completion of the project

Actions and activities necessary to accomplish the goals of the project

A desired or required sequence for performing the activities

358

One of the better known and more widely used techniques for project control is the *Gantt chart*. There are also other techniques based on network analyses. We shall look at the Gantt chart, the *critical path method* (CPM), and the *program evaluation* and *review technique* (PERT).

GANTT CHART[7]

The thrust of the Gantt chart is the graphic portrayal of the progress of a project. In the original charts the governing factor was availability and capacity of personnel and equipment. Extensions of the Gantt chart concept have recognized that the most important factor in scheduling problems is time. These newer charts are called "project planning charts."

Figure 11.2 is an illustration of a project planning chart dealing with the development of an electronic device. The elements of the project are listed on the left-hand side, and units of time in workdays appear at the top. The horizontal bars represent the schedule for the project elements, with specific tasks, activities, or operations written above the schedule line. The time duration of each activity is an estimate provided by people familiar with the activity. Work accomplished is denoted by a solid line in the schedule line. Progress is posted at regular intervals, and the time of each posting is marked by a large V at the top of the chart. A whole range of visual scheduling aids, from self-stick tapes to magnetic scheduling board kits, are now available commercially.

The time the project is behind schedule is usually determined by the maximum delay of any element from the schedule. At the last posting in Fig. 11.2, the project was 6 days behind schedule because of the receiver video amplifier. Some elements, such as the display and antenna units, were ahead of schedule. Other elements were behind schedule, but not as much as the video amplifier. Therefore, bringing the video amplifier up to schedule should have priority and, if the operations and personnel situation permit, the manager should transfer personnel temporarily from display and antenna units—the two farthest ahead of schedule—to the video amplifier.

One disadvantage of the Gantt chart is that is provides no means for assessing the impact of an element's being behind or ahead of schedule. As Cleland and King observe, "simply because a project element is behind schedule does not necessarily mean that the project is behind schedule by that amount. For most projects, only a few dates are critical in the sense that any delay in them will delay the project by a corresponding amount. The impact of slips in

---

[7]This discussion is adapted from D. I. Cleland and W. R. King, *Systems Analysis and Project Management*, McGraw-Hill, New York, 2d ed., 1975, chap. 15.

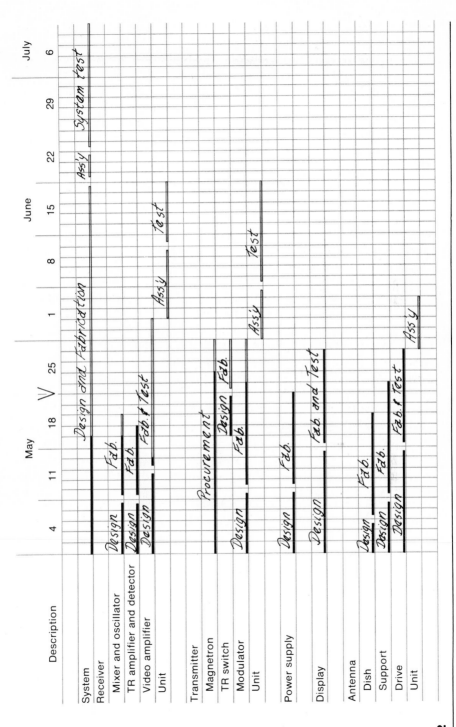

11.2 Project planning chart.

360   schedule dates depends upon the interrelationships between elements, which are not easily portrayed.[8]

Networks and network analyses are now being used quite frequently in scheduling management. They address the problem of critical activities alluded to in the above paragraph. We will examine only the two fundamental techniques, CPM and PERT.

CPM AND PERT

PERT and CPM were developed in the late 1950s and currently are being applied widely in contemporary organizations. An integral part of both techniques is the use of networks, called "arrow or precedence diagrams," to depict the precedence relationships of the elements of a project from its inception to its completion.

The construction of the precedence diagram requires the decomposition of the project into activities. Then three questions are answered about each activity of the project: What immediately precedes the activity (its *immediate predecessor*)? What can be done concurrently? What immediately follows or succeeds it (its *immediate successor*)? The diagram has two elements: *arrows* or *arcs* representing the project's activities, and *nodes* or *events* indicating the intersections of activities. Each activity or arrow has an *initial node* and a *terminal node.* The first initial node is called the *source* and the last terminal node is the *sink.* No activity can start from a node until all activities entering the node have been completed.

Consider a project with six activities, *A, B, C, D, E,* and *F,* with the following precedence relationships:

1   *B* and *C* can be started simultaneously, but each can start only after *A* is completed.

2   *D* can be started only after *B* is completed. Similarly *C* must be completed before *E* can start.

3   *F* can begin only after both *D* and *E* are completed.

Additionally, the time durations have been estimated to be *a, b, c, d, e,* and *f* hours, respectively, for activities *A, B, C, D, E,* and *F.* Figure 11.3 illustrates the precedence diagram. Node 2 is the terminal node for activity *A,* and the initial node for both *B* and *C.* Node 1 is the source and node 6 is the sink. Complications sometimes arise in the development of precedence diagrams, but discussion of these is beyond the scope of this book. The reader may consult books on networks mentioned in the Bibliography for a more detailed treatment of the subject.

[8]Cleland and King, op. cit., p. 347.

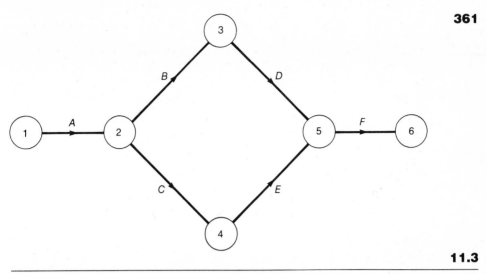

11.3

Example of an arrow or a precedence diagram.

Wiest and Levy[9] have summarized the characteristics that are essential for analysis by CPM or PERT as:

1   The project must consist of a well defined collection of jobs, or activities, which when completed marks the end of the project.

2   The activities may be started and stopped independently of each other within a given sequence.

3   The activities are ordered, that is, they must be performed in technological sequence.

Both CPM and PERT find the longest path or longest sequence of connected activities through the network. Such a path is called the *critical path* of the network and its length determines the duration of the project. A path is a set of nodes connected by arrows that begins at the initial node of a network and ends at its terminal node—in other words, its source and sink, respectively. With CPM, the amount of resources employed (cost) and the time needed to complete a project can be predicted within narrow ranges; they are assumed known with near certainty. Therefore, CPM addresses time-cost tradeoffs and is used where there is experience with similar work, such as in the construction industry. On the other hand, PERT is concerned with uncertainty in activity times and hence

[9]J. D. Wiest and F. K. Levy, *A Management Guide to* PERT/CPM, Prentice-Hall, Englewood Cliffs, N.J., 1977, p. 3.

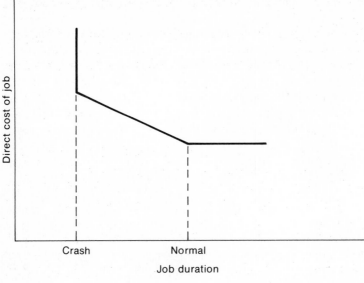

11.4

Time-cost tradeoff relationship for a typical job.

uncertainty in the project duration. It is used extensively in research and development projects.

    We now take a closer look at each of the techniques.

CPM   There are two kinds of costs involved with a project. The first kind are the direct costs associated with individual activities, which usually increase if the activities are expedited, shortened, or crashed. The second kind are indirect costs which decrease if the project is shortened. Figure 11.4 illustrates a possible, if simple, piecewise linear relationship between direct cost and job duration. The crash option was mentioned above. The normal option is doing the job at its normal and most efficient pace, with a lower level of resources than with the crash option. Over a certain range, the time can be reduced by using additional resources. There usually would be a minimum duration that cannot be reduced in spite of the addition of extra resources. This corresponds to the vertical portion of the line (crash). Similarly, slowing down the job will lower costs only up to a certain value below which no additional savings can be achieved. This is the horizontal portion of the line (normal). Thus, the crash option represents the *minimum possibe time* to complete an activity with the added costs. The normal option provides the duration for completing the job at the *lowest direct cost.*

    As the durations and costs of the activities are varied, the duration and cost of the project also vary. The CPM model specifies a

method for finding the optimum point representing the least-cost **363**
schedule.

The model begins with the identification of the activities, followed by the drawing of the precedence diagram. Then durations and costs are assigned to the activities in the diagram. There are many methods for determining the various time and cost estimates for the project. We will describe one such method using two time-cost estimates—the normal option and the crash option—and will illustrate the procedure with a simple example.

Example 11.1 Consider a project with five activities as shown in the precedence diagram of Fig. 11.5. The parameters are as indicated in the key of the figure.

Recall that the critical path represents the longest path of the network. The critical path using the normal times is *ACE* with a project duration of 15 days. Using the crash times, the critical path becomes *AD* with a duration time of 10 days. Thus our least-cost schedule will require between 10 and 15 days, inclusive.

To shorten the project's duration, we begin with the normal-time critical path, *ACE*. Activity *A, C,* or *E* will have to be expedited. To do the crashing least expensively, we choose the activity with the lowest *activity cost slope.* The activity cost slope is defined as

$$\frac{\text{Crash cost} - \text{normal cost}}{\text{Normal time} - \text{crash time}}$$

The activity cost slopes are calculated as shown in Table 11.4

Network of CPM example.

11.5

Key:

Activity name
_____→
Normal time, crashtime (days)
[Normal cost, crash cost ($)]

Activity Times and Costs for CPM Example

11.4		NORMAL		CRASH		ACTIVITY COST
ACTIVITY		TIME, DAYS	COST, $	TIME, DAYS	COST, $	SLOPE, $/DAY
A		4	270	2	670	200
B		6	190	4	310	60
C		7	210	3	610	100
D		8	300	8	300	——
E		4	180	1	540	120
			1150		2430	

along with the cost and time parameters of Fig. 11.5. Notice that the activity cost slope is inapplicable to activity *D*.

From Table 11.4, we see that activity *C* has the lowest slope among the critical path activities, *A, C,* and *E.* So we begin with expediting *C* and proceed as follows:

15-day schedule: Normal duration for the project.

Direct cost = $1150

14-day schedule: The least expensive way to gain 1 day is to expedite activity *C* by 1 day for $100.

Direct cost = $1150 + $100 = $1250

13-day schedule: Expedite activity *C* by another day for an additional $100.

Direct cost = $1250 + $100 = $1350

12-day schedule: Expedite activity *C* by yet another day for another $100.

Direct cost = $1350 + $100 = $1450

There are now two critical paths: *ACE* and *AD.* We cannot crash activity *C* any more unless there is an equal reduction on *AD;* in other words, both critical paths must be reduced.

11-day schedule: Activity *A* is the only one common to critical paths. It is the only activity that can be expedited, since activity *D* cannot be crashed by design and activity *E* cannot be crashed unless path *AD* is also reduced. Thus, we expedite activity *A* by 1 day at a cost of $200.

Direct cost = $1450 + $200 = $1650

10-day schedule: Expedite activity *A* by another day at an additional cost of $200.

Direct cost = $1650 + $200 = $1850

Total Costs of CPM Example

11.5

COSTS, $	PROJECT DURATION, DAYS					
	15	14	13	12	11	10
Direct	1150	1250	1350	1450	1650	1850
Indirect	2400	2240	2080	1920	1760	1600
Total	3550	3490	3430	3370	3410	3450

As we already noted, 10 days is the minimum duration for the project, so we cannot consider shorter periods. We now introduce the indirect costs. Suppose they amount to $160 per day. Then the total costs for the different project durations are summarized in Table 11.5. Figure 11.6 presents a graphical illustration of the same data. In either case, we observe that the optimum or least-cost schedule is 12 days at a cost of $3370.

PERT The PERT model considers uncertainties in time estimates of activities. As with CPM, we begin by identifying the activities and then draw the precedence diagram. Next we calculate the expected value of each activity duration as a weighted average of three time estimates. The three time estimates are:

1   The optimistic time t_o, with a low probability of occurring
2   The most likely time t_m, with the highest probability of occurring

Total costs of CPM example.

11.6

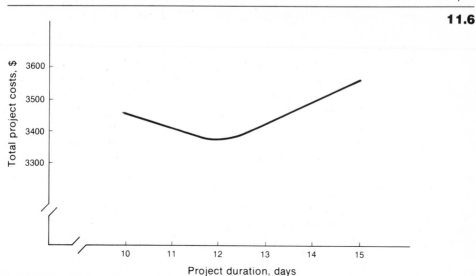

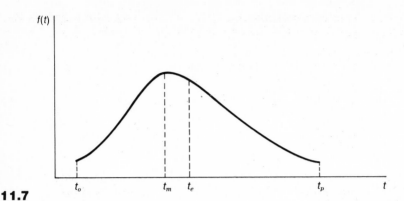

11.7

PERT beta distribution.

3   The pessimistic time t_p, also with a low probability of occurring

With PERT, we assume that t_0 and t_p are about equally likely to occur and that t_m is four times as likely to occur as either t_o or t_p, on the assumption that the three time estimates describe a beta distribution,[10] as illustrated in Fig. 11.7.

We then calculate t_e, the expected time for the activity; S_{t_e}, the standard deviation of t_e; and V_{t_e}, the variance of t_e:

$$t_e = \frac{t_o + 4t_m + t_p}{6} \tag{11.1}$$

$$S_{t_e} = \frac{t_p - t_o}{6} \tag{11.2}$$

$$V_{t_e} = S_{t_e}^2 = \left(\frac{t_p - t_o}{6}\right)^2 \tag{11.3}$$

As with CPM, the critical path is identified. Since we have calculated expected activity durations, the sum of the expected durations of the critical path activities gives the expected project duration T_e. Assuming that the times for the individual activities are statistically independent, we find that V_t, the variance of T_e, is equal to the sum of the variances of the corresponding t_e. S_t, the standard deviation of T_e, is calculated as the positive square root of V_t. Both S_t and V_t give us an idea of the degree of uncertainty associated with T_e. The higher the S_t, the more likely it is that the actual project duration will differ from T_e. We will illustrate some of these concepts with an example.

---

[10] The weights are based on an approximation of the beta distribution, which is unimodal, has finite, nonnegative limits, and is not necessarily symmetrical, unlike the more popular normal distribution.

Example 11.2 Let us redraw the network of Fig. 11.5 as Fig. 11.8 with **367**
the indicated values for the time estimates. The expected times and
variances are calculated using Eqs. (11.1) to (11.3). For instance,
for the activity from node 1 to node 2, we have

$$t_e = \frac{5 + 4(14) + 17}{6} = 13$$

$$S_{t_e} = \frac{17 - 5}{6} = 2$$

$$V_{t_e} = S_{t_e}^2 = 4$$

The critical path or the longest path of expected times includes activities from node 1 to node 2 and from node 2 to node 4. For this critical
path, $T_e = 13 + 13 = 26$. The variance of the critical path is
$V_t = 4 + 9 = 13$. Thus $S_t = \sqrt{13} = 3.6$.

After calculating the expected times for the activities and the
expected project duration T_e, as well as the corresponding variances
and standard deviations, we go on to compute the probability of
meeting a scheduled time T. We will not go into that except to point
out that the standard deviations for the activities and for the entire
project provide the manager with additional information about variations in job and project times which he may find helpful.

SIMULATION AND THE CRITICALITY INDEX   With the availability of
the digital computer, cases in which the activity times have uncertainties can be handled readily via simulation. Activity times are randomly

Network for PERT example.

11.8

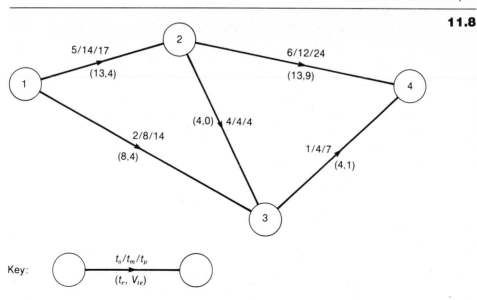

Key:
$t_o/t_m/t_p$
(t_e, V_{te})

368   selected for each activity for some suitable frequency distribution by using the Monte Carlo sampling technique. After several cycles of the procedure, the average project duration and standard deviation are calculated from the simulated data.

Additionally, a count is kept on the number of simulation runs in which a given activity is on the critical path. The criticality index of an activity is the proportion of the number of runs during which it is critical to the total number of simulation runs. For instance, an activity that is critical in 2,400 out of 10,000 runs will have a criticality index of 0.24 for the project, even though it may not be critical in the one set of calculations made by the regular PERT method. The criticality index provides a measure of the degree of attention that a manager should give an activity. The higher the criticality index, the more likely an activity is to cause serious delays in project completion if not given adequate attention or control.

To conclude this section on project control, we quote from Whitehouse and Wechsler:

> The control phase is needed to provide the project manager with information for managing his project. This monitoring is achieved by a comparison of the actual status of the project against the CPM or PERT schedule. The outgrowth of his comparison is a revised strategy. The monitoring phase is usually performed using a computer, although limited success can be achieved by using hand calculations. By periodically feeding the computer such data as (1) completed activities, (2) changes in the durations of activities, and (3) changes in the detailed structure of the network, the project manager obtains from the computer a monitor report illustrating the current status of the project. The monitor indicates those activities which are ahead of or behind schedule. Using new and improved data, the computer predicts when the project will be completed. It identifies critical activities and those about to become critical. The manager can thus see how the total project is progressing and can determine where maximum effort should be placed.[11]

POSTPROJECT EVALUATION

When a major project is close to completion, one of the final tasks is the evaluation of the overall efficacy with which the resources were used and the project efforts were managed up to that point. This is called the "postproject evaluation." Considering time, cost, and performance as the three dimensions of a project, we can generally use

---

[11]G. E. Whitehouse and B. L. Wechsler, *Applied Operations Research,* Wiley, New York, 1976, pp. 239 and 240.

objective, quantitative measures to evaluate the first two dimen- **369**
sions, but we need subjective data for the third dimension. Several
approaches can be used for this purpose. Here we will present a
brief summary of four methods which, taken together, provide the
basis for a comprehensive process for postproject evaluation.[12]

EVALUATION OF PROJECT RECORDS
The financial records and time schedules are the best sources of objective information. A detailed study of such records shows where the
cost overruns have occurred and when the delays have become critical throughout the project.

CLIENTELE ASSESSMENTS
The conduct and outcome of projects always affect a multitude
of clientele, such as the project manager and his staff, the functional
managers and their staffs, the engineers and scientists involved in the
project, the contractors and subcontractors, and the ultimate users of
the project's outputs. Questionnaires and personal interviews with
these clientele provide valuable information about all aspects of the
project that would be difficult to obtain from other sources. Each clientele group is asked to respond to a questionnaire specially
designed to probe into the areas that particularly affect that group.
The number of questions can vary from a few to several dozen,
depending on the issues involved in each area. The typical areas
evaluated by this method include preproject planning and organization, project management process (planning, organizing, directing,
and controlling), project accomplishments, and project technology
transfer. The questionnaires are generally designed with a 5- or
7-point response scale ranging from "strongly disagree" to "strongly
agree" such that a statistical analysis can be performed. In addition
to the questionnaires, the narrative responses obtained in personal
interviews with key clientele are also analyzed for their contents.
When these two are combined, an image profile for the overall performance of the project can be developed in each of the areas mentioned above.

EXPERT-PANEL EVALUATIONS
The two methods described above provide us with information about
the conduct of the project but do not necessarily indicate how successful the project has been with respect to the organizational goals
and objectives. This can be measured by quantifying the judgments
of a group of experts who have a high degree of knowledge and understanding of the pertinent issues involved in the technical aspects

---

[12]See David I. Cleland and Dundar F. Kocaoglu, *Program Evaluation of the Combined Forest Pest Research and Development Program*, U.S. Department of Agriculture, January, 1981, for the details on how such a postproject evaluation might be carried out.

370   of the project. First, a panel is formed from experts who do not have a vested interest in the project. These experts are then provided with detailed information on the project goals, outputs, and the relevant criteria that can be used as the measures of effectiveness of the goals. Finally, the experts are brought together to a meeting and presented with these goals, criteria, and outputs as the hierarchical levels of project decisions. They express their subjective opinion on the relative values of each element at each level, using the hierarchical decision model approach. The meeting essentially becomes a forum for reaching consensus among the experts. After the measurements are taken and the group consensus is reached at each level, the relative values of the project outputs are expressed in terms of their contribution to the organizational objectives. This method gives us a measure of the level to which the project has fulfilled its goals and contributed to the overall objectives.

BENEFIT/COST ANALYSIS

An economic analysis of the project outcomes and their costs is called "benefit/cost analysis." This method is especially useful for evaluating engineering projects involving the public sector, such as highway, dam, and bridge projects. The basis for the B/C analysis is the comparison of the present value of benefits with the present value of the costs. Projects with high B/C ratios are considered successful (of course, a project whose benefit/cost ratio is less than or equal to 1.0 is, by definition, too costly to be accepted as a success).

The difficult aspect of the B/C analysis is the determination of the benefits, since they are hard to define and hard to measure. One of the best methods for determining the benefits is to use a panel of experts. Once the benefits are defined and the costs delineated as described in the first method above, the analysis is performed using the familiar present value calculations.[13]

These four methods can be used by themselves or in combination with each other for postproject evaluation purposes. It may be sufficient to use only one or two of them for simpler projects. However, for the more complex projects, an integration of the four is usually needed.

SUMMARY

Projects need constant monitoring as they go through their life cycle. At the initial phase the prospective projects are evaluated for their economic, technical, and market feasibility in order to select the most promising ones. This is called the *preproject evaluation.*

---

[13]Present-value concept is discussed in any good book on engineering economy.

Once a project is under way, its day-to-day operation in **371** terms of time, cost, and performance is evaluated via project control techniques and project review meetings. These activities constitute the *ongoing project evaluation.*

Finally, as the project winds down, a careful assessment is necessary to see what happened in it, how well it was managed along the time, cost, and performance dimensions, and what it produced throughout its life cycle. This activity is called *postproject evaluation.*

DISCUSSION QUESTIONS

1 The authors state that projects are building blocks of an engineering organization's strategy to achieve overall goals and objectives. Why is this so?

2 The evaluation of a project can be done at least three times during the project's life. What are these times?

3 Why does the engineering manager need a rational methodology for the selection of engineering projects?

4 Project selection methodologies range from simple to complex. What are the different methodologies in these categories?

5 What is the advantage of using the simple rank-ordering method?

6 Why is the development of criteria for the evaluation of projects so important?

7 Who should develop the criteria for the evaluation of projects?

8 What is the essence of the hierarchical model in the evaluation of projects?

9 What are the basic questions that should be addressed during an ongoing project evaluation process?

10 How can the Gantt chart be used in project evaluation?

11 An integral part of both CPM and PERT is the use of networks. Demonstrate that you understand the use of networks.

12 What are some of the techniques that can be used for postproject evaluation?

KEY QUESTIONS FOR THE ENGINEERING MANAGER

1 What is the policy in this organization concerning the evaluation of ongoing projects?

2 Has provision been made for relating project evaluation to the strategic planning in the organization?

372
3   Who makes the decisions on the selection of a project in this engineering organization? Is this the right person to make the decision?

4   What are the techniques being used in preproject evaluation in this organization? Can these techniques be improved?

5   Considering the range of alternatives available in project selection methodologies, which are best suited to the cultural ambience in this organization?

6   Has somebody been identified as the focal point for the development of project selection methodologies in the organization?

7   Do the engineering project managers understand the various methodologies that are available for the selection of projects?

8   How are projects evaluated in this organization? Is there room for improvement in this evaluation process?

9   Do the technical people in this organization understand the advantages (and limitations) of using the Gantt chart as a project evaluation technique?

10   What is the current method for doing a postproject evaluation? Does this evaluation take into consideration the strategic objectives of the organization?

11   Who are the clientele of the key engineering projects currently being conducted in this organization?

12   Is there any opportunity for the application of expert panel evaluations in the projects currently being managed in this organization?

BIBLIOGRAPHY

Barish, Norman N.: *Economic Analysis for Engineering and Managerial Decision Making,* McGraw-Hill, New York, 1962.

Churchman, C. West, Russell L. Ackoff, and E. Leonard Arnoff: *Introduction to Operations Research,* Wiley, New York, 1957.

Cleland, D. I., and W. R. King, *Systems Analysis and Project Management*, McGraw-Hill, New York, 2d ed. 1975.

———, D. F. Kocaoglu, et al.: *Program Evaluation of the Combined Forest Pest R&D Program,* U.S. Department of Agriculture, Research Report Agreement no. OS-78-07, January, 1981.

Delbecq, Andre L., Andres H. Van de Ven, and David H. Gustafson: *Group Techniques for Program Planning,* Scott, Foresman, Glenview, Ill., 1975.

Karger, D. W., and R. G. Murdick: *New Product Venture Management,* Gordon and Breach, New York, 1972.

Kocaoglu, D. F.: "Hierarchical Decision Model under Multiple Criteria," paper presented at the TIMS/ORSA National Meeting, Washington D.C., May 5-7, 1980.

————: "Hierarchical Decision Process for Program/Project Evaluation," paper presented at the 6th INTERNET Congress, Garmisch, West Germany, September 1979.

Layard, Richard (ed.): *Cost-Benefit Analysis,* Penguin, Baltimore, 1972.

Moore, J. R., and N. R. Baker: "An Analytical Approach to Scoring Model Design: An Application to Research and Development Project Selection," Graduate School of Business, Stanford University, Stanford, Calif.

Poister, Theodore H.: *Public Program Analysis: Applied Research Methods,* University Park, Baltimore, 1978.

Scriven, M.: "The Methodology of Evaluation," in *Perspectives of Curriculum Evaluation,* R. W. Tyler, R. M. Gagne, and M. Scriven (eds.), AERA Monograph Series on Curriculum Evaluation, no. 1, Rand McNally, Chicago, 1967.

Weiss, C. H.: *Evaluation Research,* Prentice-Hall, Englewood Cliffs, N.J., 1972.

Whitehouse, G. E., and B. L. Wechsler, *Applied Operations Research*, Wiley, New York, 1976.

Wiest, J. D., and F. K. Levy, *A Management Guide to PERT/CPM,* Prentice-Hall, Englewood Cliffs, N.J., 1977.

PART FOUR

THE BROADER SYSTEMS IN ENGINEERING MANAGEMENT

TECHNOLOGICAL 12
PLANNING AND
FORECASTING

In this chapter, we will first discuss planning in general then focus on technological planning, and finally present technological-forecasting concepts.

Erich Jantsch defines planning as "an attempt at anticipating and controlling the future."[1] This definition has two components: "Anticipation of the future" refers to an assessment of the future conditions. This is the forecasting component. "Controlling the future" is the decision-making and implementation component. Actions taken without forecasting would be random, unrelated to past experiences and existing information, and not directed to future environment. Forecasting by itself is of limited value if it is not used in a planning context for evaluating the future implications of current decisions. Forecasting is meaningful only when viewed as an integral part of planning.

Technological forecasting is the activity dealing with the assessment of future technologies and their development by the engineering and scientific community. It can be done in an exploratory effort, starting with existing knowledge and moving toward the future; or it can be normative, starting with the future missions and working backward to the present.

The attempt to control the future may sound absurd. But that is what planning is all about. We first develop a perspective on the future, as we see it; then we take actions to change that future to our liking and to prepare ourselves for what we see several years ahead. We can do this by observing the pattern of emerging technologies, identifying the missing links in that pattern, and charting a route that will take us to the desired future using the existing and developing technologies as they contribute to our mission. Along the way, we concentrate on the missing links and develop them as intermediate steps to take us to the future that we like to see for ourselves. This is the essence of technological planning.

[1]Erich Jantsch, *Technological Forecasting in Perspective,* OECD Publications, Paris, 1967.

378 PLANNING

"Luck is a residue of careful planning," according to Jantsch.[2] This statement explains the whole idea behind planning.

Before we discuss the planning process, it is important to understand what planning is and is not. Peter Drucker, in one of his earlier articles on strategic planning, describes several misconceptions about planning. The following paragraph summarizes his views.[3]

First, planning is not forecasting. The result of a planning effort is a series of recommended strategies that appear to be reasonable to follow, based on the available information. A forecast, on the other hand, is a statement of what is likely to happen in the future. *Second,* planning deals with the futurity of present decisions but not with the future decisions themselves. We cannot decide on what actions to take in the future. Certain future actions may appear plausible with the information available at present. However, by the time an action is taken, we may receive additional information that could reverse our thinking process and lead to a different decision. Even though the decisions are made *in* the present, they cannot be made only *for* the present. The consequences of our decisions will remain with us well into the future. The basic idea behind planning is to take into account these future consequences. *Third,* planning does not eliminate risk. Every activity, regardless of whether it deals with technology, the economic system, or society in general, has an implicit risk associated with it. That risk cannot be eliminated through planning or other means. Planning helps us recognize and take the right risks, the acceptable risks.

Drucker points out that the decision maker can, in fact, take greater risks through effective planning. By taking the right risks, he can maximize the return on his resources.

Having listed some points of what planning is not, we now will discuss what it is. We define planning as *a systematic process of making current decisions and assessing their future impacts toward the achievement of goals and objectives of the organization.* Let us study this definition.

Planning is a continuous *process,* not a one-time activity. In contrast to a plan, which is merely a document, planning is the continuous act of making decisions. The plan is one of the outputs of this process and contains the results of planning and the implementation strategies.

Planning is a *systematic* process. It follows a series of steps and feedback loops. The steps are sometimes taken sequentially, sometimes in parallel with each other. The product of each intermediate step should be recognizable and traceable.

[2]Ibid., p. 83.
[3]Peter F. Drucker, "Long-Range Planning," *Management Science,* vol. 5, 1959, pp. 238–249.

Planning is a *decision making* activity. The decisions are al-   **379**
ways made at present, even though they are based on the information
from the past and assumptions about the future.

Planning involves the *future impacts* of decisions. These
impacts are constantly evaluated to assure that they will lead the organization toward a future of our liking.

Planning is directed to the *achievement of goals and objectives.* It is a mission-oriented activity. The general direction of the organization is first translated into broad statements concerning the
objectives of the organization. These are not operational statements,
but they provide the basis for planning. The next step is the development of explicit goal definitions from the ambiguous terms used in
the broad objectives. Once the goals are defined, strategic decisions
are made and actions are specified to achieve the goals and objectives in a hierarchical fashion.

In summary, planning is done to achieve the organizational
goals and objectives in a systematic way. An important feature of
this activity is the rationality behind it. We try to control the future by
shaping it to our liking, and we want to do so with the optimum use
of limited resources.

Since planning is a mission-oriented activity, its backbone is
the set of objectives toward which the organization moves. We can
realistically raise the question as to whose objectives they are. It can
be argued that the broad objectives are set by a group of people or
sometimes by one person, and therefore they are the personal objectives of those decision makers. This argument assumes, of course,
that when the decision makers who have set the objectives leave the
organization for some reason, the mission of the organization will be
completely different. This assumption can be justified only to a certain extent, not completely. In reality, even after the decision makers
are replaced by others, organizations may continue in basically the
same directions they followed before. Even when fluctuations take
place, it is not common to see a drastic change in the direction of
established organizations because of the culture of the organization.

On the other hand, if we extend the idea of organizational
inertia to an extreme, we can argue that the objectives are determined by the organizations taken as entities, not by the individuals
temporarily running those organizations. Of course, in this case, the
implicit assumption is that no matter who the decision makers are,
the organization will always have the same objectives. This is not
quite true either. The objectives are, by necessity, influenced to a
great extent by the value structures of the decision makers in control
of the organization.

By combining these two extremes we see that the objectives
are defined by the decision makers under personal and organizational constraints. On the personal side, the decision makers form

380

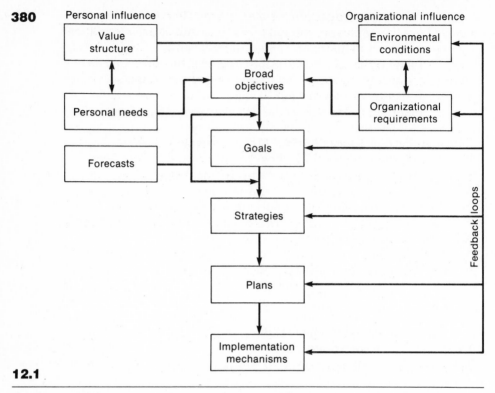

12.1

Planning process.

opinions based on their value structures and personal needs. On the organizational side, their decisions are affected by the organizational requirements and the prevailing environmental conditions.[4]

The overall planning process is depicted in Fig. 12.1.

Value structure is a personal characteristic developed through the individual's social, religious, and educational upbringing. It is acquired gradually from childhood to early adulthood. After that, it is relatively stable. However, depending upon the individual's personal needs and the level to which those needs are satisfied, the value structure may fluctuate and change its direction from time to time. On the organizational side, the needs of a fast-growing organization are quite different from the needs of a mature organization. A high-technology developer does not have the same viewpoint as a manufacturer in a stable industry. Needs change in time as organizations establish their missions and as the external system goes through major changes. All these factors are combined in the development of the organization's objectives.

---

[4]"Environment" is used in a broad sense here. It refers to the external environment consisting of the social, legal, technological, economic, and political systems.

Objectives are generally stated in broad terms to establish the **381** overall direction of the organization. Some examples are "leadership in automotive industry," "high profitability," and "social responsibility. Goals are much more specific. They establish recognizable targets to support the organizational objectives. For example, the "industry leadership" objective can be translated into specific goals, such as:

Development of a diesel engine that has a 15 percent better gas mileage than the existing models

Development of a diesel engine that has low emissions of "soot"

Strategies are a set of actions specified in support of goals. They could include the following:

Increased engineering force

Accelerated technical education for engineers to update diesel technology knowledge of existing engineering force

Acquisition of capital equipment to support diesel technology

Acquisition of increased plant capacity

After various combinations of these strategies are evaluated, the best alternatives are selected in terms of cost and effectiveness. Plans are prepared to specify the recommended actions for each strategy. Then the organizational mechanisms are developed to implement the plans. This process involves both forward and backward steps, as indicated by the feedback loops in Fig. 12.1. It is a continuous and dynamic process. It never stops. Changes are made in the process as new information becomes available. The plans are revised to reflect those changes. The process continues.

Success of the planning process largely depends on the support it receives from management. Unless the decision makers are involved and enthusiastic about it, planning does not produce much more than a few volumes of written documents. When planning is left to a few staff persons without the input and tangible support from the key decision makers, those documents containing the plans are generally left out of the decision making process in the organization.

Organizations that have developed and implemented successful planning processes see it as part of life and as a prerequisite for their survival. In such organizations, it is a continuous function performed at all levels from top to bottom. The future of an organization can be evaluated only through the use of strategic planning philosophies and techniques.

382   STRATEGIC PLANNING

Strategic planning, the capstone of all organization planning, deals with the following questions:

What has been our business in the past?

What is our current business?

What should our business be in the future?

What business should we not be following in the future?

Getting the answers to these questions demands a conscious evaluation of future opportunities and problems facing an organization. Strategic planning is a process that predicts probable future events, examines alternatives, then chooses among them. Technological planning is an inherent part of this process. The relationship of technological planning to other functional planning is illustrated below, where strategic planning documentation is portrayed.

Planning documentation

Strategic plan		To satisfy
Business-area plans	Support	customers,
Technology plan	corporate	stockholders,
Market plan	objectives,	employees,
Financial plan	goals,	community,
Production plan	strategies	etc.
Project plans		

The various plans developed in the planning process support corporate objectives, goals, and strategies in order to satisfy customers, stockholders, employees, and other groups who have some claim to the corporate resources.

Any strategic planning has to be carried out based upon a reasonably accurate forecast of the future. The awesome changes that have impacted on contemporary organizations in the past few years were not generally foreseen by those who professed an ability to forecast the future. Some of these major changes which have affected all organizations include severe recession, severe shortages in raw materials in many basic industries, sharply rising costs for energy, a crippling inflation, and abrupt shifts in customer preferences. In hindsight, these changes came from obvious events, yet there is little evidence that these changes were included in any planner's forecast.

One might conclude from this that forecasting does not really make too much difference in an organization's ability to deal with **383** future programs and opportunities. But, on the other hand, if forecasting is not done, the planner is left without any model of the future at all on which to develop his plans.

A major portion of the planning effort in technology-based organizations is directed toward the planning for the technological future of the organization. This is called technological planning.

TECHNOLOGICAL PLANNING

An engineering manager operates in an environment in which he is constantly confronted with the opportunity to make operational, short-range decisions, such as allocating resources, assigning people to emerging engineering projects, selecting design alternatives, and so on. These kinds of operational decisions are usually well known and visible to the engineering manager. They are part of the mainstream of life in an engineering organization.

A successful organization is always in motion, particularly one that is growing and coping with the environmental change that alters its chances of continued growth—or survival—as the case may be.

The seriousness of missing technological changes in the environment should be readily apparent to the perceptive engineering manager. For their well being, all managers must develop a philosophy and process for anticipating technological changes that cause existing technology to become obsolete, or open new opportunities in the organization's marketplace. The engineering organization that neglects this role runs a serious risk of becoming obsolete. There are legions of examples of organizations with impressive technological experience that have missed the potential of an emerging technological opportunity. An engineering manager who has guided a product design and development effort through its life cycle is only too aware of how changing technology can threaten the survival of this project. Technological innovation poses a challenge to general managers who ultimately must make the strategic choices for the entire organization. In this role the general manager balances technical, marketing, financial, and other issues so as to optimize the selection of a strategic role for the organization. To select this role he needs strong support from his engineering manager. The engineering manager takes a customer's needs and translates them into a proven workable design which meets all of the product requirements for price, performance, reliability, durability, dependability, etc. At the same time the engineering manager must constantly question the relevance of the state of the art embodied in the organization's products and in the manufacturing of these products. When a general manager selects a strategic objective for the organization, the engineering man-

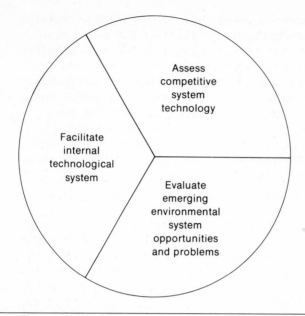

12.2

Technological planning responsibilities of the
engineering manager.

ager has the responsibility for evaluating the technological impact of
that objective on the internal technical system of the organization. In
evaluating this impact, he must assess strengths and weaknesses of
the internal system and the competitive system. He must also develop
assumptions about the probable strategies of the competitors. A
simple model, such as Fig. 12.2, can illustrate the engineering manager's responsibility for technological planning.

That responsibility includes all actions necessary to support
product design and development, and operational support to the
stream of products flowing through the organization. Emerging technologies in the state-of-the art for operational support of products
are also the responsibilities of the engineering manager. The manner in which this responsibility is carried out depends on the size of
the company. A large company may have its own research and development (R&D) laboratory where emerging technology is identified
and evaluated. R&D opportunities emerging in a technological area
are analyzed, usually through a project team composed of representatives from the appropriate disciplines. The engineering discipline
is represented in these teams to appraise the technology in terms of
its impact on product design, development, production and support.

In smaller companies an R&D effort as a separate entity may
not exist or a company may elect to perform no research and devel-

opment but rather pursue a policy of imitation.[5] In these circum- **385** stances the broad assessment and application of technology typically rest with the engineering manager. In any of the situations just described, the engineering manager's responsibility always includes an ongoing evaluation of how product design, development, production, and operational support are to be done in the organization. He may not necessarily do all of the technological planning, but he is always constrained by it and helps to provide the knowledge base for it. Planning for products to be designed and developed involves starting with long-range market and technological forecasts, determining those opportunities of interest to the company, and planning strategies to put the product into the marketplace.

THE PROCESS OF TECHNOLOGICAL PLANNING

Technology is the key to advancements within a company's product lines. A company that hopes to have long-term growth and profitability encourages the identification of new product and market ideas which will permit innovative breakthroughs into new business areas. The primary intent of overall planning is to identify new products and markets within a given technology base. A broadening of that technology base may be necessary to meet market opportunities or competitive threats. The role of technology in an organization's planning may be categorized as follows:

1   New markets for current and derivative products

2   New products for existing markets *within or outside* existing technology base

3   New products for new markets *within or outside* existing technology base

The key role of technological planning should be evident from items 2 and 3, where understanding technological change is required for an organization to determine what its business should be and what the technological strategy should be to get there. Technological planning is required to ensure that a company maintains a competitive advantage. After market requirements are translated into technical objectives, goals and strategies, a series of steps are undertaken to bring technology into line with other functional planning in the organization, such as market, finance, production, etc. A series of typical steps would include:

1   Identification of required technology with evaluation of payoff potential

---

[5]See Chap. 6.

386

2   Assessment of technology gaps based on market timing or state-of-the-art difficulties

3   Product planning to meet market requirements within financial constraints and corporate strategies

The technology plan should pull together all engineering, research, and development necessary to maintain a technological course supportive of market needs. A typical flow of technological planning is illustrated in Fig. 12.3. In essence, the purpose of such planning is to define the technological advancements required to achieve success in the market.

The engineering manager is at the interface of current technology and strategic assessment for the application of future technologies, as portrayed in Fig. 12.4.

We thus postulate that the engineering manager's strategic responsibilities are those necessary to facilitate technological survival and growth for the enterprise as a whole in a dynamic competitive and environmental system. He is the architect of technological strategy in the organization. Once the technological strategy has been

The flow of technological planning.

12.3

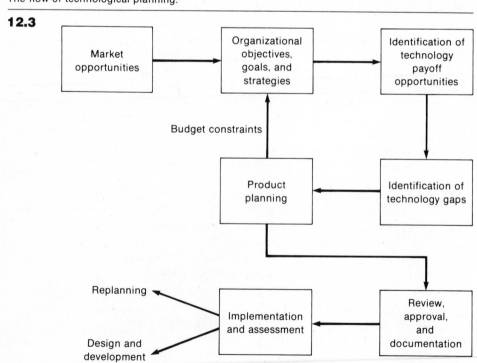

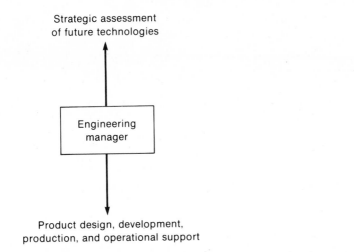

12.4

The engineering manager's interface role.

designed, he becomes the fiduciary for its implementation in the organization.

NEW PRODUCT TECHNOLOGY
The development of new products usually embodies the advancement of an existing technology. Consider the product and market opportunities that would be unleashed if coal gasification and liquidation were commercially viable. The storage of electric energy is one of the technological breakthroughs needed to develop an efficient and useful car. According to *Fortune* magazine electronics technology is on the threshold of introducing revolutionary weapons which will bring about far-reaching changes in the tactics of warfare.[6] The continuing miniaturization in the computer industry is an example of technology that almost defies the imagination. Consider the ability to provide an entire information storage and programming system consisting of 35,000 separate transistors crammed onto a silicon sliver no higher or wider than the first word of this paragraph. This microprocess, or "computer on a chip," was developed by Intel Corporation and promises to change computer technology with awesome reverberations in information processing and communication technology.

One recent technological innovation developed in companies is fiber optics. For nearly two decades the idea of transmitting information with bursts of light through glass fibers showed promise. By 1978, all the installed fiber-optic cables in the world added up to

---

[6]Juan Cameron, "Bill Perry Keeps the Pentagon's Barrels Loaded," *Fortune,* March 10, 1980.

388   some 600 miles. For many years people have been predicting that fiber optics would catch on. Finally, thanks to steady improvements in technology, fiber optics appear to be competitive with conventional copper cable in some applications. It may be that future products, in such companies as American Telephone & Telegraph and General Telephone & Electronics, using fiber-optic devices may catch on. The application of fiber-optic devices could have enormous application in the automobile and the wiring systems of aircraft, as well as in the communication industries. This is an example of a new technology that has moved out of the development stage and into the commercial mainstream of business.[7] In agricultural technology the improvement of the nitrogen-fixation properties of certain plants and the increased efficiency of the photosynthesis process in some farm crops could open the way for many new products.

In today's technology-based society the development of new products depends for the most part on an improved technology. Even with the improved technology, the development and introduction of a new product may be hampered by environment considerations that require further technology to solve. For example, the diesel engine has improved average mileage per hour over comparable gasoline-engine cars by about 25 percent. But concerns about the diesel emissions have recently emerged that will require a new technology for the solution.

There is an old saw in strategic planning literature which says: "There is nothing more important to an organization than its future." We would add: "There is nothing more important to an organization's future than its technology."

TECHNOLOGICAL FORECASTING

Martino defines technological forecasting as "the prediction of the future characteristics of useful machines, techniques and procedures."[8] As this definition implies, forecasts give us an assessment of future characteristics, not an explanation of how those characteristics are to be achieved. That explanation is given by the strategies developed for future technologies using the information obtained from forecasts.

A common error in technological forecasting is caused by overconservatism. This error could particularly affect the forecasts made by panels, where there has been a great deal of interaction among people. The strong personality of some members of the panel, or the fear of appearing too innovative or adven-

---

[7]For further information, see Gene Blinsky, "Fiber Optics Finally Sees the Light of Day," *Fortune,* March 24, 1980.
[8]Joseph P. Martino, *Technological Forecasting for Decisionmaking,* American Elsevier, New York, 1975.

turous, could force the others to agree with conservative ideas. Thus many panels have given conservative forecasts, even when they had the information to end up with drastically different conclusions. An example of a conservative forecast is the weight-to-power relation for the gas-turbine. The forecast made in 1940 was 13 to 15 pounds per horsepower; the real figure turned out to be 0.4 pound per horsepower.[9] Another error is the failure to recognize developments in other areas, or the natural limits in technology. The failure to consider these limits and to recognize the development of simultaneous technologies could result in nonsensical forecasts.

Although the nature of the forecast depends on the planning horizon and the technology involved, there are certain elements that are common to all technological forecasts:

The valid time period for the forecast

A statement of the nature of the technology

Characteristics of that technology for the appropriate time period

Probabilities associated with those characteristics

The absence of any of these elements could reduce the value of the forecast. The precision demanded for each one of these elements depends largely upon the use to which the forecast will be put, and the availability of resources for forecasting.

Forecasts are made on the basis of existing information. It must be remembered that the origin of the forecast is the past; it does not tell us anything about the future per se. It studies the interrelationships among the present and past data and explains their implications for the future.

The value of a forecast should not always be measured by how accurate it is. Other factors are just as important. We should evaluate them by the extent to which they contribute to decision making. Let us consider the case when one alternative dominates the others under all states of nature. For example, let us assume that the automobile manufacturer we discussed in Chap. 10 finds a new alternative Suppose that it gives the highest payoff under all conditions. Regardless of the state of nature (fuel shortage or not, EPA regulations or not), the decision will always be to select this new alternative. In such a case, why should we measure the value of the forecast by how well it predicts the future? If the states of nature are not a factor in the decision, forecasting those states is meaningless. Such a forecast clearly has no value regardless of its accuracy. There is a tendency to look at the past events with hindsight and

[9]Robert V. Ayres, *Technological Forecasting and Long-Range Planning*, McGraw-Hill, New York, 1969.

390   criticize forecasts that did not predict them. Can we say that a forecast that turned out to be wrong has no value at all? It depends on why that forecast was wrong. Sometimes the forecasts show conditions that are not favorable to the decision maker. If the decision maker takes appropriate steps, he can change the environment to his liking, thus proving the forecasts to be wrong. When we look back in many cases we see that the gloomy forecasts were highly inaccurate and therefore worthless. But were they? If we consider their impact on decisions, they were invaluable. In fact, the favorable conditions would probably never have materialized if the forecasts had not pointed out the danger signs.

For example, after the Three Mile Island incident, there were strong indications that we might be seeing the end of the nuclear power era. Public reaction was very strong and the prevailing assumption was that we did not have the technical capability to develop "safe" nuclear energy. Let us suppose that one company at the frontier of nuclear technology looks at the problem differently. Suppose they have a strong conviction that this is not a technical issue but a communications problem. They realize that the risks are there, but the public has never been given an accurate picture of the nature and extent of these risks except at a very emotional level. Defining the problem this way, they develop a strategy to inform the public of the tradeoffs between the risks and benefits of nuclear power. Suppose that after doing this effectively and on a massive scale they generate a realistic awareness for the assessment of the pros and cons of this technology and the public reaction is reversed. If we think about this scenario, we see that the evaluation of forecasts in terms of their accuracy alone is meaningless. We should judge them not only by how accurate they turn out to be but, more important, by how valuable they prove to be to the decision maker.

EXPLORATORY TECHNOLOGICAL FORECASTING

The methods used in technological forecasting will be discussed in this section. A wide variety of techniques have been developed over the years with different names given to them. We will classify the methods as exploratory or normative and then describe the major techniques that fall into each category, as summarized in Table 12.1.

The starting point of exploratory forecasting is the currently available knowledge base. We gather all the information available on the state of the art and the historical developments in technology up to the present. This information is used as the basis for assessing future changes that could take place in technological environment. The methods summarized below are presented in the order of analytical sophistication—from the qualitative level to the quantitative level.

Technological Forecasting Methods

EXPLORATORY	NORMATIVE	**12.1**

Intuitive methods:
  Expert forecasts
  Brainstorming
  Scenarios
  Delphi method
	Relevance trees
	Morphological analysis
	Mission flow diagrams	

Analog methods:
  Historical analogies
  Growth curves		

Causal methods		

Operations research methods:
  Gaming
  Simulation		

Aggregate methods:
  Input-output analysis
  Cross-impact analysis		

INTUITIVE METHODS

When we hear the word "intuition," we think of almost anything but rationality. However, some people are capable of grasping the multiple effects of the interrelationships in a given situation with intuition. These interrelationships escape the perception of most of us and thus put those able to perceive them into a special category—sometimes called "geniuses." In this sense, it is clear that intuition is not the opposite of rationality. Unfortunately, a genius is not always available or not always guaranteed to work well and yield perfect outputs. An alternative to the genius idea is to form a panel of experts and bring them to a consensus.

EXPERT FORECASTS   In expert panels, the personal bias is usually offset and smoothed out by group interaction except when some members of the panel have stronger convictions than others and the so-called "bandwagon effect" exists in the expert forecasts. This happens when some experts in the panel are intimidated by a few outspoken members and accept their ideas even when they do not agree. The key to the success of expert panel forecasts is in the formation of the panel. Extreme care should be exercised in selecting the panel members. The following four criteria are recommended to ensure a balanced group:

1   A high level of competence and knowledge in the field under study

2   Absence of vested interest

3   Balanced representation of opposing viewpoints

392   4   Absence of friction or intimidation in the group

When the experts meet, they are presented with the issues, given all the available information, and asked to have an open discussion for the assessment of future technological changes in the area under discussion. A moderator conducts the meeting and facilitates the discussions. An off-site location is usually more productive than the office environment for these meetings. A disadvantage of expert-panel forecasts is the tendency toward overconservatism.

BRAINSTORMING   The overconservatism of the expert-panel meetings can be overcome by using a slightly different approach called the brainstorming method. It is used specifically to stimulate a different form of thinking, something adventurous and extraordinary. For a successful brainstorming session, a well trained, highly skilled leader is crucial. He is to establish a set of rules, keep the participants from breaking the rules, and encourage the inhibited persons to participate in the discussion.

The following rules improve the probability of success of brainstorming sessions:

> Concentrate efforts only on well-defined problems and avoid wandering from topic to topic.
>
> Consider all the ideas that are in some way related to the problem.
>
> Do not criticize any idea, even if it looks outrageous.
>
> Do not start exploring the detailed implications of any idea, until several ideas have been selected and a particular interest has emerged in developing one or more of them.

Brainstorming sessions are conducted in a sequence. They usually begin with the statement of a problem, or the presentation of a specific topic. Typically five to eight people participate in these sessions. They can be from different disciplines and have different backgrounds; they may or may not know each other.

The first session is normally dedicated to suggesting ideas about the proposed topic under the direction and encouragement of the leader. It usually takes about 2 to 4 hours to complete the initial session, ending with a list of ideas. In the next few sessions, which are usually shorter than the first one, consensus is sought to select the best ideas. After the best ideas have been chosen, the implications of those ideas are studied and acceptable ways to implement them are discussed. The result of a well conducted brainstorming session is either a set of solutions to specific problems, a set of alternatives to achieve a particular goal, or an assessment of future conditions.   **393**

SCENARIOS   Scenarios are the descriptions of a series of events that could be expected to occur simultaneously or sequentially. They are developed to reflect the logical interactions that can take place among the major forces shaping up in the technological, economic, social, and international subsystems. The branch points are critical in scenarios. These are the points where small changes could have major impact on the eventual outcome of the situation. When a scenario is written, it can be tested under various conditions by changing the assumptions and conditions at branch points. This way, one scenario typically leads to a large number of related scenarios with slight modifications.

Herman Kahn is a well-known authority on this subject. He and his team at the Hudson Institute develop and test scenarios on a wide range of socioeconomic, technological, national, and international topics. Kahn points out two major advantages of scenarios:[10]

1   Scenarios counteract "carry-over" thinking and force the policy maker to consider situations other than the "surprise-free projections."

2   Scenarios help the analyst to look at the crucial details in the context of their impact on the total picture. Without a scenario, the analyst might be satisfied by an abstract definition of the big picture, but miss the dynamic interactions that could lead to significant changes in it.

Scenarios can be generated by combining historical events or by a role-playing exercise in a simulated situation. As an example, a family of scenarios for an energy crisis can be generated by various combinations of the following elements:

Parties involved ($p_1, \ldots, p_{n1}$):   OPEC nations, Western Europe, Japan, communist bloc, the United States, developing countries

Actions ($a_1, \ldots, a_{n2}$):   Oil embargo, continuous price increases, sudden reduction in oil production

Counteractions ($c_1, \ldots, c_{n3}$):   Military intervention, food embargo, international cooperation

[10]Herman Kahn and Anthony J. Wiener, *The Year 2000: A Framework for Speculation*, Macmillan, New York, 1967.

394   Results $(r_1, \ldots, r_{n4})$:   Reduced industrial output, mass transportation, change in life styles, worldwide conflict

Effects $(e_1, \ldots, e_{n5})$:   High inflation, unemployment, economic crisis

Strategies $(s_1, \ldots, s_{n6})$:   Conservation, coal research, development of renewable energy sources

DELPHI This technique, developed by T. J. Gordon and Olaf Helmer,[11] differs from the other interactive methods in that it eliminates the face-to-face interactions, and the problems are usually stated as questions. The face-to-face interaction is replaced by a questionnaire, which the anonymous participants answer and return by mail.

The questionnaires are prepared by a director, who filters, analyzes, and processes the answers. The processed answers are sent back to the participants, keeping all the pertinent data still anonymous. The basic idea is to lead the participants, through a controlled questionnaire, toward a consensus. There is still a good deal of interaction among the participants, but this time in a completely anonymous environment. The results are expressed in terms of means, standard deviations, etc. This way, most of the disadvantages of the other methods, especially inhibition and bandwagon effects, are avoided. However, there are still some unsolved problems. First, a bias may be introduced, due to the general willingness to participate. Second, the director may guide the discussions in a certain direction at each iteration, and those participants whose answers were far away from the mean may be forced to shift toward the consensus. Third, all participants are given equal weight, which in certain cases may not hold true because of special interests of someone in the problem, too much or too little expertise in that field, etc. Fourth, if there is a time pressure, or if there is an implicit imposition of an answer, some participants may tend to agree with the majority. This kind of bandwagon effect may be controlled to an extent by the director, by properly changing the nature of the questions. We must point out that this control is unavoidable in face-to-face interactions.

There can be modifications of the Delphi technique, which involve the removal of some of its characteristics, usually one at a time. For example, the anonymity of the participants could be removed, the feedback can be reduced, or the process can start with no questions but only the statement of the problem.

---

[11] Theodore J. Gordon and Olaf Helmer, "Report on a Long-Range Forecasting Study," RAND Corporation, Paper 2982, September 1964.

ANALOG METHODS

Another approach to technological forecasting is to compare technologies by looking at historical parallels between them or by applying the models developed in the past for a technology analogous to the current one.

HISTORICAL ANALOGIES When two situations are shown to be analogous, we can use the known outcome of the one that has already been observed to estimate the outcome of the unknown situation. The key to this approach is systematic, detailed comparisons. Martino compares the space program with the railroads in terms of technological, economic, management, political, social, cultural, intellectual, religious or ethical, and ecological similarities and differences, between the decline of railroad construction in the 1870s and the termination of the space program in the 1970s. He shows the validity of historical analogy along these dimensions and points out that the transfer of space technology to nonspace applications was predictable through such a systematic comparison.[12]

GROWTH CURVES When we fit curves through technological phenomena, the growth pattern generally follows an S-shaped exponential curve, always with an upper limit and sometimes also with a lower limit. These observations suggest the existence of a general expotential law governing the technological and social growth process. Special curves have been developed to explain that process.

The best-known growth curves are the Pearl curve and the Gompertz curve.

The Pearl curve is symmetrical; the Gompertz curve, unsymmetrical. They were developed originally to describe biological phenomena and mortality rates but were subsequently applied to a wide range of technological processes.

CAUSAL METHODS

All the models described in previous sections have a common weakness. They try to forecast technological change and explain a specific phenomenon without capturing the essence of that phenomenon. They do not attempt to develop cause-effect relationships, but merely study the historical observations or unrelated historical events and draw conclusions from such observations without explaining the reasons for those conclusions. If we want to understand the causal relationships underlying the technological phenomena, we need entirely different types of models.

Several models have been developed for this purpose. Ridenour's explains the behavior of technological trends by an ex-

[12]Op. cit.

396   ponential model.[13] Hartman and Isenson propose differential equations to explain the accumulation of knowledge as a function of scientists, available information, and the rate of information generation.[14]

A more sophisticated model developed by Acey Floyd explains technological growth on the basis of attempts for improving existing capabilities.[15]

OPERATIONS RESEARCH METHODS
Sometimes also called "operational models," these methods use the operations research approach in large-scale multivariate models. We will briefly discuss gaming and simulation in this section. The major difference between the two methods is in the degree of involvement of the decision makers in the actual development and use of the model.

GAMING   In gaming, the participants are given a set of initial conditions and a set of rules. Within these rules and initial conditions they are free to play the game as they see fit. According to the specified rules, a scenario is developed by defining logical and plausible events sequentially as well as simultaneously. The scenario is simulated as the participants specify parameter values by their role playing. Since the participants are assigned specific roles, but are given enough independence to play their roles, the bandwagon effect is greatly reduced in games.

The gaming technique has been found especially useful in studying future environments, developing business strategies, and setting objectives.

The so-called war games of the military are well-known applications of the gaming method. The military forces of the whole nation or of a region play hypothetical war situations, usually against a neighbor country or one that represents a threat. The roles played by the enemy simulate risks that had not been considered before.

The Boeing Corporation has developed a special gaming method known as "dynamic contextual analysis." Two teams assume different objectives and philosophies in the game. They try to achieve their objectives by exploring special technical requirements that contribute to their goals. When certain moves result in a technological demand, the corporation considers it a forecast and takes appropriate steps to use it as part of its technological planning effort.

---

[13]Louis Ridenour, "Bibliography in an Age of Science," *2d Annual Windsor Lectures,* University of Illinois, Urbana, 1951.
[14]Lawton Hartman, "Technological Forecasting," and Raymond Isenson, "Technological Forecasting: A Planning Tool," both in G. A. Steiner and W. Cannon (eds.), *Multinational Corporate Planning,* Crowell-Collier, New York, 1966 (cited by Jantsch, op. cit.).
[15]Acey Floyd, "A Methodology for Trend-Forecasting of Figures of Merit," in James R. Bright (ed.), *Technological Forecasting for Industry and Government,* Prentice-Hall, Englewood Cliffs, N.J., 1968.

SIMULATION   Large scale simulation models serve a purpose similar to that of the games. The difference is the absence of competitive role playing in simulation models. Industries and societies are modeled. The socioeconomic behavioral and technological subsystems are defined in terms of their interactions. The mathematical model is used for testing policies and studying the effects of technological changes. These continuous simulations borrow heavily from control theory concepts. The feedback loops, exponential time delays, and system fluctuations are embedded in the model. The systems-dynamics approach is an example of such models.[16]

AGGREGATE METHODS

The interactions among technologies and technological forecasts are important elements in assessing the future environment. There may be a large number of such interactions when several technological forecasts are considered at any given time. Comparison of the forecasts for these interactions can be tedious and unproductive if not done systematically. Two approaches to systematize the process of comparison and evaluation of the interactions among forecasts and their impacts on each other are summarized in this section.

INPUT-OUTPUT ANALYSIS   Leontieff's input-output analysis is a useful tool for technological forecasting.[17]   It considers changes in productivity, impacts of innovation, capital expansion, and the diffusion of technology. All these factors influence technological change. Input-output analysis has already proved to be a valuable technique for manpower and unemployment forecasts. The U.S. Bureau of Labor Statistics has made forecasts for the labor force, using the coefficients of technological change derived from various tables for the United States economy. A RAND Corporation study has used input-output analysis to forecast the establishment of new industries in the aerospace sector. Arthur D. Little has identified suitable R&D programs in oceanology for the state of Hawaii by using the input-output approach.[18]

CROSS-IMPACT ANALYSIS   The cross-impact analysis is performed on a square matrix whose rows and columns list technological events in chronological order. The upper half of the matrix indicates the impact of each particular event on other events when that particular event occurs. The lower half shows the impact if a particular event fails to occur. The diagonal is left blank. Cross-impact analysis assumes that an event can have an impact on only those other events that will occur later in time. The objective of this technique is to study

---

[16]Jay Forrester, Industrial Dynamics, MIT Press, Cambridge, Mass., 1961.
[17]Wassily Leontieff, *Input-Output Economics*, Oxford University, New York, 1966.
[18]Jantsch, op. cit.

398 the interactions among the various technological events and to obtain better estimates for the probabilities of their occurrence.

A detailed discussion and an example of cross-impact analysis can be found in Martino.[19]

NORMATIVE TECHNOLOGICAL FORECASTING

The forecasting methods we have described up to now are all exploratory methods. They start with the current level of knowledge and try to assess the future of technology. There is a different category of methods, that tries to answer the same question from the opposite end by asking: "What should be in the future?" Since these methods try to establish a norm and a direction of technological advance, they are known as "normative methods." They are oriented toward the definition of functional capabilities of a given technology, the capabilities that must be accomplished before going further in the pursuit of the final goals. In this sense, the normative methods set the pace in relation to what must be looked for, and when. They are used to establish the relative importance of research and development in one area of the technology as compared to other areas.

The three most commonly used methods of normative forecasting are described below.

RELEVANCE TREES

If we can distinguish different levels in a problem that we are analyzing, we can use relevance trees to visualize the problem in a systematic way. The idea behind the relevance tree is to start with a very broad statement of the problem and then disaggregate it into several components of similar importance. As we keep disaggregating these components into more and more branches, we finally arrive at the bottom of the decision hierarchy in the problem, where further disaggregation is impractical or infeasible. The number of branches originating at each node does not have to be the same for every node. However, care must be taken to ensure that all the possibilities for each branch are covered and that there is no overlap among the branches starting at the same node. Each node represents a goal for its branches. When all branches reach their targets, the node becomes a subgoal in terms of the upper branches and nodes. If we have stated a problem and defined the branches as alternative solutions to that problem, then we can find a path from the bottom to the top of the tree. This path is one of the many possible ways to solve the problem.

The structure of the relevance tree depends on the nature of the problem and, to some extent on the person or group of persons

[19]Martino, op. cit., pp. 271–281.

performing the analysis. In general, the relevance-tree method is **399** used more as a description tool than as a forecasting method. However, if we develop a relevance tree to define something not yet achieved, the nodes represent subgoals in the context of the complete hierarchy, and each one of the branches represents an alternative solution or a subgoal. All the subgoals of a given level must be met in order to achieve the next level subgoal, and so on. In this sense, by stating what is needed for the solution, relevance trees form a good basis for normative forecasting.

The different degrees of complexity, difficulty, urgency, and importance among the branches of a given node can be accounted for by allocating different weights to each branch. These weights are normalized (they add up to 1.0) and are known as the "relevance numbers." The importance of each branch is represented by its relevance number. By multiplying the relevance numbers of each branch across every path in the tree, we come up with the relative order of importance of every approach.

The methodology for the relevance-tree approach is similar to that of the hierarchical decision models discussed in Chaps. 10 and 11 and the Appendix.

MORPHOLOGICAL ANALYSIS
Morphological analysis first breaks a problem into as many parts as possible. All the solutions for every part of the problem are then outlined. Since there is usually more than one solution for every part, the total number of solutions is determined by the combinations of partial solutions. In a problem with four parts, each having 3, 2, 3, and 1 solutions respectively, the total number of combinations is $3 \times 2 \times 3 \times 1 = 18$. Some of these solutions can be eliminated because of infeasible or incompatible conditions, thus reducing the solutions to a smaller set.

Since many of the feasible solutions require a certain technology, and can be stated as technological goals, this method is used as a normative approach to technological forecasting. It is similar to relevance trees in this respect.

Neither of these methods is applicable to situations where a sequence of events is being considered. A mission flow diagram is more appropriate for such problems.

MISSION FLOW DIAGRAMS
This method was originally developed for military use but it is equally applicable to technological forecasting. It starts with the mapping of possible ways, routes, and means through which a given mission can be accomplished.

The identification of all important, significant steps in each of the routes is an important requirement. Once this identification is

400   done and the diagram is completed, the difficulties of each path, the associated costs, and the development needs are stated.

Paths are weighted in mission flow diagrams in a way similar to the relevance trees. The weights represent not only the importance of certain steps but also the relative importance of investigating a given technology.

SUMMARY

An engineering manager develops technical strategies and manages engineers and scientists to implement those strategies in a dynamic environment. The technological subsystem in which he operates is undergoing constant changes. He has to assess the future technological environment and take actions to take advantage of that environment. His decisions are made in the present but affect the future. Therefore, he needs to probe into the future before he can develop the technical strategies to take his organization from where it is today to where he wants it to be in the future. He can use *technological forecasting* to develop a data base for his decisions.

Various approaches and methodologies of technological forecasting are available today. After the engineering manager makes an assessment of the future conditions, he makes decisions with future implications. This decision-making process is called *planning.*

Planning requires time, resources, conceptual skills, and organized thinking. It is costly. However, it is not a matter of whether an organization can afford planning; it is whether it can afford not to do planning. If an engineering manager does not plan his organization's future, his competitors will do it for him to *their* liking.

DISCUSSION QUESTIONS

1 What are the essential elements found in a definition of planning?

2 The authors describe forecasting as an integral part of planning. Why is this so?

3 What are the main misconceptions about planning that Peter Drucker discusses?

4 "Luck is a residue of careful planning." What is the *real* meaning of this statement?

5 What are the key elements in the planning process? How do the steps in this process relate to the question: "What business are we in?"?

6   Define and relate the notion of objectives, goals, and strategies.   **401**

7   What are some recent technological changes that could affect the "business" that a college or university is in?

8   What are the key technological planning responsibilities of an engineering manager?

9   How might the role of technology in an organization's planning be categorized?

10   Design and develop a model which depicts the flow of technological planning.

11   Identify and show examples of the main technological forecasting methodologies.

12   A good forecast necessarily results in good technological planning. Defend or refute this statement.

KEY QUESTIONS FOR THE ENGINEERING MANAGER

1   What are some of the key technologies that have been used successfully by this organization that have been the result of careful planning?

2   Can the planning process depicted in Fig. 12.1 be adapted to planning in this engineering organization?

3   Has the engineering manager assumed the technological planning responsibilities as portrayed in Fig. 12.2?

4   How does the flow of technological planning in the organization compare with the "theoretical" model of Fig. 12.3?

5   What are the leading technological factors affecting product planning in this organization?

6   How are the competitors doing in maintaining an advantage in the technological factors affecting their product planning?

7   How is strategic planning currently being carried out in this organization? Is there any room for improvement?

8   What forecasts are currently being accomplished to facilitate strategic planning in this organization?

9   Have the technological forecasting methodologies suggested in the chapter been tried in this organization? Why or why not?

10   How would one rate the effectiveness of strategic planning in this organization on a scale of 1 to 10? What are the problems (or opportunities) that currently exist with respect to strategic planning?

11   What have I done to encourage the key people in this engineering organization to develop a long-range perspective in their work? Do they perceive that I have a long-range perspective in managing this engineering organization?

12   What might be the significant technological improvements in our key products over the next 5 to 10 years?

BIBLIOGRAPHY

Ayres, Robert V.: *Technological Forecasting and Long-Range Planning,* McGraw-Hill, New York, 1969.

Drucker, Peter F.: "Long-Range Planning," *Management Science,* vol. 5, 1959, pp. 238–249.

Floyd, Acey: "A Methodology for Trend-Forecasting of Figures of Merit," in James R. Bright (ed.), *Technological Forecasting for Industry and Government,* Prentice-Hall, Englewood Cliffs, N.J., 1968.

Gabor, Dennis: *Inventing the Future,* Secker & Warburg, London, 1963.

Hartman, Lawton: "Technological Forecasting," in G. A. Steiner and W. Cannon (eds.), *Multinational Corporate Planning,* Crowell-Collier, New York, 1966.

Isenson, Raymond: "Technological Forecasting: A Planning Tool," in G. A. Steiner and W. Cannon (eds.), *Multinational Corporate Planning,* Crowell-Collier, New York, 1966.

Jantsch, Erich: *Technological Forecasting in Perspective,* OECD Publications, Paris, 1967.

Kahn, Herman, and Anthony J. Wiener: *The Year 2000: A Framework for Speculation,* Macmillan, New York, 1967.

King, William R., and David I. Cleland: *Strategic Planning and Policy,* Van Nostrand Reinhold, New York, 1978.

Leontieff, Wassily: *Input-Output Economics,* Oxford University, New York, 1966.

Martino, Joseph P.: *Technological Forecasting for Decisionmaking,* American Elsevier, New York, 1975.

Ridenour, Louis: "Bibliography in an Age of Science," *2d Annual Windsor Lectures,* University of Illinois, Urbana, 1951.

THE 13
ENGINEER
AND THE
ENVIRONMENT

The purpose of this chapter is to alert the engineering manager to his responsibility in managing the engineering function so as to respect the environment. The application of technology has created environmental problems—and opportunities. No one would deny this. The solution of environmental problems entails technological, economic, social, political, and legal considerations that need to be studied and evaluated. The application of technology is the prime concern of the engineering manager; this chapter can provide that manager with insight into the challenges he faces in considering the environmental impact of his work.

Today's citizens feel a general discontent with the quality of our environment. Information on the quality of the air in our cities is broadcast reqularly to keep the citizens informed. Many public bathing beaches are so contaminated that they have been closed for an indefinite period of time. Global weather changes are being blamed on the growing contamination of the earth's atmosphere. Privacy from congested population is becoming a precious luxury that only a few people will be able to afford in the future. These and many other problems growing out of our burgeoning global population portend environmental challenges of awesome proportions.

THE ENVIRONMENTAL CHALLENGE

All too frequently the media remind us of the ever-present potential for environmental contamination. In 1979, for example, *The New York Times*[1] reported two incidents that could have become environmental

---

[1] *The New York Times*, July 22, 1979, p. 1.

404 disasters. Two oil supertankers collided off the coast of Venezuela on July 19, 1979. An aerial reconnaissance revealed an oil slick around one of the tankers some 30 square miles in size. In the second incident a runaway Mexican oil well 50 miles off the coast of Yucatan had been pouring crude oil into the Bay of Campeche at the rate of 1.2 million gallons a day for several weeks. As oil from this well was carried by the Gulf currents north to Texas, a serious threat to the Texas coast ecosystem became a real possibility.

Another serious pollution of the environment occurred in the case of the Allied Chemical Company, which "blundered into trouble when it contracted out the production of a chemical called Kepone."[2] Kepone is a pesticide ranked with the deadliest of substances, developed by the Allied Chemical Company in the 1950s. It is used mainly against potato bugs in Europe and was manufactured for years without any known harmful effects on the workers producing it. Allied's troubles began with Kepone when the corporation decided to have the pesticide produced by another company. The company that had won the contract operated a small plant in Hopewell, Virginia, which according to *Fortune* magazine turned out to be incredibly lax about safety precautions. When employees began coming down with severe nervous and other disorders, the plant was promptly closed by health authorities. It was subsequently learned that Allied itself had previously discharged the pesticide into the James River. The river was closed to sport and commercial fishing and the Allied Chemical Company became embroiled in numerous costly lawsuits as well as a celebrated water pollution trial. The company paid a staggering $20 million in fine settlements and legal fees.[3]

The workers in the places where Kepone was manufactured have recovered from the ill effects of this chemical, but neither Allied's chairman nor his fellow executives will ever be the same after the beating they took in the courts of law and public opinion.

A different serious situation is the threat of "acid rain," the result of airborne pollutants, sulphur and nitrogen oxides, combining with atmospheric moisture, as reported by *The Wall Street Journal.*[4] Already the acid rain and snow have rendered "several hundred lakes in eastern Canada and the U.S." fishless. Thousands of additional lakes are threatened, as are crops, forests, buildings, and the public's health.

These kinds of environmental contamination incidents serve to remind us of the awesome challenge faced by those people who manage the application of technology. How to maintain a research and production capability and at the same time protect the environment is a challenge where engineers and scientists play a key role.

[2]*Fortune,* Sept. 11, 1978, p. 6
[3]Ibid.
[4]*The Wall Street Journal,* Sept. 21, 1979, p. 18.

We usually think of pollution as the concern of modern times; our **405** modern concern does, however, have a historical precedence.

EVOLUTION OF ENVIRONMENTAL CONCERN

One early reference to environmental pollution is found in an essay written by John Evelyn on air pollution in London in 1661. This essay, sent to King Charles II, states in its foreword:

That this Glorious and Antient City, which from Wood might be rendred Brick (like another Rome) from Brick made Stone and Marble; which commands the Proud Ocean to the Indies, and reaches the farthest antipodes, should wrap her stately head in Clowds of Smoake and Sulphur, so full of Stink and Darkness, I deplore with just Indignation.[5]

Unfortunately, the essay no more commanded the attention of the public than did some of our environmental crises of the late 1960s.

The air pollution problem in Los Angeles is well known, but such pollution has historical corollaries. Four centuries ago Juan Rodriguez Cabrillo recorded in his diary that smoke from Indian fires in the area went up for a few hundred feet and then spread out to blanket the valley with haze. Historians report that because of this phenomenon he named what is today called San Pedro Bay "the bay of smokes." Cabrillo was observing the effect of a thermal inversion (a layer of warm air above a layer of cold air). The wind patterns in the eastern Pacific and the ring of mountains surrounding the Los Angeles basin are ideal conditions for the formation of inversions, usually at about 2000 feet above the floor of the valley. They occur on about 7 days out of every 16.

Even before the environmentalism of the 1960s and the early 1970s, there was ample evidence of the disastrous effects of air pollution. In Donora, Pennsylvania, on October 26, 1948, a thermal inversion trapped fog and smoke, producing a lethal situation in the town. Fifteen men and five women died during a weekend when the polluted air was trapped over the city. In all probability the lives of many others were shortened as a result of this. London offers another example; during early December, 1952, a combination of fog and thermal inversion was accompanied by severe weather. London homes were heated by coal, resulting in heavy smoke. The atmospheric content of sulphur dioxide rose to double its usual level. The London smog in 1952 was blamed for the death of approximately 4000 people.

Environmental concern in this country is a relatively recent

[5]I.R. McIntyre and L. W. Woodhead, "Engineering and Environmental Management," *The Journal of the Institution of Engineers, Australia,* March-April, 1976, p. 33.

406   phenomenon, evolving from a complex interplay of people, institutions, critical events, mass media, and a growing heightened public perception of some of the environmental problems that exist in our world. Any detailed discussion of these forces goes beyond the scope of this chapter. Nevertheless, it is useful to touch upon three key events that provided the basis for today's environmental awareness: first, a few highly visible environmental accidents; second, partly as a result of these accidents, a change in public priorities on environmental issues; finally, the influence that was exerted by prominent conservationists and the mass media.

During the early and mid-1960s most people were preoccupied with the threat of nuclear war with the Soviet Union, heightened by the Cuban missile crisis and the Vietnam war, communism, inflation, unemployment, racial tension, and crime in the streets. However, two environmental accidents happened that caused a change in the public sense of priorities.

The first was the oil spill of the *Torrey Canyon* in March, 1967. The tanker, carrying in excess of 119,000 tons of crude oil, broke apart in rough seas off Lands End, England. In spite of frantic efforts to prevent the oil from doing damage to the environment, pollution of unprecedented magnitude occurred. Television audiences in this country witnessed the use of everything from detergents to Napalm to prevent the flow of the oil onto the beaches—all of which proved largely unsuccessful. Ultimately, great quantities of oil polluted wide expanses of English beaches, killing countless shore birds and crippling much of the tourist trade. The subsequent testimony by the British investigation team was typical of the worldwide concern about the potential probability and liability of such future disasters.

Another oil spill that caused public confidence to be shaken out of its complacency occurred in January, 1969, when an offshore drilling rig in the Santa Barbara channel struck a large oil deposit and set off an environmental catastrophe. The resultant blowout caused a crack in the ocean floor and several million gallons of crude oil escaped to pollute the beaches around Santa Barbara, a well-known garden spot of California. Despite herculean efforts to contain the oil, miles of coastal waterways and beaches became polluted with crude oil. Water fowl and other aquatic life were killed. The long-term effects of that oil spill are still not fully known today.[6]

The mass media attracted national attention to the pollution of the beaches in Santa Barbara. There is no question that television was instrumental in stimulating public interest in the need to develop a strategy to reduce the probability of this sort of disaster occurring again.

[6]James Ridgeway, *The Politics of Ecology*, Dutton, New York, 1971, pp. 118–119.

During this same period, the 1960s, several concerned **407** conservation groups (for example, the Izaak Walton League) sponsored a "clean air week." President Johnson had spoken of the importance of beautifying America in 1965 and doubtless helped to bring about some changes in public attitude. In the same year, 43 percent of a Harris poll sample expressed concern about the pollution of rivers and streams.[7] Another index of increasing public concern was the publication of 350 articles on pollution by *The New York Times,* more than twice the number that had been published the previous year. Also, four pieces of important environmental legislation were enacted in 1965: the Water Quality Act, the Water Resources Act, the Rural Water Sewage Act, and the Highway Beautification Act.

Another major sign of public interest during this same period was the increasing membership of the Sierra Club from some 15,000 to more than 85,000. More significantly, its eastern membership went from 750 to 19,000, according to Trop and Roos.[8]

Perhaps the greatest contribution of all to the growth of an environmental awareness in this country was Rachel Carson's best seller *Silent Spring,* which brought public attention to the dangers of pesticides. The popular appeal of *Silent Spring* marked the beginning of a cooperative alliance between conservationists and the mass media. From 1960 on, the reading public in this country has been provided with environmental literature of one sort or another. As public pressure, supported by the conservation organizations and public opinion itself, began to mount during the late 1960s, a number of senators and congressmen became involved in the environmental crusade. In the wake of the 1968 elections, the competition heightened as the White House took an active part in doing something to protect the environment.

The year 1970 marked the milestone in the start of a new era in environmental management. On January 1, 1970, President Nixon signed into law the National Environmental Policy Act (NEPA); on January 9, he sent to Congress Reorganization Plan No. 3 of 1970, creating the Environmental Protection Agency (EPA). The National Environmental Policy Act made explicit the responsibility of the federal government for the quality of the American environment. It did this in explicit language and with institutional arrangements and procedures for the operational implementation of environmental policy. Organizationally, the Environmental Protection Agency consolidated ten federal pollution control programs, going to a single organization based upon the concept of the environment as a single interrelated system.

[7]Cecelia Trop and Leslie Roos, "Public Opinion and the Environment," *The Politics of Ecosuicide,* Leslie Roos (ed.), Holt, New York, 1971, p. 54.
[8]Ibid., p. 62.

408   In defining the roles and relationships of this legislation in environmental management, President Nixon declared that the Council on Environmental Quality, created by the National Environmental Policy Act, "focuses on what our broad policy in the environmental field should be; the Environmental Protection Agency would focus on setting and enforcing pollution control standards." Both these measures, NEPA and EPA, were the outcomes of a public consciousness for the state of the environment in which we live; this awareness grew from the early 1960s with a speed and scope utterly unanticipated by most of the leaders in our society.

The passage of the National Environmental Policy Act established the need for environmental assessment. This act was passed with little guidance established by way of ground rules on how it would be implemented. The act established the intent of the law and left such matters as interpretation, procedure, scope, etc. to be implemented at a later time. Since then there have been opportunities to apply the National Environmental Policy Act and a great interest has developed in the assessment process, not only in government circles but in management and technical communities as well. The evaluation comes to focus in a document called an *Environmental Impact Statement.*

THE ENVIRONMENTAL IMPACT STATEMENT

The key part of the National Environmental Policy Act of 1970 is the requirement for an environmental evaluation of the major federal actions that could significantly affect environmental policy. This process is known as the NEPA assessments process, and the resulting document is called an Environmental Impact Statement. The act covers projects and programs initiated by government and those in the private sector which require federal involvement. The language of the act extends the scope of actions requiring an Environmental Impact Statement to an extremely large proportion of federal activities. In addition, state governments and regional and local agencies now typically require some kind of an environmental review. Engineers and scientists, as well as those who manage them, should recognize that the number of projects and programs requiring an environmental impact type of assessment will doubtless increase in the future. As the requirement for environmental evaluation expands, it will be necessary for more people to become familiar with the procedures of preparation and review.

Basically, the Environmental Impact Statement is an official document, prepared by a federal or state agency proposing to undertake or fund an action or project. In essence, the Environmental Impact Statement defines and evaluates the effects on the environment of a proposed project or action and its alternatives. It further attempts to determine the possibility of reducing negative impact by

reviewing alternative ways of accomplishing the project. If the envi-   **409**
ronmental impacts prove to be sufficiently serious, the project may
be abandoned. Under certain conditions, however, depending on
other considerations, it may still be decided to proceed with the
project. The best example of this was the oil shortage created by the
oil embargo of 1973, which resulted in approval by the Congress of
the United States of the construction of the Alaskan pipeline in spite
of serious environmental impact questions that were not answered.

The Environmental Impact Statement should be placed in
its proper perspective, as explained by Rosen:

> It is a tool prepared to assist the decision maker in making
> sound and rational decisions regarding the environmental
> effect of various alternatives. The statement itself does not
> make that decision. It is only part of the data that are used to
> make a final determination. Other factors include such con-
> siderations as need, cost, public benefit, etc. The statement
> serves to assist in the planning for specific projects and in
> the development of regional and local land use policies.[9]

The mandate of the National Environmental Policy Act to
require appropriate consideration of environmental values is intended to condition the technical and economic decisions that have
an impact upon the environment. In the past these decisions have
tended to ignore environmental costs and benefits. They have been
made almost exclusively in terms of narrow economic efficiency and
technical feasibility. Neither the value of free resources such as air
and water nor the social costs of polluting them have significantly
affected decision making for either public or private investment. Unfortunately, in any environmental assessment the consequences of
alternative actions cannot always be put into a common denominator for measurement, such as in dollar terms for a direct comparison
with economic and technical costs and benefits. Quantitative assessments are not always possible during the early stages of a program, and qualitative judgments are sometimes the most that can be
accomplished.

The environmental impact decision requires the structuring
of the analysis such that each alternative action can be assessed in
terms of:

1   Types of environmental and social impacts

2   Whether the impacts are likely to be beneficial or
detrimental based upon present knowledge

---

[9]Sherman J. Rosen, *Manual for Environmental Impact Evaluation*, Prentice-Hall, Englewood Cliffs, N.J.,
1976, p. 5.

410   3   Who and what is affected and in what ways

A typical series of steps taken in the environmental analysis can best be illustrated by citing two examples in which an analysis was done.

PORT JEFFERSON HARBOR

The first example concerns the environmental aspects of dredging Port Jefferson Harbor for delivering petroleum products to Long Island, New York.[10] The environmental aspects of this problem were investigated under a contract to the New York District U.S. Army Corps of Engineers as part of a larger study in which the Corps also evaluated the economic and engineering feasibility of the alternatives. The environmental analysis in this situation consisted of the following:

1   Establishing a valid cost-effective framework considering, in particular, the extent of the geographical area affected

2   An identification of the environmental parameters that were to be affected

3   A review of the available literature to prepare a qualitative summary of the effects of the alternative plans and the environmental parameters

4   An evaluation of the nature of the environmental impacts as beneficial or detrimental, taking into account countervailing effects and uncertainty

5   Tabulation of these evaluations by alternative, geographical area, and type of impact

6   A summary of the analysis to help determine the patterns of environmental impacts

During this environmental analysis, the following parameters were identified and examined: water quality, shell fish, fin fish, water fowl, other wildlife, oil pollution, ground water, air quality, oil disposal, pleasure boating, road traffic, wave action, storm surge, fire hazard, aesthetics, and land use. These parameters represent the major categories of environmental concern that had been expressed by advocates and opponents of the various alternatives. The parameters are not completely independent. For example, oil

---

[10]The following information is taken from U.S. Army Engineer District, New York, Preliminary Assessment of Alternative Plans of Supplying Petroleum to the Port Jefferson, New York Service Area, Feb. 13, 1973, as cited in James Wells and Douglas Hill, "Environmental Values in Decision Making: Port Jefferson as a Case Study," *Journal of Environmental Sciences*, July-August 1974.

pollution could be considered an aspect of water quality, fire   **411**
danger, and aesthetics. Nor do they infer equal importance. The list
was chosen to catalog the issues as completely as possible and to
minimize the chance that any major environmental concern might
be omitted.

EAGLE CREEK DAM AND RESERVOIR

Another example of an environmental statement is contained in the
evaluation of the Eagle Creek Dam and Reservoir in Ruidoso, New
Mexico. This statement, prepared under the auspices of Public Law
91-190, evaluates the environmental impact of the construction of an
earthen and rock-fill dam to provide domestic water storage to the
village of Ruidoso, New Mexico. The reservoir created by the dam covered approximately 70 surface acres. The statement presents the alternatives and decisions coming out of planning the construction
project with public and governmental involvement. An outline of this
environmental statement is shown in Table 13.1. Note the breadth of
the assessment done as a prerequisite to initiating this construction
project.

Engineering managers will doubtless come in contact with
some facet of an Environmental Impact Statement, although the
statement is the responsibility of federal agencies. Statements are
prepared under the jurisdiction of state and local agencies and
in some cases by privately owned firms. This allows for the use of outside practitioners when workloads preclude the meeting of schedules or when specialized expertise is required. The engineering manager must recognize that this is an area in which he will need
specialized assistance, and he should know how to look for it.

In dealing with the subject of the effect of technology on the
environment, it is clear that we need a more definitive formulation
and resolution of problems that can affect the environment
in which we live. In the past, this has been the responsibility of the
architect, the landscape architect, the urban planner, and other professional people. Today it is clear that a much wider range of skills
is required to deal with the complex environmental problems and
opportunities that face us. More important, the engineering manager
should develop a deep sense of commitment to manage the application of technology so as to protect the environment as much as possible. To do this the manager must be aware of the major environmental threats with which he must deal. A brief summary of these
threats is contained in the next section.

ENVIRONMENTAL THREATS

In many ways, technology has been responsible for making our environment much more hospitable than it was in the past. In some ways,
however, technology has also made it a more hostile place. Over-

Outline: USDA Forest Service Environmental
Statement
Eagle Creek Dam and Reservoir, Ruidoso, New
Mexico*

13.1

  I. Description of proposed action
    A. Proposal
    B. Background and existing situation
    C. Present environmental setting and land-use activities
        1. Geology
        2. Soils
        3. Vegetation
        4. Timber
        5. Air
        6. Water
        7. Climate
        8. Land ownership
        9. Special uses and easements
       10. Wildlife
       11. Recreation
       12. Wilderness
       13. Range
       14. Historical and archeological
       15. Fire
       16. Minerals
       17. Visual quality
       18. Insects and diseases
 II. Environmental impacts
    A. Soils
    B. Water
    C. Land ownership
    D. Special uses and easements
    E. Range
    F. Air
    G. Wilderness
    H. Minerals
    I. Fire
    J. Timber
    K. Recreation
    L. Wildlife
    M. Insects and disease
    N. Downstream values
    O. Visual
    P. Transportation alternatives and impact
    Q. Socioeconomic
    R. Mitigation of environmental impacts
III. Adverse environmental effects which cannot be avoided
    A. Resources and activities
    B. Social
IV. Relationship between local short-term uses of environment and maintenance
    and enhancement of long-term productivity
 V. Irreversible and irretrievable commitment of resources
VI. Alternatives to proposed action
VII. Preferred alternatives
    A. Dam and spillway location
    B. State highway relocation
    C. Land ownership
    D. Powerline relocation
    E. Reservoir clearing
    F. Recreation development

VIII. Consultation with others: To initiate public and other agency involvement, a let- **413**
ter, data sheet, and maps describing the project proposal and requesting
thoughts on the project were distributed to approximately ninety addressees. A
sample of the addressees is indicated below:
Bureau of Indian Affairs
State Planning Office
Albuquerque Wild Life and Conservation Association
U.S. Corps of Engineers
Ruidoso Realtors Association
Sierra Club Office
Federal, state, county, and local agencies

*SOURCE: Environmental Statement proposal for Eagle Creek Dam and Reservoir, Lincoln National
Forest, USDA, Forest Service Southwestern Region, Albuquerque, N.M.

population and industrialization have contributed in many ways to the
general deterioration of the environment in which we live and on
which we are completely dependent for sustenance. Only in recent
times have we become more aware of the harmful effects of the
seemingly ever-increasing number of chemical and biological substances that we produce and to which we expose ourselves. These
are substances for which no evolutionary experience exists, and
against which human cells have evolved no natural defenses.

Perhaps the most serious threat of all is the threat to human
health because of the deterioration of the environment. Pollutants of
one form or another reach us through the air we breathe, the water
we drink, the food we eat, and the sounds we hear. These direct
threats are not the only ones. There are other threats that are not so
obvious—those which impinge on the complex environmental system upon which all human existence ultimately depends.[11]

AIR POLLUTION
Probably the form of pollution with which we are most familiar is air
pollution. It comes from many sources—from motor vehicles, industrial plants, residential centers, to name just a few. Those of us who
live in or near cities can see it, and we can feel it when it burns our
eyes and irritates our lungs. Virtually every major metropolis of the
world has serious air pollution problems. Travelers often can see a
city first by the pall of smog that hangs over it. Pollution at times
cuts down the amount of sunlight that reaches the city. It has been
estimated that the pollution in New York City reduces the amount of
sunlight by nearly 25 percent and in Chicago by about 40 percent—a grim harbinger of what we can expect in the world if current
trends are allowed to continue. Even the entire planet itself is affected to some degree. Meteorologists speak of a nebulous circle of

[11]The ensuing material is paraphrased from Paul R. Ehrlich and Ann H. Ehrlich, *Pollution Resources Environment,* 2d ed., Freeman, San Francisco, 1972, pp. 166–169.

414   air pollution around the earth. Smog has been observed over oceans, over the North Pole, and in other unlikely places.

WATER POLLUTION

In many of our communities, there is also a direct threat to human health through the pollution of the water. Even the drinking water that flows from the tap in some localities has already passed through seven or eight people. In some cities the water has been claimed to be unfit to drink. For example, in the late 1960s, the water in Wilmington, Delaware, New London, Connecticut, Chattanooga, Tennessee, and Eugene, Oregon, was not adequately protected from its source to the faucet. Although chlorination may help, there is growing evidence that high content of organic matter in water can somehow protect viruses from the effect of chlorine. In recent years there have been serious outbreaks of infectious hepatitis in the United States. A major suspect for the route of transmission of infectious hepatitis is the "toilet to mouth pipeline" of many metropolitan water systems that are not made safe by chlorination.

Water pollution in some of our lakes has become so serious that all swimming is banned and people are advised not to eat any of the marine life from those waters.

Water pollution through sewage disposal provides one of the classic examples of pollution coming about by virtue of population growth. In situations where raw sewage is pumped directly into a river, the natural purification capability of the water is not enough to decontaminate the sewage. As population increases, the waste cleansing ability of a river becomes overstrained and either the sewage or the intake water must be treated if the river water is to be safe for drinking. Should the population of the river area increase further, more and more elaborate and expensive treatments are required to keep the water safe for human use and to maintain a desired level of marine life.

SOLID WASTE

A serious problem that has yet defied any real solution is the accumulation of solid waste in open dumps and land fills. Such repositories of solid waste are not just aesthetic disasters. If these wastes are burned they can contribute to air pollution. Water from rainfall seeping through the land fills pollutes ground water supplies. The solid waste repositories also serve as a breeding ground for such disease-bearing organisms as rats, cockroaches, and flies. Disposal of metal and plastic containers, the junking of some 7 or 8 million automobiles each year, and the collection of urban kitchen wastes reach an astronomical number of tons in a year.

It is becoming universally recognized that the current way of dealing with the solid waste problem is completely inadequate, and if there is any area in which technology can help in the whole matter of **415** improving our environment, it is in the design of alternatives for the safe and aesthetic disposal of solid waste.

PESTICIDES AND RELATED COMPOUNDS

Some substances, such as chlorinated hydrocarbons, lead, mercury, and fluorides, reach us in so many ways that they must be considered as general pollutants. Chlorinated hydrocarbons—perhaps the best example being DDT—were put to use late in World War II. DDT was the most commonly used and most thoroughly studied of all synthetic insecticides. After many years of heavy usage, we now realize that we were dangerously polluting the environment with DDT. Other chlorinated hydrocarbon insecticides, including Dieldrin benzine hexachloride and polychlorinated biphenyls (PCBs) have also been found to be pollutants.

These compounds are used in a variety of industrial processes and become a part of our environment through vaporization from storage containers, emittance from factory smokestacks, or the dumping of industrial wastes in waterways. It is difficult at this time to evaluate the magnitude of the threat to human health represented by the level of various chlorinated hydrocarbon uses. Most of the analysis has been done with DDT, but humans have been exposed to a wide range of related compounds, some of which are believed to have considerably higher immediate toxicity. There are indications that Dieldrin, perhaps four times as toxic as DDT, may be involved in cirrhosis of the liver, and that benzine hexachloride may contribute to liver cancer. The critical question concerning all these is not so much their immediate toxicity but the long-term effects that they may have.

HEAVY METALS

Pollution by heavy metals—lead, mercury, cadmium, arsenic, and others—is a real concern. There is no lack of understanding on the part of the technical community regarding acute lead poisoning. It manifests itself through such symptoms as loss of appetite, weakness, awkwardness, apathy, and miscarriage. Ample evidence exists for what lead can do, although we do not yet know what the results of chronic low-level exposure to lead are, particularly to that which exists in the atmosphere from automobile exhausts.

We are constantly exposed to lead in our environment in the form of air contamination from gasoline containing tetraethyl lead —although the introduction of lead-free gasoline promises to reduce this health hazard. There are other ways in which we can be exposed to lead, such as through insecticides, paints, solder used to seal food cans, lead piping, abraded particles of lead contained in ceramics and glassware, etc.

Mercury is a prime ingredient of agricultural fungicides that have been used extensively for treating seeds. Small amounts of mercury are found in fossil fuels and are released when the fuels are burned. Even medicines flushed down drains and broken medical thermometers may make significant contributions to the environmental mercury levels.

FLUORIDE POLLUTION
Fluoride treatment of water supplies is an emotion-charged subject, but it is linked with a potentially serious health hazard. The scientific evidence supporting the safety of mass fluoridation is not as good as it ought to be, but neither is there convincing evidence that it is harmful. Increased fluoride concentration has been detected in the foods and beverages processed in communities supplied with fluoridated water.

RADIATION
Ionizing radiation is everywhere. It comes from cosmic rays, from radioactive substances and the earth's crust, and from certain natural radial isotopes. At the moment, the bulk of man-made radiation exposure comes through the use of medical and dental x-rays for diagnosis and therapy. However, another source of radiation exposure is fallout from nuclear weapons. A potentially important source of radiation exposure is the use of nuclear reactors as a source of electrical supply. The accident at the Three Mile Island nuclear power plant at Harrisburg, Pennsylvania, in March, 1979, serves as a grim reminder of the risk of radiation exposure from such plants.

CHEMICAL MUTAGENS
Biologists have recently become concerned about the risks of mutations other than those caused by radiation. An awareness has developed that humans are being exposed to many thousands of synthetic chemicals whose mutagenic potential is not known. Such chemicals as caffein and LSD had been alleged to be dangerously mutagenic in humans, but the results of tests on experimental animals have been variable and therefore inconclusive. Among the other chemicals that have been shown to be capable of causing mutations are nitrogen mustards from which organophosphate insecticides are derived, hydrogen peroxide, formaldehyde, cyclohexylamine, and sodium nitrate.

GEOLOGICAL HAZARDS
Geological hazards include those caused by earthquakes and those caused by human activities such as landscaping. The latter include changing the land for some purpose, such as the building of large

dams to meet the needs of growing population for additional water **417** supplies.

NOISE POLLUTION

People everywhere have recently become aware of a new kind of pollution—noise pollution. This problem has been thrown into sharper focus by the discovery that some teenagers have suffered permanent hearing damage following long exposures to amplified rock music and by public concern about the effects of sonic booms. Anyone who has been caught in a traffic jam in a major city and has time to ponder that situation can surely appreciate the uncomfortable feeling that noise pollution creates.

ENVIRONMENTAL DETERIORATION IN CITIES

The deterioration of the environment both physically and aesthetically is apparent in many of our cities. The dehumanizing effects of life in the slums and ghettos, particularly where there is little hope for improved conditions, has often been cited as the cause for urban rioting and disturbances. Crime rates usually reach their zenith in such neighborhoods. Such symptoms of general physiological maladjustments suggest that modern cities provide a less than ideal environment for human beings.

THE EPIDEMIOLOGICAL ENVIRONMENT

This environment is not with us as yet, but considering the size of our population today, its density and its mobility, the potential for a worldwide epidemic has never been greater and it will probably continue to grow. Our awareness of this potential threat, however, seems to be practically nonexistent. There is much that we do not understand about the behavior of viruses, but we do know that the spontaneous development of a highly lethal strain of human virus from an extremely dangerous animal virus is possible. We also know that overcrowded conditions and our mobility could help to spread such a virus. This is an area that will bear close watching in the future. Along with this is the possibility of accidents resulting from chemical biological warfare research. The inability to completely preclude any such accident from occurring has been made clear by the Skull Valley, Utah, chemical biological warfare disaster of 1968 in which thousands of sheep were poisoned when a chemical agent escaped.

In 1969, President Nixon announced the unilateral renunciation by the United States of the use of biological warfare and directed that the United States stocks of biological agents be destroyed. Destruction of biological warfare materials was accordingly carried out in 1970 and 1971. We do not really know, however, whether other nations have destroyed any of their chemical biological warfare

418   agents—not a comforting thought on which to close this section of our chapter.

The reader should have gained an appreciation of the systems nature of these environmental threats—that indeed everything is related to everything else. In the environmental sense we are dealing with the concept of a "system of effects."

THE SYSTEM OF EFFECTS

The application of technology may set in motion systems effects that were not anticipated. Systems effects are those that appear subsequent to or as a result of, the immediate and obvious consequences of an action. The automobile provided a means of transportation for people, but it contributed to pollution and social change and preempted the growth of mass transit. The automobile also contributed to the current energy crisis. Thus, the solution of one problem created other problems. Literally, the notion that everything is related to everything else is quite true. When technology is introduced, other changes may result, such as social and economic effects. Consider the example of the building of the Aswan Dam in Egypt, a civil engineering project designed to provide an improved power generation capability to Egypt. If one can visualize the total environmental area of the dam as a system into which technology has been introduced, an important environmental question can be raised: How does the Aswan Dam fit into the appropriate Egyptian system—to include the economic and social subsystems thereof?

Unfortunately, the planning for the Aswan Dam appears to have neglected any real evaluation of some of the potential side effects. After the dam was completed significant systems effects appeared:

1   An excessive use of water by the farmers has caused a water logging of much of the farmland;

2   With a greater exposure of water surface to the atmosphere, salts have built up in the soil through increased evaporation. The Dam has also ended the Nile River's annual floods that had helped to flush out the salts from the land;

3   The still waters of the reservoir and the irrigation canals have become fertile breeding ground for insects that cause intestinal diseases and malaria;

4   The surplus water resulting from too much irrigation will have to be carried away by some artificial draining system—an expense of massive proportion for Egypt;

5   Certain nutrients are being denied to the soil

because the Nile no longer floods each year. This requires **419**
replacement with commercial fertilizers; the cost of commercial fertilizers has gone beyond the wherewithal of the
farmers even with government subsidies.[12]

Some of these problems would be solved if the excessive irrigation by the farmers could be reduced. But this would be difficult
to bring about, because such a change would require a massive
reorientation of social tradition. For centuries the waters of the Nile
have been regarded as a divine gift—and their free use as a basic
right in the culture of Egypt.

The total systems effect of the Aswan Dam may lead to a
high price of long-term degradation of the soil at the expense of a
short-range power generation capacity. Were the total systems effects—economic, social, and technological—taken into account
when the planning and decision making for the dam was carried
out? The evidence suggests that they were not.

We are, of course, guilty of Monday morning quarterback-
ing, yet there is value in such postmortem analysis, for it helps us to
avoid making the same mistakes again. It also serves to remind us
that engineering management has come to encompass a much
broader systems application of technology than has been practiced
in the past. There are no longer simple engineering systems to be
built and operated without an assessment of what they do to the
total system.

A system is characterized not only by its parts but by the interaction of these parts as well. The interactions and the interfaces of
these parts make it impossible to try to introduce a new or modified
part without an assessment of the effect on the other parts and the
total system. By concentrating on only one part and seeking to optimize it, we may find that the rest of the system responds in an un-
suspected manner. The introduction of DDT provided an effective
pesticide for control of malaria through its effectiveness in killing
the mosquito. Today we also know the harmful side effects of DDT;
its use has been banned or restricted. The early developers of DDT
apparently did not anticipate the side effects that it might have.
Indeed, almost every day one hears about some project or product
that has created an unanticipated systems effect and has become a
problem for the user.

We may not be clever enough to anticipate all the systems
effects, but the first step in discovering them is to be aware of and
accept the viewpoint of the potential impacts when technology is introduced into an environment. Having such an awareness facilitates
the design of a strategy that increases the probability of anticipating

[12]*The Wall Street Journal*, Sept. 24, 1976.

420   the systems effects. The engineering manager's role is a key one in helping to anticipate these effects.

THE ENGINEERING MANAGER'S ROLE

In recent years people have complained that many of the sins of modern society are attributable to what the engineer has accomplished. Certainly there are visible and sometimes disturbing evidences of engineering projects that have destroyed some of the aesthetic qualities of an environment. However, people apparently forget the facts that through the engineer's work we live longer, eat better, are better sheltered, and have more leisure than any other people ever before in the history of humankind. Civilization quite clearly has made significant gains in technology, and there have been some accompanying vices. Today the whole field of technology is being watched to see if we take care to preserve those environmental things of value.

The engineer's lot today is not a happy one. He is always in the business of disturbing nature and creating ecological imbalances. In protecting people against ever-present disease and bad weather, the engineer uses tools and structures that undoubtedly modify the environment. When he provides access across a river to make it possible for people to reach each other, he changes the skyline. Every time he builds a farm-to-market road, often through forest and jungle untouched for centuries, he has tampered with the balance of nature.

Consciously or subconsciously, people have been dependent upon science and technology to protect themselves against their universally hostile surroundings. However, public attitudes today reflect a disenchantment with technology in general. This is reflected in the opposition of the public to the construction of new power plants, smelters, roads, the supersonic transport, and many other projects. It is evidenced also in the rising public concern with the pollution of our land, air, and water. It is found in a public wish for more efficient transportation, new means to recycle waste design of ways to prevent oil spills at sea, restoration and reforestation of land that has been strip-mined, and other similar consumer-oriented and environment-oriented issues.

This new concern with the environment may very well have been precipitated by the disenchantment of the public with the way in which technology has been applied in the past. The solutions to a better environment lie in more, not less, technology. According to Kellogg: "New technology can help us; indeed new technology will play an essential role in the battle to reverse the environmental crisis."[13]

[13]H. H. Kellogg, "Engineers and the Environment," *Journal of Metals*, June 1972, p. 13.

Engineers are caught at the crux of the environmental crisis. **421** Technology has to be developed by engineers' assuming the initiative in the battle to both conserve resources and protect the environment. If the engineering community remains aloof, indifferent, or neutral to this task, it will be impossible to accomplish.

As stated by H. H. Kellogg:

[Engineers] must lead a campaign requiring both technologic and social efforts to conserve mineral and fuel resources and to minimize pollution. On the social front we must publicly urge the adoption of new methods for collection and recycling waste materials, new laws restricting the use of scarce metals for purposes that preclude recycling, and new public priorities which will reverse the trend towards use of products that involve severe environmental stress. On a technologic front we must work hard to develop new processes and products that lessen the environmental crisis, new processes to recover a greater fraction of the metal value in a given ore, new processes that create less pollution or reduce pollution from existing processes, new processes that utilize less fossil fuel or less electrical energy, new processes for recycle of metal waste, new metal products that last longer and are more readily recycled, new products that use materials of less environmental sensitivity.[14]

Any effective program to protect and enhance the environment would of necessity be complex and diverse, since the protection of the environment, given today's life style, has to be dealt with through the application of science and technology. But the protection of the physical environment requires social and economic sanctions, brought to bear through a political process through which the will of the people is expressed, decisions made, and appropriations authorized to implement environmental programs. All this has to come to focus in the engineering and scientific disciplines.

Engineering societies have taken the lead in adapting environmental policy statements. For example, the American Society of Civil Engineering has expressed professional awareness in regard to the ecological balance in its preamble, which states:

Environmental goals must emphasize the need for assuring a desirable quality of life in the context of the expanding technology necessary to sustain and improve human life. Implicit in these goals is the need to develop resources and facilities to improve the environment of man and the need to abate deleterious effects of technology on the environment.

This environmental policy statement expresses

[14]Ibid., p. 16.

422

professional awareness and regard for ecological balance. It is clear, therefore, that neither returning the environment to its pristine form, on the one hand, nor allowing technology to develop to the total disregard of its long- and short-range environmental effects, on the other hand, is acceptable.[15]

All this means that the engineering manager's role has been broadened. Engineering organizations have also recognized the broadened role of the engineering manager. A case in point is the manner in which Kaiser Engineering views the role of a con-struction project manager. The manager is provided various services from the company to support the project, including engineering, design, procurement, and construction. Environmental impact and control is an important part of his responsibility.

In addition, the project team also includes many other special services, such as economics, market analysis, financing, social awareness, and others as shown in Fig. 13.1. The environ-mental impact and control aspects of a project in Kaiser Engineer-ing have always been considerations on construction projects. Today they constitute a substantially larger portion of the work ef-fort to comply with environmental impact programs.

We see the role of the engineer—the technologist—as being important to environmental protection, and the managing of an envi-ronment is probably not much different than managing something else. In essence it means the application of management knowledge, skills, and attitudes to design strategies to integrate technological, social, and economic resources to achieve a desirable environmental setting.

How can this be done? We suggest the following processes for the development of an adequate environmental awareness on the part of engineers and engineering managers:

1 Accept the idea that ecologists, economists, sociologists, and other nontechnical specialists will become involved in the planning and design of engineering projects. These other disciplines will be represented not as status symbols, but rather as equal partners.

2 Accept ecological, economic, and social expertise as it becomes more widely available to the engineering com-munity. Much of this expertise can be provided gratis by government agencies through their review and participation in the development of an Environmental Impact Statement.

3 Develop a commitment to the human need for open space and an aesthetic environment.

[15]See Maxwell Stanley, "The Engineer and the Environment," *Civil Engineering*, July 1972.

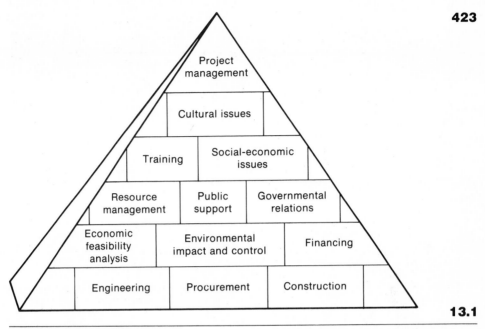

Building blocks of decision-making considerations. *(From V. E. Cole, "Managing the Project," Realities in Project Management, Proceedings of the Ninth Annual International Se-
minar/Symposium, October 22–26, 1977, Chicago, Ill.)*

13.1

4   As part of a code of conduct, stimulate an environmental consciousness in organizations by asking questions about each project such as: (*a*) Has this project been
augmented properly with the skills necessary to determine
total environmental impact? (*b*) Has this project thoroughly
analyzed feasible alternative ways of accomplishing end
purposes that could enhance, or at least not detract from,
environmental considerations? (*c*) Has the client—that is,
the one paying for the engineering knowledge—been given
a clear choice of environmental alternatives in accomplishing end purposes?

This changing role in dealing with environmental considerations can
be illustrated by viewing the philosophy used in making decisions.
The general approach to engineering decisions before the onset of
the current environmental awareness can be portrayed in a model
similar to that in Fig. 13.2. In this model, economy is the primary
goal and other goals, such as environmental considerations, are not
considered. This model makes the decision process unresponsive to
human factors and public goals or attitudes. The introduction of en-

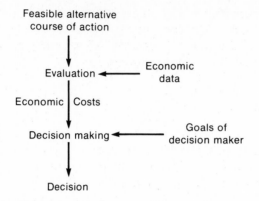

13.2

Traditional decision-making model. *(Adapted from I.R. McIntyre and L. W. Woodhead, "Engineering and Environmental Management," The Journal of the Institution of Engineers, Australia, March–April, 1976, p. 33.*

vironmental factors as a constraint influences the feasible alternatives presented to the decision maker and may alter the conclusion that would have been reached through only economic rationale.

Today's environmentally concerned engineering manager would be more apt to use a decision model similar to the one in Fig. 13.3. This is a multidimensional approach that integrates social and environmental factors along with the environmental considerations.

In Fig. 13.3 one can see the clear potential for conflict—and the need to trade off these conflicting dimensions in decision mak-

Conflicting dimensions in decision making. *(Adapted from I. R. McIntyre and I. W. Woodhead, "Engineering and Environmental Management," The Journal of the Institution of Engineers, Australia, March–April, 1976, p. 36.)*

13.3

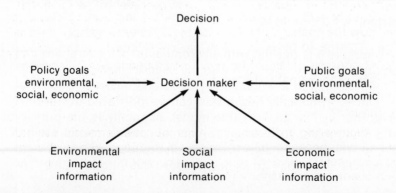

ing. The challenge is to ensure that the conflicting dimensions are **425** clearly defined in the decision process and decisions are made only in the full light of considerations about all relevant factors of environmental, technological, social, and economic significance.

SUMMARY

The purpose we have hoped to achieve in this chapter is to give the reader a summary exposure to the complex subject of environmental considerations. If we have succeeded in attracting the engineering manager's attention to this important area of his responsibility, then we have accomplished our purpose.

The brief historical perspective at the outset was intended to present an experience base for the engineer. This material should make it clear that environmental pollution is nothing new. What is new is the attention that society is giving to legislation mandating the environmental assessment of any effort that could impinge on the environment.

Engineers and engineering managers should be aware of the many ways by which our environment is contaminated. The concept of a system—a relationship between sub-systems in a particular ambience—should be an important part of the engineer's lexicon. Today's engineer must understand the systems approach—the notion that everything is related to everything else. Armed with these important ideas, the engineer is in a good position to take a positive stance in protecting the environment.

DISCUSSION QUESTIONS

1   A systems approach to the responsibilities of corporations may have helped reduce the current levels of environmental ills. Comment.

2   The concept of an Environmental Impact Statement is embedded in a cost-benefit analysis approach to decision making. Can you suggest how the costs and benefits of environmental impacts can be measured relative to other germane considerations?

3   Some sources of air pollution are not generated by humans. Name some of these natural sources of air pollution.

4   What are some other types and sources of pollution not covered in this chapter?

5   The engineering profession may well be viewed critically by the general public because of ecological disruptions. What strategy should engineers evolve to meet this challenge? How must they redefine clientele and responsibilities?

426 6 How does the growing public concern over the environment fit into the discussion on managing change as presented in Chap. 6?

7 What roles do the engineer's knowledge, skills, and attitudes play in influencing the environmental considerations of a project?

8 Why do the authors believe that more and not less technology is needed to overcome environmental ills?

9 For what reasons are we more concerned about our environment today than in previous times?

10 What moral responsibility does an engineer manager have toward protecting the environment?

11 A burgeoning population is often described as an environmental hazard. What is the rationale for such a statement?

12 What is meant by the idea of systems effects? Select an engineering project and trace through some of the most probable systems effects.

KEY QUESTIONS FOR THE ENGINEERING MANAGER

1 What opportunity have you provided for the engineers assigned to your organization to assess the potential environmental impact of their engineering work?

2 Have you ever conducted a dialog with your engineers on the subject of the engineer and the environment?

3 Have you encouraged your engineers to do professional reading on the subject of environmental awareness?

4 What assistance have you provided to your engineers to help them in making sound and rational decisions regarding the environmental effects of various engineering alternatives?

5 What degree of commitment do you have to protect the environment as you manage the application of technology in your organization?

6 Are the engineers assigned to your organization aware of the major environmental threats with which they must deal?

7 Do you and the engineers with whom you work understand the notion of the "system of effects"?

8 In reviewing engineering projects, are you alert for evidence of the effects that appear subsequent to or as a result of the immediate engineering option under study?

9 Have you discussed with your engineers the development of an organizational strategy to increase the probability of anticipating the side effects of key engineering decisions?

10 Have the engineering societies to which you belong taken a positive position toward the environmental question? Have you assisted these societies in this respect?

11 Have you encouraged your engineers to seek ecological, eco-   **427**
nomic, and social expertise when it becomes available to assist them
in evaluating engineering alternatives?

12 Has the material in this chapter given you some insight into the
complex subject of environmental considerations?

BIBLIOGRAPHY

Andover, James J.: "EE's and the Environment," *IEEE Spectrum,*
April 1973.

Bungay, H. R., 3d: "The Status of Engineering in Environmental Programs," *Water and Sewage Works,* February 1973.

Coppoc, W. J.: "Challenges in Utilizing Fossil Fuels," *Chemical Engineering Progress,* vol. 69, no. 6, June 1973.

David, E. E.: "An Energy Policy from the Federal Standpoint," *Civil
Engineering Progress,* vol. 69, no. 6, June 1973.

"Educating the Environmental Designer," *Civil Engineering,* January
1969.

Ehrlich, Paul R., and Ann H. Ehrlich: *Population Resources Environment,* Freeman, San Francisco, 1972.

"Environmental Engineering Stars at ASCE Chattanooga Meeting,"
Civil Engineering, June 1968.

Foster, William S.: "Environment and the Engineer," *American Water
Works Association Journal,* vol. 63, September 1971.

Friedlander, Jacob, and Merril Eisenbud: "Engineering a Better Environment," *Mechanical Engineering,* November 1970.

Galler, Sidney R.: "Environmental Enhancement: The Challenge to
Engineers and Scientists," *Journal of WPCF,* vol. 45, no. 5, May 1973.

A Guide to Models and Governmental Planning and Operations, Office of Research and Development, Environmental Protection
Agency, August 1974.

Jensen, Eugene: "Engineering and the Environment," *Water and
Sewage Works,* June 1972.

Kahne, Stephen: "Contribution to Decision Making in Environmental
Design," *IEEE Proceedings,* March 1975.

Kellogg, H. H.: "Engineers and the Environment," *Journal of Metals,*
June 1972.

428   Kennard, D.C., Jr.: "Environmental Technology: How Are We Doing?" *Journal of Environmental Sciences,* June 1967.

Koenig, Herman E.: "Ecology, Engineering, and Economics," *Proceedings of the IEEE,* March 1975.

——— et al.: "Engineering for Ecological, Sociological and Economic Compatibility," *IEEE Transactions on Systems, Man and Cybernetics,* July 1972.

McIntyre, Ian R., and R. W. Woodhead: "Engineering and Environmental Management," *Journal of the Institute of Engineers, Australia,* March-April 1976.

MacLaren, James W.: "Engineering the Metropolitan Environment," IC-70-CIV4, paper presented at the Annual Meeting of the EIC, Vancouver, B. C., September 10-12, 1969.

Managing the Environment—Socio-Economic Environmental Studies, EPA-600 5-73-010, Office of Research and Development, Environmental Protection Agency, November 1973.

Marini, Robert C.: "Cost Effectiveness of Current Environmental Engineering Practices," *Water and Sewage Works,* April 30, 1973.

Platts, J. R.: "An Approach to Environmental Engineering," *Electronics and Power,* May 1972.

"Professional Collaboration in Environmental Design," *Civil Engineering,* July 1966.

Rosen, Sherman J.: *Manual for Environmental Impact Evaluation,* Prentice-Hall, Englewood Cliffs, N.J., 1976.

Santos, M. A.: "Ecological Systems versus Human Systems: Which Should Be Supreme?" *Journal of Environmental Systems,* vol. 4, no. 4, Winter 1974.

Shefer, Daniel: "Optimal Use of Natural Resources—The Choice between Preservation and Development," *Journal of Environmental Systems,* vol. 4, no. 4, Winter 1974.

Skinner, John E.: "Can Development and Ecology be Reconciled?" *Civil Engineering,* April 1971.

Slayton, William L.: "Environment by Design," *Civil Engineering,* October 1971.

Stanley, C. Maxwell: "The Engineer and the Environment," *Civil Engineering,* July 1972.

"Technology and the Environment: A New Concern on Capitol Hill," *Science,* vol. 157, August 1967.

Tipton, William R.: "Environmental Considerations in Project Management," *Journal of Metals,* May 1977.

Walker, E. A.: "Engineers and the Environment; Being Heard and Being Right," *Journal of Metals,* April 1973.

Wells, James, and Douglas Hill: "Environmental Values in Decision Making: Port Jefferson as a Case Study," *Journal of Environmental Sciences,* July-August 1974.

Wolman, Abel: "The Civil Engineer's Role in the Environmental Development," *Civil Engineering,* October 1970.

Zimmerman, R. L.: "Environmental Improvement from a Product Standpoint," *Journal of Paint Technology,* vol. 45, no. 580, May 1973.

LAW AND THE ENGINEER 14

Never before in its history has the engineering profession faced the legal exposure that it does in modern society. The rapid expansion of technology has stimulated equally rapid developments in the law. The proliferation of goods through mass production, advertising, and marketing has given rise to militant consumerism. Concern for the safety of the public is generating a flow of legislation aimed directly at the engineer. The Boat Safety Act,[1] the Consumer Product Safety Act,[2] the Nuclear Regulatory Act[3] and the Motor Vehicle Safety Act[4] are a few examples of the federal laws recently enacted.

Merely keeping abreast of developments in the law that have a direct effect on you as an engineer would be a full-time task. Lack of time will prevent you from developing familiarity with all the laws that may affect your profession. However, should you act without basic knowledge of the law, you will act at your peril. "Ignorance of the law is no excuse" is still a valid maxim.

The engineer will not be permitted to excuse a violation of the law by claiming that he lacked awareness, nor that he acted in good faith and with the best of intentions. And, while at first it seems unfair, the principles underlying this apparent callousness apply equally to everyday life. Imagine the chaos that would exist if ignorance of the traffic laws excused drivers. It is sound logic to demand conformity with the law, even of those lacking familiarity.

As an engineer, you may face many legal problems but suffer no adverse effects. It is only in the breach of a law that problems arise. However, the danger of acting without a basic understanding of the laws relating to your profession is obvious.

*This chapter was written by Ronald Leslie, B. S., J. D., a practicing attorney and member of various state and federal bar associations. He currently heads the litigation section of Rockwell International Corporation where, as a result of the technical aspects of his practice, he enjoys numerous contacts with engineers at all levels.
[1]Federal Boat Safety Act of 1971, 46 U.S.C.A. §§1451–1479.
[2]Consumer Product Safety Commission Improvements Act of 1976, 15 U.S.C.A. §§ 2051–2081.
[3]Energy Reorganization Act of 1974, 42 U.S.C.A. §§5801–5849.
[4]National Traffic and Motor Vehicle Safety Act of 1966, 15 U.S.C.A. §§1391–1431.

432   When considering the "law" that may apply to your profession, you must consider both the common law and the statutory law. The common law has been developing in the courts during recorded legal history on a case-by-case basis. Statutory law is codified in state and federal statutes. Statutory law, such as the safety acts previously mentioned, is relatively easy to locate, read, and comprehend. Indeed, when an act or statute is of particular concern to the engineer, a copy should be obtained and kept readily at hand.

The common law is not simply delineated. This vast body of legal principles is not static. It is constantly developing and growing as new cases are added to those previously recorded. In many state jurisdictions, the common law is held to include not only decisions of the American courts but also English decisions recorded prior to the American Revolution. Because common law is in effect a living fabric, its principles are subject to reinterpretation and continual growth. There is thus a problem in keeping abreast of new decisions refining the principles of common law as they are handed down by the courts. But while there may be difficulty, there are methods you can learn that will give you knowledge of key legal concepts and theories, in essence a "feel" for the law.

A question that may be foremost in your mind, assuming that you cannot achieve full knowledge of the law, is how best do you protect yourself and how do you avoid the criminal and civil penalties that might result from even an innocent breach? The answer is relatively simple: by developing and nurturing a sixth sense that will warn that an undertaking may have legal ramifications; by recognizing the potential of a legal problem and seeking legal advice before it becomes critical. Developing this sixth sense or feel is essential in today's business environment. It need not be either a time-consuming or painful task. It will be a rewarding and enjoyable experience.

In developing this sixth sense, the first step to be taken is to strive to gain an awareness of the major statutes or governmental regulations that directly affect your field of concentration. Trade journals, which usually highlight laws and proposed legislation of particular interest to their subscribers, are most informative in this regard. Public law libraries, usually in close proximity to the local courthouse, are readily available sources for federal and state statutes and the explanatory texts accompanying them.

In addition to developing an awareness of pertinent legislation, you must develop familiarity with the major theories and trends in the common law. Understanding basic legal principles and theories will serve to alert you to situations that may require in-depth legal consideration. Seminars, trade journals, and analytical news articles that are directed to reviewing pertinent legal concepts or theories are most helpful.

Of current concern to engineers is the trend to afford

consumers and users quick relief for damages occasioned by defective products. The problems, principles, and theories underlying this major trend have been collectively referred to as "product liability." An awareness of this area of the law would be most helpful to you and, by example, it will serve to give you a "feel" for other areas of the law.

433

PRODUCT LIABILITY[5]

Claims and lawsuits against manufacturers are now so common that it is difficult to conceive that only a decade ago they were relatively rare. In recent years, the number of suits filed against manufacturers by those allegedly injured by defective products has been accelerating. Consequently, the financial impact of product liability has been dramatic. Insurance premiums have skyrocketed. Indications are that the increase of claims and litigation resulting from allegedly defective products will continue indefinitely. Because of the proliferation of product liability, there is the likelihood that many engineers will find themselves involved in a product liability claim or lawsuit, perhaps as an expert witness. The following legal principles apply to product liability.

EXPRESS AND IMPLIED WARRANTY
The first area of consideration under the general heading of product liability is that of breaches of express and implied warranties. Generally speaking, the word "warranty" infers that a proposition or fact is true and is treated as synonymous with "guarantee." The concept of warranty finds its genesis in the law of contracts. For legal purposes, the law concerning warranty has been summarized by the Uniform Commercial Code.[6]

The Uniform Commercial Code, which codifies or defines the essential law of commerce, has been adopted throughout the United States. It is a general statement of law concerning trade or commerce.

Under the Uniform Commercial Code, *express* warranties are defined as follows:

Express Warranties by Affirmation, Promise, Description, Sample.

(1)   Express warranties by the seller are created as follows:

(a)   Any affirmation of fact or promise made by the seller to the buyer which relates to the goods and

---

[5]A comprehensive legal review of product liability may be found in Louis R. Frumer and Melvin I. Friedman, *Products Liability*, vols. 1–4, Matthew Bender, New York, 1977.
[6]A general discussion of the Uniform Commercial Code may be found in Ronald A. Anderson, *Anderson on the Uniform Commercial Code*. vols. 1–4, Lawyers Co-operative, Rochester, N.Y., 1970.

434

becomes part of the basis of the bargain creates an express warranty that the goods shall conform to the affirmation or promise.

(*b*) Any description of the goods which is made part of the basis of the bargain creates an express warranty that the goods shall conform to the description.

(*c*) Any sample or model which is made part of the basis of the bargain creates an express warranty that the whole of the goods shall conform to the sample or model.

(2)   It is not necessary to the creation of an express warranty that the seller use formal words such as "warranty" or "guarantee" or that he have a specific intention to make a warranty, but an affirmation merely of the value of the goods or a statement purporting to be merely the seller's opinion or commendation of the goods does not create a warranty.[7]

Express warranties may be created by a seller through advertisements, catalogs, formal contract specifications, and written correspondence. Any affirmation that in effect defines the quality, quantity, or characteristics of goods or products may create an express warranty.

You may be drawn into inadvertently creating an express warranty while attempting to define or develop technical specifications or parameters for a product with a customer. You must be careful to ensure that the statements you convey to a buyer concerning the character and quality of goods forming the basis of a sale are accurate and factual. This is true whether the statements are oral or written. You must not treat any statement of fact you make relating to a product as inconsequential, for in the eyes of a buyer you as an engineer are a technical expert. Reliance may be placed on all you, as an expert, convey.

To develop the terms of a warranty, the courts may take into consideration affirmations of fact beyond the descriptive terms of a written contract between a seller and buyer. Warranty thus may not be limited merely to the guarantees set forth in a contract, but may include statements contained in brochures, catalogs, and advertisements upon which a buyer relied in purchasing the goods. The courts have held it necessary to review all the affirmations made by a seller to determine the essential character of the goods that have been warranted by the seller to the buyer. A written contract or purchase order may only partially define the quality or character of the goods that are being sold. Thus, in the absence of specific language in the contract narrowly defining them, warranties set forth in a contract may be broadened by other guarantees or promises made by the seller or his agents.

[7]Uniform Commercial Code—Sales, §2–313.

It should also be noted that under the Uniform Commercial **435** Code express warranties are created by a seller.[8] "Sellers" are further defined within the code as including not only the manufacturers of goods but also distributors and retailers. Express warranties can be created by any of those within this chain of distribution and their agents or employees.

An express warranty can also be created by means other than an expression in writing. A warranty can be created by a *sample* that is suggested to the buyer as being indicative of the type of product that will be furnished to him. Care must be exercised to ensure that a sample is truly representative of the goods to be sold.

Frequently, a seller, fearful that certain of his statements may be misconstrued, attempts to limit his responsibility by exculpatory language in a written contract. The contract may be drafted with language to the effect that all express warranties, however made, merge into the agreement and only those warranties specifically set forth in the agreement will be binding upon the seller. This exculpatory language may be helpful. However, the cautious engineer should not assume that it will be held legally binding by a court in the event of a controversy.

There are numerous cases in which sellers have been held responsible for express warranties beyond those within the contract despite contractual language limiting warranties. It is, therefore, the better approach to seek to avoid every possibility of error and to ensure that all statements are accurate and do not conflict with those contained in a written contract. For example, if by contract the seller agrees to furnish the buyer an automotive component with a useful life of 50,000 miles, it could create serious problems if the seller's engineer were to affirm in writing that the product has a life expectancy of at least 100,000 miles. Such a statement can form the basis of a warranty relied upon by the buyer, who then demands a useful life substantially beyond the 50,000-mile life set forth in the contract. The statement may become the basis for a lawsuit, if in fact the product fails to meet the warranty promised by the engineer as the agent of the seller.

Advertising is the ally of mass production. Without it, the proliferation of goods throughout society would not have reached its present extent. To some degree, the contents of advertising are controlled by agencies such as the Federal Trade Commission. However, there is great discretion allowed as to the content of advertisements. Frequently, the engineer will be requested to review the technical content of an advertisement. The engineer should ensure that the content of the advertisement accurately conveys the technical aspects of the product. As an example, a statement contained in an ad-

[8]Uniform Commercial Code, §2-103(1)(d).

436   vertisement that a component will last *"up to* 50,000 miles" conveys an entirely different meaning from one that states that a product will last *"a minimum* of 50,000 miles." In the former statement, the express warranty will be met although the product may not have lasted beyond 50,000 miles. The maximum life as guaranteed by the seller according to the advertisement is 50,000 miles. Indeed, even a life of 40,000 miles may meet this guarantee. Any mileage obtained beyond 50,000 miles is a boon to the purchaser and in excess of that warranted by the manufacturer. On the other hand, in the second example, the product will not meet the advertised warranty if it fails at less than 50,000 miles. It must exceed the guaranteed threshold of 50,000 miles in order to meet the promise made to the buyer. Should it fail to meet the threshold, the buyer has in fact been damaged, as the goods have failed to conform to the express warranty. As can be seen by these examples, the engineer should scrutinize advertising carefully.

The question of whether warranties have been met is one that forms the basis of many legal confrontations over the interpretation of technical language within a contract. An engineer asked to review the technical warranties to be set forth in a proposed contract must exercise great caution. The language used to describe guaranteed levels of quality or quantity can be critical. For instance, a printing press may be sold under a warranty that it will produce "2000 signatures [folded pages] per hour." This language, taken on its face, does not specify whether the 2000 signatures will be of a quality sufficient to permit sale by the printer, or if it is intended to convey the minimum quantity that can be produced at the maximum mechanical speed of the printing press irrespective of quality. Are the terms intended to convey that all 2000 signatures would be of acceptable quality to the printer and salable, or that only a portion of them would be of a quality acceptable for eventual sale? Unfortunately, the contract language noted does not specify the quality of the end product. Because this contract is not sufficiently clear on its face, the buyer and seller may have two entirely different interpretations of its meaning. If, in fact, the warranty or promise is to produce a given number of impressions based upon mechanical speed and is not intended to infer acceptable quality, it should so state. Warranties of quality or quantity must be objectively evaluated by the engineer. Does the language adequately convey the true intention of both parties? Should a dispute over the meaning of the language develop, would a court be able to satisfactorily define the intention of the parties from the face of the written agreement?

Another common means of conveying warranties is a catalog. The descriptions contained in catalogs are normally intended to define the characteristics of goods and may be suggestive of such matters as expected life or performance. As such, a catalog is a treasury of express warranties upon which a buyer can contend he **437** relied when purchasing the listed items. Descriptive terminology, drawings, and depictions in a catalog are express warranties and should be treated as such by the seller.

In addition to express warranties, there can be warranties relating to a sale that are not specifically documented in writing or affirmed in oral statements. These are warranties *implied* by law. Implied warranties relate to the merchantability and the fitness for a particular purpose of goods. These implied warranties of merchantability and fitness for a particular purpose are defined by the Uniform Commercial Code as follows:

Implied Warranty: Merchantability; Usage of Trade.

(1)   Unless excluded or modified..., a warranty that the goods shall be merchantable is implied in a contract for their sale if the seller is a merchant with respect to goods of that kind. Under this section the serving for value of food or drink to be consumed either on the premises or elsewhere is a sale.

(2)   Goods to be merchantable must be at least such as

(a) pass without objection in the trade under the contract description; and

(b) in the case of fungible goods, are of fair average quality within the description; and

(c) are fit for the ordinary purposes for which such goods are used; and

(d) run, within the variations permitted by the agreement, of even kind, quality and quantity within each unit and among all units involved; and

(e) are adequately contained, packaged, and labeled as the agreement may require; and

(f) conform to the promises or affirmations of fact made on the container or label if any.

(3)   Unless excluded or modified ... other implied warranties may arise from course of dealing or usage of trade.[9]

Implied warranties arise as a matter of law and do not require specific action on the part of a manufacturer or seller or their agents. They are unlike express warranties, which can only be created through actions on the part of the seller that affirm the character or quality of goods to be sold. The bulk of claims and litigation involving warranties relate to breaches of implied warranties.

[9]Uniform Commercial Code, §2–314.

438   Implied Warranty: Fitness for Particular Purpose.

Where the seller at the time of contracting has reason to know any particular purpose for which the goods are required and that the buyer is relying on the seller's skill or judgment to select or furnish suitable goods, there is unless excluded or modified under the next section an implied warranty that the goods shall be fit for such purpose.[10]

The implied warranty of fitness for a particular purpose can prove to be a pitfall to the engineer. Frequently, a buyer relies upon the seller's engineers as technical experts to advise or assure him that the product being purchased is properly suited for his intended use. The buyer may advise the seller's engineers of the application intended for a product, perhaps giving the parameters of performance that he intends to achieve. He relies upon the expertise of the engineer to confirm that the particular product will meet his requirements. The builder of a logging vehicle may leave the selection of the most efficient type of drive axle for use in logging operations to the manufacturer from which he purchases component axles. It is his belief that the axle manufacturer is an expert and that he is justified in relying upon the manufacturer's expertise to determine the best axle for the intended use. If the axle manufacturer, acting through its engineers, assumes the responsibilities of selecting the proper axle, it assumes the responsibility of choosing an axle fit for the particular purpose of off-highway logging operations.

Assume a buyer relying upon a seller's expertise advises the seller that he intends to purchase a particular product for a marine application and that the product will be frequently submerged in salt water. It thus becomes incumbent upon the seller to select a product that will withstand the ravages of salt water and that will otherwise function satisfactorily in a marine environment. To furnish a component consisting of high carbon steel, which would be unusually susceptible to corrosion in this type of application, may be a breach of the warranty of fitness for a particular purpose. Such action by the seller could give rise to a claim for damages on the part of the buyer.

Under the Uniform Commercial Code, a manufacturer may be held responsible for a breach of express and implied warranties irrespective of his knowledge that a specific defect existed. The liability of a manufacturer or seller will not be negated because of ignorance of a defect. A manufacturer of a completed product is also directly responsible for a defect in a component that he assembled into his product. The law is clear that the manufacturer is responsible for a breach of an express or implied warranty relating to the finished product even though the component that caused the failure was not

---

[10]Uniform Commercial Code, §2-315.

manufactured directly by him. In practice, the manufacturer can join **439**
the producer of the component in a suit commenced against him. If
no lawsuit is commenced but a claim is made, he can tender that
claim to the component manufacturer and demand that it save him
harmless from the costs and expenses involved. The engineer called
on to analyze a failed product should keep in mind the possibility of
tendering the matter to the manufacturer of a defective component.
Frequently, the ultimate responsibility of the component manufacturer may be overlooked during the technical analysis of a failed
product.

NEGLIGENCE
The next area of consideration under the broad spectrum of product
liability is that of negligence. *Negligence* is defined as the omission to
do that which a reasonable person guided by ordinary considerations
that regulate human affairs, would do under the circumstances. Or it
can be the doing of that which a reasonable and prudent person
would not do under the circumstances. It is the failure to use ordinary
care or exercise the degree of care that a prudent person would exercise under a given or like circumstance. Negligence is a breach of the
duty of due care that is owed to those within the scope of the risk. It is
the legally culpable omission of a duty owed by one to another.

Negligence is a theory of law that has particular application
to the engineer, for there are many duties that rest upon the
shoulders of an engineer because of his involvement in the design,
manufacture, testing, inspection, and packaging of products, and finally, in the instruction in their use or warning to the end user.
Because of his exposure throughout the development, manufacture,
and sale of products, the engineer is unusually susceptible to claims
of negligence. It is essential that the engineer understand the basic
concepts underlying the theory so as to minimize his exposure.

While the term "negligence" is easily defined by the common law cases, questions that will arise in the mind of an engineer
regarding the responsibilities he carries in today's dynamic business
environment will be many and troublesome. Questions will arise as to
what is "reasonable" action. What is "prudent" engineering under
the circumstances? What are the design criteria that must be considered in determining whether or not the test of "reasonable prudence"
has been met?

In attempting to establish a criteria against which he may
gauge his actions, the engineer may be tempted to look to the conduct of others within his profession. He may look to the established or
customary practice within the industry to determine whether or not
the actions he intends to take will be construed as "reasonable."
Custom and usage within the industry are certainly evidence as to
that which may be considered reasonable and prudent under a given

440   circumstance. However, while it is evidence of the established practice of other engineers, it is not absolute proof of "reasonableness." Whether the custom and usage are reasonable may be questioned, although they may have existed for an extended period of time. In an early case,[11] a question was presented to the court by the owner of barges that had been lost at sea when they went down because of the lack of radio equipment aboard the ship towing them. A radio would have given warning of the approach of severe weather. At the time of the catastrophe, it was not the ordinary practice of shipowners to have radio equipment aboard ships. The defendant shipowner argued that it was not an industry custom and therefore it could not be considered negligent for failure to have a wireless aboard the towing vessel. The court acknowledged that it was not the custom within the industry to have radio equipment aboard vessels, but the court went on to state that the shipowner was negligent in failing to recognize the technological advances that had occurred in radio communication and in failing to meet a readily achieved degree of safety by carrying radio equipment. Because of this legal precedent, which has been repeatedly affirmed in subsequent cases, the engineer cannot rely upon merely meeting the practice of others in the industry, unless such practice is reasonable and in keeping with current technology.

A term that is frequently raised in discussing questions of negligence is "state of the art." Ordinarily, an engineer may defend against a claim of negligence by contending that he met the standards of his profession and the state of the art. State of the art is essentially the level of technological development of an industry at a given period of time. The state of the art can be relied upon as evidence that reasonable care and prudence were exercised by the engineer whose actions were in conformance with it. However, it is not the ultimate criterion for determining whether or not the engineer's actions were reasonable and prudent. Reliance on state of the art is somewhat akin to reliance upon the custom of an industry. The law maintains state of the art as an open-ended concept. This approach permits gradual technological improvement. If the state of the art were a static standard, there would be no incentive for improvement and growth in the technological area. Thus, reliance on the state of the art as indicative of due care must be objectively evaluated to determine if the level of technological development achieved by an industry is in keeping with that which could be achieved without artificial restraints that inhibit growth.

In evaluating his actions, the engineer can look to the established standards of his colleagues, the custom and usage of his industry, and the state of the art for his industry as being evidence that he exercised due care in fulfilling his duties.

[11] *The T. J. Hooper*, 60 F. 2d. 737 (1932).

STRICT LIABILITY

441

The third and final theory to be considered under the broad concept of product liability is that of strict liability. *Strict liability* has been legally defined as follows:

Special Liability of Seller of Product for Physical Harm to User or Consumer.

(1) One who sells any product in a defective condition unreasonably dangerous to the user or consumer or to his property is subject to liability for physical harm thereby caused to the ultimate user or consumer, or to his property, if

(a) the seller is engaged in the business of selling such a product, and

(b) it is expected to and does reach the user or consumer without substantial change in the condition in which it is sold.

(2) The rule stated in Subsection (1) applies although

(a) the seller has exercised all possible care in the preparation and sale of his product, and

(b) the user or consumer has not bought the product from or entered into any contractual relation with the seller.[12]

The theory of strict liability can prove the most troublesome of the three theories that are comprised in the broad concept of product liability. It is a difficult concept for the engineer to comprehend, because in practice it affixes responsibility for a product that gives rise to harm or injury even though the highest degree of care and skill have been exercised in its design and manufacture. Strict liability is not based upon a finding of fault. If a transaction meets the criteria set forth in the definition of strict liability, the seller will be held responsible for a loss or injury occasioned by a product despite the fact that he has taken every step to produce a safe product.

Historically, the theory of strict or absolute liability on the part of a seller found its birth in early common law cases dealing with the liability of sellers of food that had proved harmful or dangerous. The tendency of the law in those early cases was to affix liability upon the purveyor of food regardless of any defense of a lack of intent to cause harm, lack of negligence, or lack of any direct contract between the seller and the injured party. The trend that developed in these early cases gained momentum and, eventually, decisions were recorded permitting not only recovery against the sellers of food but also the manufacturers and sellers of nonfood products. The underly-

[12]*Restatement of the Law, Second, Torts 2d,* vol. 2, American Law Institute, St. Paul, Minn., 1965, pp. 347—348, §402A.

442 ing rationale that was taking form in these early cases was to the effect that the seller had marketed a product upon which the general public placed reliance and it was therefore proper to require the seller or manufacturer to stand behind that product. Moreover, there was an inclination toward assuming that as between the injured party and a manufacturer, the latter was in a better economic position to absorb the financial impact resulting from the loss. Underlying the current application of strict liability in the courts is the assumption that manufacturers stand in the shoes of insurers for their products, upon which the public has placed great reliance. Should an injury or loss occur, the manufacturer or seller is better able to stand the loss than the individual consumer.

While the theory of strict liability may be construed by some as placing an undue burden upon the manufacturers and sellers of goods, it is considered by others to be basically humanitarian in its approach. In any respect, strict liability will apply only if the basic requirements of the definition previously set forth are met. The engineer should not believe it to be an impossible theory under which to practice his profession. To the contrary, the current trend in the courts is to require strict conformity to the definition before it will apply.

Fundamental to the application of strict liability is the requirement of a "seller." A *seller* is any entity engaged in the "business" of selling specific products for use or consumption. It may include a manufacturer, a distributor, or a retailer. However, a casual seller, one not engaged in the ordinary business of selling specific goods, will not meet the definition. By way of example, if an entity is in the business of manufacturing forgings or castings, and in remodeling its plant sells used equipment, it would not be construed as a "seller" under the definition. The sale was in effect a casual transaction and not one undertaken in the usual course of its business of selling castings. The ordinary business of the casual seller is that of manufacturing castings or forgings, not selling equipment. In order for the theory of strict liability to apply, it is necessary that the actor be usually engaged in the business of selling products such as that which gives rise to the injury.

Second, the product must reach the user or consumer without substantial change in the condition in which it left the hands of a manufacturer or seller. Minor alterations that could be reasonably anticipated by the seller will not absolve him of responsibility. However, a substantial modification or alteration to the product is a sufficient basis for negating the strict liability doctrine. The seller is not responsible for loss occasioned by a product which he had delivered in a safe condition and which thereafter, because of a substantial modification, gave rise to an injury. The seller is responsible only

if the product at the time of injury is substantially in the same condi- **443**
tion as when it left his hands.

Lack of adequate warning makes a product defective. In order to avoid being construed as defective, the product must be properly packaged. If warning labels are required because of the nature of the product or its probable use, they must be affixed. To be legally sufficient, warning labels may have to not only warn of dangers but also instruct in proper use of the product.

A product is not in a defective condition when it leaves the seller in a condition safe for the use anticipated. Strict liability will not apply if the end user or consumer uses the product in some unanticipated or unforeseeable manner so as to cause injury to himself. If an individual uses a glass vase as a convenient tool to hammer a nail into a wall and the vase shatters, causing injury, it would not be construed as a defective or unsafe product under the theory of strict liability. The use of a vase as a hammer is not to be reasonably anticipated or foreseen by its manufacturer or seller. It is a use inconsistent with the basic nature of the product and therefore any injury that is occasioned by the unreasonable use would not be compensable under the strict liability doctrine. However, if a use can be reasonably anticipated even though inconsistent with the intended purpose of a product, then the seller may be strictly liable. A manufacturer must anticipate reasonably foreseeable misuse. The manufacturer of pliers should anticipate that on some occasions they will be used as a substitute hammer for tapping a troublesome nut. This is a reasonably foreseeable misuse of the product and one that the reasonably prudent engineer will take into account in his design.

The product must also be an unreasonably dangerous one in order for strict liability to apply. There is no hard or fast definition of "unreasonably dangerous." Experience undoubtedly is the best criterion. In order to be unreasonably dangerous, a product must exhibit characteristics in use that were not contemplated by the general public at the moment of purchase. For example, if a purchaser buys an ordinary carpenter's hammer and, during normal usage, the head of the hammer shatters, striking him in the eye and giving rise to an injury, then it can be construed or defined as an unreasonably dangerous product. In some instances, a seller may be required to furnish a warning label and instructions in the use of the product in order to reduce the possibility of having it held as unreasonably dangerous.

We have now considered breaches of warranty, negligence, and strict liability, which are the three theories of seller's liability collectively referred to as "product liability." Each may serve as the basis for holding a supplier, whether a manufacturer, wholesaler, or retailer, liable for injuries and damages occasioned by a defective prod-

444   uct. Ordinarily, when a suit is commenced alleging product liability, the plaintiff or party bringing the action will base the claim collectively on each of the three concepts.

PREVENTIVE MANAGEMENT

It is appropriate to end our review of product liability on a positive note by giving consideration to actions that may be taken to reduce the exposure of an engineer to product liability. Positive measures may be generally referred to as "preventive management". Awareness of the potential liability that might arise is only half the battle. Developing a positive attitude that will avoid exposure to a liability situation is the other half. We have noted the unusual susceptibility of an engineer because of his involvement in the functions of design, manufacture, testing, inspection, labeling, and development of instructions or warnings for products.

As fundamental preparation to undertaking the initial design of a new product, the engineer must evaluate the current state of the art as it affects the safety or integrity of the proposed design. The engineer must be cognizant of recent technological advances that may enhance the safety of his product. It is incumbent upon him to ensure that the design incorporates up-to-date safety advances. It is also necessary to ensure that due consideration is given to governmental design standards and pertinent industry standards. Custom and usage within the industry should be considered but not taken as the sole criterion. Thought should be given to the nature of the materials selected to ensure that they do not adversely affect the safety of the final product. Although it is impossible to make any product "idiotproof," it is necessary for the engineer to anticipate some misuse of the product. Reasonably foreseeable misuse should be considered and steps should be taken to reduce the likelihood of an injury or loss.

Inspection procedures are of prime importance. Inspection during and particularly following manufacture will reduce many claims for defective products. Inspection may be proper on a random basis, depending upon the nature of the goods involved.

Testing of the product during its design stage and random testing during manufacture can prove a major defensive device should a question of product integrity arise. Recorded testing data may support the engineer's assertions that design integrity was properly verified. Records of tests during production may be evidence that proper quality control standards were met during manufacture.

The care and consideration that is given to the areas of design, manufacturing, testing, and inspection are ineffectual if records documenting the approach and the results are not retained. Therefore, as a basic necessity, a good record retention program must be established and adhered to in order to retain records that are essential to the defense of a product liability claim or litigation. An **445** engineer should not be willing to trust to his memory to recollect the factors that were considered in projects undertaken and completed many years prior to the date an incident or loss concerning the product may occur.

It is also most helpful to retain copies of catalogs, brochures, and other advertising for indefinite periods of time as evidence for defense of a product liability claim or litigation.

THE ENGINEER AS AN EXPERT WITNESS

Because of his education, training, and career experience, an engineer has developed special knowledge and skill in the area of his discipline. The expertise that he has developed will qualify him to act as an expert witness should litigation arise under theories of product liability. Engineers may be called upon either to support or to attack the integrity of a particular product. Whether he is called upon to evaluate a product either before a dispute develops or during the course of a lawsuit, the engineer should be careful to ensure that he has the requisite qualifications to pass judgment upon the product. The engineer should feel comfortable in the task requested of him. If he believes that he lacks the requisite qualifications, whether educational or technical, he should refuse to evaluate the integrity of a product. This will avoid embarrassment or difficulties that may arise if his conclusions are disputed.

In litigation, the qualifications of an engineer as an expert witness are open to cross-examination by the adverse party. Indeed, it is the duty of counsel representing an opposing party to impeach the qualifications of an expert witness if given a reasonable basis for so doing. Thus, it is imperative that the engineer have the requisite background to permit him to testify as an "expert."

In addition to having the broad educational and background experience to qualify as an expert witness, the engineer must also be able to support the grounds upon which he has based his opinion. Of necessity, this requires a careful and documented approach to the evaluation and conclusion that is drawn.

In dealing with a specific product failure, it is incumbent upon the engineer to secure an opportunity to inspect personally the failed product. If the underlying circumstances merit it, the failed component should be subjected to laboratory testing. Of necessity, the testing may require partial destruction of the product being inspected. Approval must be obtained before destructive testing is undertaken. The approval of trial counsel should be obtained, since destructive testing may have an adverse effect at trial in regard to the admissibility into evidence of a product that has been altered from the time it occasioned a loss.

446   When the engineer is working with trial counsel in developing the technical aspects of a case, he should make certain that there is a complete and candid understanding between himself and counsel. The engineer must not assume that trial counsel is competent to fully grasp the technical aspects of the case. The engineer must be prepared to translate technical terminology and theory into readily understandable terms for the benefit of counsel, judge, and jury.

SUMMARY

The engineer must develop a sixth sense that will alert him to possible legal implications and permit him to seek competent advice before undertaking any project. This sixth sense can be developed by acquiring an awareness of major legal concepts that affect the engineering profession. Of current significance to all engineers is the concept of product liability. "Product liability" is a collective term describing three legal theories. These theories, which form the basis for recovery against the manufacturer and seller of a defective product, are: (1) breaches of express and implied warranty; (2) negligence; and (3) strict or absolute liability. Because of his relationship to the design and manufacture of products, there is a strong likelihood that the engineer will become involved in a product claim or litigation as an "expert." The engineer called upon to render technical assistance as an expert should ensure that he has the requisite educational background and practical knowledge regarding the subject of the controversy. The engineer can also prove instrumental in avoiding product liability exposure by instigating and pursuing preventive management techniques.

DISCUSSION QUESTIONS

1   Why is it important that today's engineer have an appreciation of the legal implications of the engineering process?

2   What is meant by the statement "Ignorance of the law is no excuse"?

3   What is the difference between common law and statutory law?

4   Common law might be defined as an evolving system of principles for living in an organized society. Defend or refute this statement.

5   When an engineer develops a sense of discomfort about the legal ramification of a course of action he is taking, what should he do?

6   What is meant by "product liability"? Why are there more claims **447** and lawsuits against manufacturers today than in previous times?

7   If an engineering manager determines that one of the engineers assigned to his department has been designated to become a technical advisor in a litigation, what action should be taken?

8   What is an implied warranty? What is an express warranty?

9   What is the Uniform Commercial Code?

10   How is negligence defined? Why does the legal theory of negligence have particular application to the engineer?

11   In the context of negligence, can an engineer's reliance on the custom of an industry be considered reasonable and prudent?

12   How are the notions of the "state of the art" and "negligence" related?

KEY QUESTIONS FOR THE ENGINEERING MANAGER

1   What action have I taken to acquaint the engineers assigned to this organization with their legal liabilities and responsibilities?

2   Do I fully realize my responsibility for providing the assigned engineers an awareness of their legal position in practicing the engineering profession?

3   An engineering manager should have enough appreciation of engineering and law to know when he might be headed for trouble. Do I have such an appreciation?

4   Common law is somewhat akin to organizational policy. Do I understand the relationship? When might an organizational policy become a consideration in a legal action?

5   Have I reviewed the record retention policy in this organization to determine its value (or potential liability) in a legal action? Have I reviewed this policy with the legal department?

6   Am I aware of the major statutes and government regulations that have direct effect upon my engineering organization? Have I made these available to the engineers assigned to the organization?

7   Am I familiar with the major theories and trends in the common law that affect the engineering work of this organization? Are the assigned engineers also familiar with these major theories and trends?

8   Have there been any product liability cases against this engineering organization? Have these cases been analyzed to determine if engineering can reduce the risk of future exposure?

9   Do the engineers in this organization have an appreciation of the nature of express and implied warranty?

10   Have I ever asked the legal department of this organization to present a workshop on the subject "Engineering and the Law"?

448   Might such a workshop be of interest to the members of this engineering community?

11   Do I realize that negligence is a theory of law that has particular application to the engineer? Do the assigned engineers realize this?

12   What actions can I take to reduce the exposure of the assigned engineers to product liability?

BIBLIOGRAPHY

American Jurisprudence, Second, vols. 57 and 58, "Negligence," Lawyers Co-operative, Rochester, N.Y., 1971.

American Jurisprudence, Second, vol. 63, "Product Liability," Lawyers Co-operative, Rochester, N.Y., 1972.

Anderson, Ronald A.: *Anderson on the Uniform Commercial Code,* Lawyers Co-operative, Rochester, N.Y., 1970.

Frumer, Louis R., and Melvin I. Friedman: *Products Liability,* vols. 1–4, Matthew Bender, New York, 1977.

Products Liability Reporter, vols. 1 and 2, Commerce Clearing House, New York, 1974.

Restatement of the Law, Second, Torts 2d, vol. 2., American Law Institute Publishers, St. Paul, Minn., 1965.

APPENDIX
HIERARCHICAL DECISION MODEL

Engineering managers are frequently faced with multilevel decisions under conflicting objectives and criteria. The hierarchical decision model is an approach to analyze such decisions. It can be particularly effective for project planning, project evaluation, and resource allocations. We referred to such a model in Chaps. 10 to 12 in the book. This appendix illustrates the model for a project selection decision. It is developed and applied in three stages for this purpose:

1   Priorities of the decision elements at the impact level (organizational objectives)

2   Relative impact of each element at the target level (goals) on each organizational objective

3   Relative contribution of each project to each goal

The measurements required at each stage involve the quantification of subjective values implicit in decision-makers' judgments. This is done by the constant-sum method, using a series of pairwise comparisons among the decision elements.[1] First, the $n(n-1)/2$ pairs are randomized for n elements. Then the decision makers are asked to distribute a total of 100 points between the elements of each pair in the same proportion as the relative value of the two elements with respect to each other. For example, if one element is viewed to be four times as high as the other one, the decision maker allocates 80 and 20 points, respectively. If one is considered slightly higher than the other, the allocation may be 52 and 48. If they are equal in value, each is given 50 points. After all the comparisons are made, the sub-

---

[1]For a description of the constant-sum method, see Kocaoglu (1976) and Guilford (1954).

jective values implicit in the decision-makers' judgment are recovered, using the three matrices discussed for subjective probabilities in Chapter 11.[2]

HIERARCHICAL MODEL

At stage 1, pairwise comparisons are made for the relative priorities
of the objectives. If there are four objectives, O1, O2, O3 and O4, six
pairwise comparisons are needed. Let us assume that the decision
maker is presented with these pairs in a randomized order and he
provided the following values:

O3:	75	O1:	60	O2:	30	O3:	67	O1:	45	O4:	60
O2:	25	O4:	40	O1:	70	O4:	33	O3:	55	O2:	40

These responses are evaluated in matrices *A, B,* and *C* as follows:

Matrix A:

a_{ij} = allocation to element *j* when compared with element *i*.
For example:

$$a_{12} = 30$$

$$a_{21} = 70$$

are the values obtained in the third pairwise comparison (O2 versus
O1).

$$
\begin{array}{c}
\ \ O1\ \ O2\ \ O3\ \ O4 \\
\begin{array}{c} O1 \\ O2 \\ O3 \\ O4 \end{array}
\begin{bmatrix}
- & 30 & 55 & 40 \\
70 & - & 75 & 60 \\
45 & 25 & - & 33 \\
60 & 40 & 67 & -
\end{bmatrix}
\end{array}
$$

Matrix B:

$$b_{ij} = \frac{a_{ij}}{a_{ji}}$$

For example:

$$b_{12} = \frac{30}{70} = 0.43$$

$$b_{21} = \frac{70}{30} = 2.33$$

---

[2]Saaty (1980) approaches this problem with a different solution methodology, using the eigenvectors
derived from the comparison data.

Cell ij in this matrix contains the ratio of the implicit values assigned **451** to objectives O_j and O_i when they were compared with each other at stage 1.

$$
\begin{array}{c}
 \quad O1 \quad\ O2 \quad\ O3 \quad\ O4 \\
\begin{array}{c} O1 \\ O2 \\ O3 \\ O4 \end{array}
\left[
\begin{array}{cccc}
1.0 & 0.43 & 1.22 & 0.67 \\
2.33 & 1.0 & 3.00 & 1.50 \\
0.82 & 0.33 & 1.0 & 0.50 \\
1.50 & 0.67 & 2.00 & 1.0
\end{array}
\right]
\end{array}
$$

The diagonal elements of matrix B are equal to 1.0 because they represent the comparison of each objective with itself.

Matrix C:

$$
c_{ij} = \frac{b_{ij}}{b_{i,j+1}} \qquad \text{for } i = 1, \ldots, n
$$
$$
j = 1, \ldots, n-1
$$

For example:

$$
c_{11} = \frac{1.0}{0.43} = 2.33
$$

$$
c_{12} = \frac{0.43}{1.22} = 0.35
$$

$$
c_{13} = \frac{1.22}{0.67} = 1.82
$$

$$
\begin{array}{c}
 \quad \dfrac{O1}{O2} \quad\ \dfrac{O2}{O3} \quad\ \dfrac{O3}{O4} \\
\begin{array}{c} O1 \\ O2 \\ O3 \\ O4 \end{array}
\left[
\begin{array}{ccc}
2.33 & 0.35 & 1.82 \\
2.33 & 0.33 & 2.00 \\
2.48 & 0.33 & 2.00 \\
2.24 & 0.34 & 2.00
\end{array}
\right]
\end{array}
$$

$$
\begin{array}{rccc}
\text{Mean} = & 2.35 & 0.34 & 1.96 \\
\text{Standard Deviation} = & 0.09 & 0.01 & 0.08
\end{array}
$$

Each column of this matrix contains the ratios of the values assigned to two objectives (such as O1 and O2) when they are compared with the row elements. For the first column:

$$
c_{11} = \frac{b_{11}}{b_{12}} = \frac{O1/O1}{O2/O1} = \frac{O1}{O2} = \frac{1.0}{0.43} = 2.33
$$

$$
c_{21} = \frac{b_{21}}{b_{22}} = \frac{O1/O2}{O2/O2} = \frac{O1}{O2} = \frac{2.33}{1.0} = 2.33
$$

$$c_{31} = \frac{b_{31}}{b_{32}} = \frac{O1/O3}{O2/O3} = \frac{O1}{O2} = \frac{0.82}{0.33} = 2.48$$

$$c_{41} = \frac{b_{41}}{b_{42}} = \frac{O1/O4}{O2/O4} = \frac{O1}{O2} = \frac{1.50}{0.67} = 2.24$$

Although all elements of a column represent the same ratio (O1/O2 for column 1), they are not necessarily equal to each other. This is because of the inconsistencies inherent in judgmental measures. In column 1, c_{11} and c_{21} are equal because both of them are obtained from the direct comparison of O1 and O2. However, c_{31} is the result of a three-way comparison among O1, O2, and O3; it is not equal to c_{11}. Similarly, c_{41} gives the O1/O2 ratio from a three-way comparison of O1, O2 and O4; it again has a different value. The mean of each column is calculated to smooth out these variations. Using the column means as the average ratios perceived by the decision maker, we can calculate the relative priorities of the objectives as follows:

Assign 1.0 as the relative priority of O4.

O4 = 1.0

O3 = 1.96 × 1.0 = 1.96

O2 = 0.34 × 1.96 = 0.67

O1 = 2.35 × 0.67 = 1.57

Normalizing these values, we obtain

O1 = 30%

O2 = 13%

O3 = 38%

O4 = 19%

Sum = 100%

Notice that the calculations above were made for a specific orientation of the columns. For the O1-O2-O3-O4 orientation, we determined the relative priority of O1 as

$$O1 = \frac{O1}{O2} \frac{O2}{O3} \frac{O3}{O4}$$

If we arbitrarily change the column orientation to, let us say, O1-O4-O3-O2, the relative priority of O1 will be calculated as

$$O1 = \frac{O1}{O4} \frac{O4}{O3} \frac{O3}{O2}$$

These two values will not necessarily be equal. The reason is again the internal inconsistencies of the judgmental measures. These inconsistencies can once again be smoothed out by taking the

average of all *n*! column orientations. In our example this means **453**
4! = 24 sets of calculations for the *C* matrix. If the measurements are
relatively consistent, however, as indicated by the low values of
standard deviations in this example, the variation among the 24 sets
of values will be minimal; thus the means obtained in the first orientation can be used as the final result.

In case of multiple decision makers, the calculations above
are modified by using the sum of the allocations made by all participants in each pairwise comparison. Thus, for *m* decision makers, the
values of cells *ij* and *ji* in the matrix *A* will add up to 100*m*. The rest of
the calculations are identical to the case of individual decision
makers.

At stage 2, the goals are evaluated in terms of their relative
contributions to the objectives. This is accomplished by taking two
sets of measurements:

1   Pairwise comparisons of the contribution of
goals to a particular objective, taking the objectives one at
a time.

2   Scale correction for the relative measures among
objectives.

Let us consider three goals—G_1, G_2, and G_3—in support of
the four objectives evaluated in stage 1. Assume that the constant-
sum measurements shown in Table A.1 have been obtained in the
pairwise comparisons of goals with respect to each objective.

Pairwise Comparisons of Goals for Their
Contribution to Each Objective.

Relative contribution of goals to objective 1:   **A.1**

G1: 80	G3: 58	G2: 16
G2: 20	G1: 42	G3: 84

Relative contribution of goals to objective 2:

G1: 20	G3: 75	G2: 58
G2: 80	G1: 25	G3: 42

Relative contribution of goals to objective 3:

G1: 65	G3: 27	G2: 60
G2: 35	G1: 73	G3: 40

Relative contribution of goals to objective 4:

G1: 40	G3: 60	G2: 50
G2: 60	G1: 40	G3: 50

Relative Contribution of Goals to Objectives
(Objectives Taken One at a Time)

A.2	Goals\Objectives	O1	O2	O3	O4
	G1	0.38	0.13	0.52	0.24
	G2	0.10	0.50	0.29	0.38
	G3	0.52	0.37	0.19	0.38
	Sum	1.00	1.00	1.00	1.00

The evaluation of these measurements through the three matrices (*A, B,* and *C*) results in Table A.2.

The values in Table A.2 need some explanation. Since goals were evaluated with respect to a particular objective at a time, each column represents the relative contributions to that objective only. For example, the first column indicates that G1 contributes 3.8 times G2 to objective 1, and G3 contributes 5.2 times G2 to the same objective. These relative values are valid regardless of the level of absolute contributions to the first objective. In other words, we do not know if G1 contributes 3.8 times a high value or 3.8 times a low value. In column 2, G2 contributes 3.85 times as much as G1 to the second objective but again we do not know if it is a high contribution or a low contribution vis-à-vis the contribution of G2 to any other objective.

This calls for a scale correction to obtain a compatible set of measurements in the columns. It can be done by reversing the pairwise comparisons. We evaluate each goal's relative contribution to the four objectives by taking the goals one at a time for this purpose. Let us assume that the new pairwise comparisons are as shown in Table A.3.

Pairwise Comparison of Each Goal's
Contribution to the Objectives

A.3

Relative contribution of goal 1 to objectives:

O4: 36	O2: 22	O1: 78	O3: 64	O4: 67	O3: 50
O1: 64	O3: 78	O2: 22	O4: 36	O2: 33	O1: 50

Relative contribution of goal 2 to objectives:

O4: 76	O2: 67	O1: 20	O3: 37	O4: 45	O3: 66
O1: 24	O3: 33	O2: 80	O4: 63	O2: 55	O1: 34

Relative contribution of goal 3 to objectives:

O4: 40	O2: 69	O1: 62	O3: 30	O4: 52	O3: 22
O1: 60	O3: 31	O2: 38	O4: 70	O2: 48	O1: 78

The comparisons of Table A.3 result in the following normalized ratio **455** scale values:

	O1	O2	O3	O4	Sum
Goal 1 contributions:	.35	.10	.35	.20	1.00
Goal 2 contributions:	.10	.39	.19	.32	1.00
Goal 3 contributions:	.39	.24	.11	.26	1.00

We can now apply the scale correction by combining the two sets of measurements as follows:

For goal 1, multiply the first column of Table A.1 by (.35/.38) the second by (.10/.13), the third column by (.35/.52) and the fourth column by (.20/.24). This gives us one corrected matrix.

For goal 2, follow the same procedure by multiplying the first column of Table A.1 by (.10/.10), the second column by (.39/.50), etc. This will give us a second matrix.

Similarly, obtain a third corrected matrix by using (.39/.52), (.24/.37), (.11/.19) and (.26/.38) as the correction factors.

Normalize each of the three new matrices and take the average value of each cell.

This way, we explicitly recognize the scale differential in the relative contributions of the goals to the various objectives. The final goal/objective matrix obtained through these calculations is shown in Table A.4.

Studying Table A.4, we conclude that the contribution of goal 1 to objective 2 (0.03) is the same as G2's contribution to O1 (0.03), half as much as G1's contribution to O4 (0.06) and G2's contribution to O3 (0.06); one-third as much as G3's contribution to O2 (0.09), etc.

Short version of scale correction:

The procedure above for scale correction requires a large number of additional comparisons. An alternative is to select only the

Relative Contribution of Goals to Outputs, G_{jk}
(After Scale Correction)

Goals\Objectives	O1	O2	O3	O4	**A.4**
G1	0.11	0.03	0.11	0.06	
G2	0.03	0.12	0.06	0.10	
G3	0.15	0.09	0.04	0.10	

Note: Sum of cell values = 1.00

Largest Contribution to Each Objective Based on
Table A.1.

A.5 Goals\Objectives	O1	O2	O3	O4
G1			0.52	
G2		0.50		0.38
G3	0.52			

highest valued element from the columns of Table A.1 and compare
them with each other. The highest values in each column are shown
in Table A.5.

Suppose these four cells are compared as follows:

Contribution of G3 to O1: 56		Contribution of G2 to O4: 48	
Contribution of G2 to O2: 44		Contribution of G1 to O3: 52	

Contribution of G2 to O2: 52		Contribution of G3 to O1: 58	
Contribution of G1 to O3: 48		Contribution of G1 to O3: 42	

Contribution of G3 to O1: 60		Contribution of G2 to O2: 55	
Contribution of G2 to O4: 40		Contribution of G2 to O4: 45	

These comparisons give the normalized ratio scale measures for the
four cells shown in Table A.6.

The combination of Tables A.1 and A.6 again results in the
corrected goal/objective matrix as follows: Multiply the first column
of Table A.1 by (.31/.52), the second column by (.25/.50), the third
column by (.23/.52), and the fourth column by (.21/38). Normalize
the resulting matrix.

Consistent comparisons were used in all the computations
above. The two versions of scale correction give identical results
even though we use much less information in the short version. In

Short Version of Scale Correction

A.6	O1	O2	O3	O4
G1			.23	
G2		.25		.21
G3	.31			

Relative Contribution of Projects to
to Goals, P_{ij}

A.7

Projects\Goals	G1	G2	G3
P1	0.02	0.05	0.03
P2	0.10	0.01	0.12
P3	0.04	0.08	0.10
P4	0.11	0.02	0.03
P5	0.09	0.10	0.10

Note: Sum of the cell values = 1.00

reality, such a high consistency cannot be expected. However, the computational ease of the short version could make it an attractive alternative in many cases. It is particularly useful when we are dealing with sparse matrices.

Stage 3 measurements are identical to stage 2. In this case, the proposed projects are evaluated with respect to their contribution to the goals. The result of this stage is the project/goal matrix obtained in the same way as the goal/objective matrix.

Let us assume that measurements have been taken for five projects and Table A.7 has been obtained as the project/goal matrix.

We can now combine the project/goal and goal/objective matrices to obtain the project/objective matrix as follows:

$$P_{ik} = \sum_{j=1}^{3} P_{ij} \cdot G_{jk}$$

For example,

$$P_{11} = (0.02, 0.05, 0.03) \begin{bmatrix} 0.11 \\ 0.03 \\ 0.15 \end{bmatrix} = 0.0082$$

Calculating all other values similarly, we can develop the project/objective matrix shown in Table A.8.

Project/Objective Matrix

A.8

Projects\Objectives	O1	O2	O3	O4
P1	8.2	9.3	6.4	9.2
P2	29.3	15.0	23.6	19.0
P3	21.8	19.8	13.2	20.4
P4	17.2	8.4	14.5	11.6
P5	27.9	23.7	19.9	25.4

Note: For ease of calculations, the cell values are multiplied by 1000

Finally, the project/objective matrix is premultiplied by the objective priorities to obtain the relative value of each project as follows:

$$V_1 = (0.3, 0.13, 0.38, 0.19) \begin{bmatrix} 8.2 \\ 9.3 \\ 6.4 \\ 9.2 \end{bmatrix} = 7.9$$

Similarly,

$$V_2 = 23.3$$
$$V_3 = 18.0$$
$$V_4 = 14.0$$
$$V_5 = 23.8$$

For ease of comparison, the project relative values are normalized to add up to 100.

$$V_1 = \quad 9.1$$
$$V_2 = 26.7$$
$$V_3 = 20.7$$
$$V_4 = 16.1$$
$$V_5 = 27.4$$

Sum 100.0

The relative values above indicate that projects 5 and 2 are very close to each other. They both have higher overall values than the other projects. In fact, each of them is about three times as desirable as project 1. Because of the subjective nature of evaluations, the difference between 26.7 for project 2 and 27.4 for project 5 should not be considered significant. Both should be viewed as being superior to the next project (project 3) in the list.

PROBABILISTIC HIERARCHICAL MODELS

The analysis above is based on a deterministic approach. Assuming all projects will succeed, we have ranked them in terms of their overall relative values as follows:

$$P5 > P2 > P3 > P4 > P1$$

Obviously, all projects are not equally likely to succeed. We can incorporate the risk involved in the projects by estimating the probability of success and determining the *expected* relative value of each one. This can be done in three ways. The three methods are explained **459** below in the order of increasing complexity:

METHOD 1
Suppose that we estimate that the relative value of a project obtained in the hierarchical model will be achieved with probability p, and the project will fail with probability $1 - p$. The expected value of that project can be calculated as follows:

$$E(\text{Project}) = (p) \ (\text{Relative value of project}) + (1 - p) \ (0)$$

$$= (p) \ (\text{Relative value of project})$$

METHOD 2
Rather than use only a success-failure dichotomy, we develop a probability distribution for the level of achievement that can be expected for each project. After determining the probability of achievement at 20, 40, 60, 80, and 100 percent levels, we calculate the expected value. In this calculation, it is assumed that the achievement level and the relative value of the project are proportional.

For this method, the decision maker is asked a series of questions for each project as follows:

What is the likelihood that this project will completely fulfill its objectives in terms of time, cost and performance?

What is the probability that it will achieve at least 80 percent of its objectives? 60 percent, 40 percent, 20 percent?

The answers to these questions provide the cumulative probability distribution for the level of achievement of the project. Let us assume that the following answers (see Tables A.9 and A.10) have been obtained for the five projects in our example.

Cumulative Probabilities for Project
Achievement Levels

CUMULATIVE PROBABILITIES				A.9
ACHIEVEMENT LEVEL, PERCENT	PROJECT 1	PROJECT 2	PROJECT 3	PROJECT 4	PROJECT 5
Full	0.7	0.3	0.6	0.9	0.4
At least 80	0.8	0.5	1.0	1.0	0.5
60	0.95	0.7	1.0	1.0	0.6
40	1.0	0.9	1.0	1.0	0.7
20	1.0	0.95	1.0	1.0	0.8

Probabilities for Project Achievement Levels

A.10	PROBABILITIES				
ACHIEVEMENT RANGE, PERCENT	PROJECT 1	PROJECT 2	PROJECT 3	PROJECT 4	PROJECT 5
Achievement: 100	0.7	0.3	0.6	0.9	0.4
80 ≤ achievement < 100	0.8−0.7 = 0.1	0.2	0.4	0.1	0.1
60 ≤ achievement < 80	0.95−0.8 = 0.15	0.2	0.0	0.0	0.1
40 ≤ achievement < 60	1.0−0.95 = 0.05	0.2	0.0	0.0	0.1
20 ≤ achievement < 40	0.0	0.05	0.0	0.0	0.1
0 ≤ achievement < 20	0.0	0.05	0.0	0.0	0.2
Sum	1.00	1.00	1.00	1.00	1.00

Probabilities of achievement within specific ranges are obtained as the differences of the cumulative probabilities shown in Table A.9.

We can now calculate the relative values at the midpoints of the probability ranges and determine the expected relative value of each project as follows:

$$E(\text{Project 1}) = 9.1 \ [0.7 + 0.9(0.1) + 0.7(0.15) + 0.5(0.05)]$$
$$= 84$$

$$E(\text{Project 2}) = 26.7 \ [0.3 + 0.9(0.2) + 0.7(0.2) + 0.5(0.2) + 0.3(0.05) + 0.1(0.05)] = 19.8$$

$$E(\text{Project 3}) = 19.9$$

$$E(\text{Project 4}) = 15.9$$

$$E(\text{Project 5}) = 18.1$$

When the risks are considered, project 5 is no longer the best choice because of a relatively high probability of failure and a low probability of full achievement. In fact, although project 3 was previously ranked third in terms of its relative value, now it becomes the best choice.

METHOD 3

The achievement level and the relative value of project were assumed to have a linear relationship in method 2. In other words, it was assumed that if a project is to achieve 60 percent of its objectives its relative value would be reduced to 60 percent of the full-achievement level. This assumption is removed in method 3.

This method is more sophisticated than the previous two methods, but it is also more difficult to implement because of the large number of pairwise comparisons required in measurements. In

this method, each project is defined at a discrete number of achieve- **461**
ment levels. The lowest level corresponds to the failure of the project,
the highest refers to full achievement. Intermediate levels are defined
in terms of partial outputs within the time and cost constraints. For
example, if the project involves the design and construction of a
power plant, the various achievement levels could be defined as
follows:

Level 1 (lowest)	No work is done
Level 2	Engineering design is complete
Level 3	Engineering design and shop drawings are complete
Level 4	Design and drawings plus half of the construction complete
Level 5 (highest)	The project is complete

The number of levels and their definition depend on the nature of the
project. They should be recognizable breakpoints in the projects.
Typically 3 to 7 levels are sufficient for evaluation purposes.

Once the levels are defined, each level is treated as a
subproject within the project for the purpose of pairwise comparisons.

Let us assume that the power plant is P1 in our previous example with a relative value of 9.1 units. This is the value of the fifth
level (completion of the project). The first level (no work done) is assigned a value of zero. Levels 2 through 5 are evaluated among
themselves by repeating the procedure for obtaining the proj-
ect/goal matrix. In other words, pairwise comparisons are made for
the measurement of the relative impact of each level within the project to the goals, treating each level as a subproject.

After the subproject/goal matrix is obtained, the relative
value of each level is determined through the goal/objective matrix
and the objective priorities. Suppose that the normalized relative values of the levels of project 1 are as shown below:

Level 1	0.0	(By definition)
Level 2	0.05	
Level 3	0.15	
Level 4	0.20	
Level 5	0.60	
Sum	1.00	

462   The relative value of level 5 corresponds to 9.1 calculated earlier. Using it as the basis, the values of other levels are calculated as follows:

$$P1(\text{level } 1) = 0.0$$

$$P1(\text{level } 2) = 0.05 \, \frac{9.1}{0.6} = 0.76$$

$$P1(\text{level } 3) = 0.15 \, \frac{9.1}{0.6} = 2.28$$

$$P1(\text{level } 4) = 0.20 \, \frac{9.1}{0.6} = 3.03$$

$$P1(\text{level } 5) = 0.60 \, \frac{9.1}{0.6} = 9.10$$

These are the relative values of project 1 at various levels of achievement obtained without assuming a linear relationship between the achievement level and its value. In other words, even though P1 (level 5) is (9.10/2.28) = 3.99 times P1 (level 3), we do not imply that level 3 represents roughly one-fourth of the total project output.

After the relevant levels of all projects are evaluated in a similar way, the probabilities of the levels are obtained as explained in method 2.

At this point, we have all the information we need for determining the expected values of the projects. Suppose that the relative values and probabilities of the various levels of achievement have been determined for each of the five projects. We evaluate Project 1 as follows.

Project 1:

ACHIEVEMENT LEVEL	1 to 2	2 to 3	3 to 4	4 to 5	FULL ACHIEVEMENT
Average relative value	$\dfrac{0 + 0.76}{2} = 0.38$	$\dfrac{0.76 + 2.28}{2} = 1.52$	$\dfrac{2.28 + 3.03}{2} = 2.66$	$\dfrac{3.03 + 9.10}{2} = 6.07$	9.10
Probability	0.0	0.05	0.15	0.10	0.70

We can now use the same approach as in method 2 to calculate the expected value of each project and select those which have the highest expected values.

$$E(\text{Project 1}) = 0.38(0) + 1.52(0.05) + 2.66(0.15) + 6.07(0.10) + 9.10(0.70)$$

$$= 7.45$$

Note that the relative value for a range of achievement levels was determined as the average of the relative values at the ends of that range. This assumes piecewise linearization of the relative values in **463** the short span defined by the specific range of achievement levels.

Method 3 is more involved than the previous two methods, but it is also a more sophisticated and realistic method for evaluating the projects by the hierarchical model in probabilistic domain.

BIBLIOGRAPHY

Bell, Thomas R.: "A Consistency Measure for the Constant-Sum Method," M.S. thesis, University of Pittsburgh, 1980.

Childs, M., and H. Wolfe; "A Decision and Value Approach to Research Personnel Allocation," *Management Science*, vol. 18, no. 6, February 1972, pp. 269–278.

Cleland, D., and D. F. Kocaoglu: "Program Evaluation of the USDA Combined Forest Pest R & D Program," USDA Research Report, 1981.

Guilford, J. P.: *Psychometric Methods*, 2d ed., McGraw-Hill, New York, 1954.

Kocaoglu, D. F.: "A Systems Approach to the Research Allocation Process in Police Patrol," Ph.D. dissertation, University of Pittsburgh, 1976.

Kocaoglu, D. F: "Disaggregation of Decisions to Allocate Patrol Resources to Police Precincts," in L. P. Ritzman et al. (eds.), *Disaggregation*, Martinus, Nijhoff, Boston, 1979, pp. 541–552.

Kocaoglu, D. F.: "Hierarchical Decision Process for Program/Project Evaluation," Proceedings of the 6th INTERNET Congress, Garmisch, W. Germany, vol. 4, September 1979, pp. 239–249.

Nagurney, F. K.: "An O. R. Approach to the Evaluation of Prehospital Emergency Medical Care," Ph.D. dissertation, University of Pittsburgh, 1978.

Saaty, Thomas, *The Analytic Hierarchy Process*, McGraw-Hill, New York, 1980.

Shepherd, J. C. "An Interactive Support System to Choose the Optimal Portfolio of Projects under Conditions of Uncertainty and Multiple Criteria," Ph.D. dissertation, University of Pittsburgh, 1980.

Thurston, L. L.: "A Law of Comparative Judgment," *Psychological Review*, vol. 34, 1927, pp. 273–286.

INDEX

PROPERTY OF
USBI